T0293278

STRUCTURED DEPENDENCE BETWEEN STOCHASTIC PROCESSES

The relatively young theory of structured dependence between stochastic processes has many real-life applications in areas including finance, insurance, seismology, neuroscience, and genetics. With this monograph, the first to be devoted to the modeling of the structured dependence between random processes, the authors not only meet the demand for a solid theoretical account but also develop a stochastic processes counterpart of the classical copula theory that exists for finite-dimensional random variables.

Presenting both the technical aspects and the applications of the theory, this is a valuable reference for researchers and practitioners in the field, as well as for graduate students in pure and applied mathematics programs. Numerous theoretical examples are included, alongside examples of both current and potential applications, aimed at helping those who need to model the structured dependence between dynamic random phenomena.

Encyclopedia of Mathematics and Its Applications

This series is devoted to significant topics or themes that have wide application in mathematics or mathematical science and for which a detailed development of the abstract theory is less important than a thorough and concrete exploration of the implications and applications.

Books in the **Encyclopedia of Mathematics and Its Applications** cover their subjects comprehensively. Less important results may be summarized as exercises at the ends of chapters. For technicalities, readers can be referred to the bibliography, which is expected to be comprehensive. As a result, volumes are encyclopedic references or manageable guides to major subjects.

All the titles listed below can be obtained from good booksellers or from Cambridge University Press. For a complete series listing visit www.cambridge.org/mathematics.

125 F. W. King *Hilbert Transforms II*
126 O. Calin and D.-C. Chang *Sub-Riemannian Geometry*
127 M. Grabisch *et al. Aggregation Functions*
128 L. W. Beineke and R. J. Wilson (eds.) with J. L. Gross and T. W. Tucker *Topics in Topological Graph Theory*
129 J. Berstel, D. Perrin and C. Reutenauer *Codes and Automata*
130 T. G. Faticoni *Modules over Endomorphism Rings*
131 H. Morimoto *Stochastic Control and Mathematical Modeling*
132 G. Schmidt *Relational Mathematics*
133 P. Kornerup and D. W. Matula *Finite Precision Number Systems and Arithmetic*
134 Y. Crama and P. L. Hammer (eds.) *Boolean Models and Methods in Mathematics, Computer Science and Engineering*
135 V. Berthé and M. Rigo (eds.) *Combinatorics, Automata and Number Theory*
136 A. Kristály, V. D. Rădulescu and C. Varga *Variational Principles in Mathematical Physics, Geometry and Economics*
137 J. Berstel and C. Reutenauer *Noncommutative Rational Series with Applications*
138 B. Courcelle and J. Engelfriet *Graph Structure and Monadic Second-Order Logic*
139 M. Fiedler *Matrices and Graphs in Geometry*
140 N. Vakil *Real Analysis through Modern Infinitesimals*
141 R. B. Paris *Hadamard Expansions and Hyperasymptotic Evaluation*
142 Y. Crama and P. L. Hammer *Boolean Functions*
143 A. Arapostathis, V. S. Borkar and M. K. Ghosh *Ergodic Control of Diffusion Processes*
144 N. Caspard, B. Leclerc and B. Monjardet *Finite Ordered Sets*
145 D. Z. Arov and H. Dym *Bitangential Direct and Inverse Problems for Systems of Integral and Differential Equations*
146 G. Dassios *Ellipsoidal Harmonics*
147 L. W. Beineke and R. J. Wilson (eds.) with O. R. Oellermann *Topics in Structural Graph Theory*
148 L. Berlyand, A. G. Kolpakov and A. Novikov *Introduction to the Network Approximation Method for Materials Modeling*
149 M. Baake and U. Grimm *Aperiodic Order I: A Mathematical Invitation*
150 J. Borwein et al. *Lattice Sums Then and Now*
151 R. Schneider *Convex Bodies: The Brunn-Minkowski Theory (Second Edition)*
152 G. Da Prato and J. Zabczyk *Stochastic Equations in Infinite Dimensions (Second Edition)*
153 D. Hofmann, G. J. Seal and W. Tholen (eds.) *Monoidal Topology*
154 M. Cabrera García and Á. Rodríguez Palacios *Non-Associative Normed Algebras I: The Vidav-Palmer and Gelfand-Naimark Theorems*
155 C. F. Dunkl and Y. Xu *Orthogonal Polynomials of Several Variables (Second Edition)*
156 L. W. Beineke and R. J. Wilson (eds.) with B. Toft *Topics in Chromatic Graph Theory*
157 T. Mora *Solving Polynomial Equation Systems III: Algebraic Solving*
158 T. Mora *Solving Polynomial Equation Systems IV: Buchberger Theory and Beyond*
159 V. Berthé and M. Rigo (eds.) *Combinatorics, Words and Symbolic Dynamics*
160 B. Rubin *Introduction to Radon Transforms: With Elements of Fractional Calculus and Harmonic Analysis*
161 M. Ghergu and S. D. Taliaferro *Isolated Singularities in Partial Differential Inequalities*
162 G. Molica Bisci, V. D. Radulescu and R. Servadei *Variational Methods for Nonlocal Fractional Problems*
163 S. Wagon *The Banach-Tarski Paradox (Second Edition)*
164 K. Broughan *Equivalents of the Riemann Hypothesis I: Arithmetic Equivalents*
165 K. Broughan *Equivalents of the Riemann Hypothesis II: Analytic Equivalents*
166 M. Baake and U. Grimm (eds.) *Aperiodic Order II: Crystallography and Almost Periodicity*
167 M. Cabrera García and Á. Rodríguez Palacios *Non-Associative Normed Algebras II: Representation Theory and the Zel'manov Approach*
168 A. Yu. Khrennikov, S. V. Kozyrev and W. A. Zúñiga-Galindo *Ultrametric Pseudodifferential Equations and Applications*
169 S. R. Finch *Mathematical Constants II*
170 J. Krajíček *Proof Complexity*
171 D. Bulacu, S. Caenepeel, F. Panaite and F. Van Oystaeyen *Quasi-Hopf Algebras*
172 P. McMullen *Geometric Regular Polytopes*
173 M. Aguiar and S. Mahajan *Bimonoids for Hyperplane Arrangements*
174 M. Barski and J. Zabczyk *Mathematics of the Bond Market: A Lévy Processes Approach*

ENCYCLOPEDIA OF MATHEMATICS AND ITS APPLICATIONS

Structured Dependence between Stochastic Processes

TOMASZ R. BIELECKI
Illinois Institute of Technology

JACEK JAKUBOWSKI
University of Warsaw

MARIUSZ NIEWĘGŁOWSKI
Warsaw University of Technology

CAMBRIDGE
UNIVERSITY PRESS

University Printing House, Cambridge CB2 8BS, United Kingdom

One Liberty Plaza, 20th Floor, New York, NY 10006, USA

477 Williamstown Road, Port Melbourne, VIC 3207, Australia

314–321, 3rd Floor, Plot 3, Splendor Forum, Jasola District Centre, New Delhi – 110025, India

79 Anson Road, #06–04/06, Singapore 079906

Cambridge University Press is part of the University of Cambridge.

It furthers the University's mission by disseminating knowledge in the pursuit of education, learning, and research at the highest international levels of excellence.

www.cambridge.org
Information on this title: www.cambridge.org/9781107154254
DOI: 10.1017/9781316650530

First published 2020

A catalogue record for this publication is available from the British Library.

Library of Congress Cataloging-in-Publication Data
Names: Bielecki, Tomasz R., 1955– author. | Jakubowski, Jacek, author. | Niewęgłowski, Mariusz, 1976– author.
Title: Structured dependence between stochastic processes / Tomasz R. Bielecki, Jacek Jakubowski, Mariusz Niewęgłowski
Description: Cambridge ; New York, NY : Cambridge University Press, [2020] | Series: Encyclopedia of mathematics and its applications | Includes bibliographical references and index.
Identifiers: LCCN 2020006654 | ISBN 9781107154254 (hardback)
Subjects: LCSH: Markov processes. | Dependence (Statistics)
Classification: LCC QA274.7 .B55 2020 | DDC 519.2/33–dc23
LC record available at https://lccn.loc.gov/2020006654

ISBN 978-1-107-15425-4 Hardback

Contents

1
Introduction

The main objective of this book is to study the structured dependence between stochastic processes, such as for example Markov chains, conditional Markov chains, and special semimartingales, as well as between Markov families. In particular, we devote considerable attention to the modeling of structured dependence.

In a nutshell, the structured dependence between stochastic processes is the stochastic dependence between these processes subject to the relevant structural requirements.

Some major practical motivations for modeling the structured dependence between stochastic processes can be briefly summarized as follows. This methodology allows one to build models for multivariate stochastic differential systems subject to the constraint that the idiosyncratic dynamical properties of the multivariate system, that is, the dynamics of the univariate constituents of the modeled multivariate system, are matched with the univariate data. Both the multivariate and the univariate dynamics can be modeled as desired. For example, in case of Markov structures, both the multivariate and the univariate dynamics are Markovian, which allows one to tap in to the rich theory and practice of Markov processes. Moreover, and quite importantly, by using structured-dependence modeling one can separate estimation of the univariate, idiosyncratic, parameters of the model from estimation of the parameters accounting for the stochastic dependence *between* the univariate constituents of the model. In particular, within the structured-dependence model one can match the dynamics of the univariate constituents of the system that are estimated from the univariate data.

Let $\mathcal{P}$ represent a certain property of stochastic processes, such as the Feller–Markov property or the special semimartingale property. Let us fix an integer $n > 1$ and let $\mathcal{P}^n$ denote a class of n-variate stochastic processes[1] that take values in $\mathcal{X} := \mathcal{X}_1 \times \cdots \times \mathcal{X}_n$ and that feature the property $\mathcal{P}$. The problem of modeling the structured dependence between stochastic processes can be summarized as follows.

Given a collection of univariate stochastic processes, say Y^i, $i = 1,\ldots,n$, taking values in $\mathcal{X}_i$, $i = 1,\ldots,n$, respectively, and all featuring property $\mathcal{P}$, construct

[1] We refer to Appendix A for the key definitions in the area of stochastic analysis that are used throughout this book.

n-variate stochastic processes, say $X = (X^1, \ldots, X^n)$, such that $X \in \mathcal{P}^n$ and the structural properties of X^i are the same as the structural properties of Y^i for all $i = 1, \ldots, n$. For example, if $\mathcal{P}$ represents the Markov property and the structural property of process Y^i is that it is a Markov chain with generator function $\Lambda^i(\cdot)$, $i = 1, \ldots, n$, then we want X to be an n-variate Markov chain and X^i to be a Markov chain with generator function $\Lambda^i(\cdot)$, $i = 1, \ldots, n$. Or, if $\mathcal{P}$ represents the special semimartingale property and the structural property of process Y^i is that it is a special semimartingale process with characteristic triple (B^i, C^i, ν^i), $i = 1, \ldots, n$, then we want X to be a special semimartingale and X^i to be a special semimartingale process with characteristic triple (B^i, C^i, ν^i). Any such process X is called an n-dimensional *$\mathcal{P}$-structure* for Y^i, $i = 1, \ldots, n$, and the processes Y^i, $i = 1, \ldots, n$, are called the *predetermined margins for X*. We will sometimes use the term *stochastic structure* if reference to a specific class $\mathcal{P}$ is not needed.

In many cases of interest the structural properties of a stochastic process can be used to determine the law of the process. For example, in case of a Markov process starting at a given time, the distribution of the process at this time and the generator of the process determine the law of this process. Also, with semimartingales, it is frequently the case, albeit not always, that the initial law of a semimartingale and its characteristic triple determine its law. In such situations, any $\mathcal{P}$-structure X for Y^i, $i = 1, \ldots, n$, provides the additional benefit that the law of X^i may be chosen so that it agrees with the law of Y^i, $i = 1, \ldots, n$. If this can be done then we refer to $\mathcal{P}$-structure X as a *$\mathcal{P}$-copula structure.*

In this book we focus on constructing $\mathcal{P}$-structures, which naturally lead to $\mathcal{P}$-copula structures.

The problem of constructing a $\mathcal{P}$-copula structure clearly is reminiscent of one of the classical problems in probability: given a family of $\mathbb{R}^1$-valued random variables, say U^i, $i = 1, \ldots, n$, construct $\mathbb{R}^n$-valued random variables, say $V = (V^1, \ldots, V^n)$, such that the law of V^i is the same as the law of U^i for all $i = 1, \ldots, n$. The elegant and satisfactory solution to this problem is given in terms of copulae (see Appendix A) and the celebrated Sklar theorem (see Sklar (1959)). Here is a statement of the Sklar theorem,

Theorem 1.1 *Let $(\Omega, \mathcal{F}, \mathbb{P})$ be the underlying probability space, and let $V = (V^1, \ldots, V^n)$ be an $\mathbb{R}^n$-valued random variable on this space. Next, let*

$$F(v^1, \ldots, v^n) = \mathbb{P}(V^1 \le v^1, \ldots, V^n \le v^n)$$

be the cumulative probability distribution function of V, and let $F^i(v^i) = \mathbb{P}(V^i \le v^i)$, $i = 1, \ldots, n$ be the corresponding marginal cumulative probability distributions. Then there exists a function C, called the copula, such that

$$F(v^1, \ldots, v^n) = C\left(F^1(v^1), \ldots, F^n(v^n)\right)$$

for each $v = (v^1, \ldots, v^n) \in \mathbb{R}^n$. Moreover, if the marginal distributions are continuous then the copula C is unique.

In general, the Sklar theorem does not extend to the case of (infinite-dimensional) vector-space-valued random variables such as are found in a stochastic process.[2] This means that, in general, there does not exist a copula that could be used to construct a $\mathcal{P}$-copula structure for Y^i, $i = 1,\ldots,n$. Thus, other methods need to be used for this purpose. A possible method, based on the theory developed in this book, would be to construct a $\mathcal{P}$-structure for Y^i, $i = 1,\ldots,n$, which, in cases when structural properties of the process can be used to determine its law, would lead to constructing a $\mathcal{P}$-copula structure for Y^i, $i = 1,\ldots,n$.

In this volume we propose methods of constructing $\mathcal{P}$-structures for various classes of stochastic processes: Feller–Markov processes, Markov chains, conditional Markov chains, special semimartingales, Archimedean survival processes, and generalized Hawkes processes.

The practical importance of stochastic structures cannot be exaggerated. In fact, stochastic structures provide great flexibility for modeling the dependence between dynamic random phenomena with the preservation of important structural features and marginal features. Moreover, these structures allow for separation of the estimation of the marginal structural properties of a multivariate dynamical system from the estimation of the dependence structure between the margins. This feature is key for the efficient implementation of stochastic structures.

It needs to be remarked that in the case $n = 2$ the concept of a stochastic structure is related to the concept of the coupling between two probability measures. In fact, if $\mathcal{X}^1 = \mathcal{X}^2$ then the law of a bivariate stochastic structure $X = (X^1, X^2)$, considered as a probability measure on the canonical space supported by $\mathcal{X}^1 \times \mathcal{X}^2$, is a coupling between the laws of X^1 and X^2.

The theory of structured dependence is closely related to the theory of consistency for multivariate stochastic processes. The theory of consistency studies the following question: which processes $X = (X^1,\ldots,X^n) \in \mathcal{P}^n$ are such that all X^i, $i = 1,\ldots,n$, feature property $\mathcal{P}$? We devote much attention in this book to consistency.

In the case of Markov processes, the study of Markovian consistency is related to the study of so-called Markov functions. The latter considers the following issue: let M be a Markov process taking values in $\mathcal{M}$, and let $\phi : \mathcal{M} \to \mathcal{Y}$. Then the question is, for which processes M and for which functions ϕ is the process Y, given as $Y_t = \phi(M_t)$, a Markov process? In the case of Markovian consistency, one asks this question for $M = X = (X^1,\ldots,X^n)$ and for functions ϕ_i given as the coordinate projections, that is, $\phi_i(x^1,\ldots,x^n) = x^i$. Markov functions were studied in Chapter X, Section 6, in Dynkin (1965), Rogers and Pitman (1981), and Kurtz (1998), among others. In this book we aim to contribute significantly to the theory and practice of Markov functions by providing an in-depth study of Markovian invariance under coordinate projection.

We tackle only the consistency of a process $X = (X^1,\ldots,X^n)$ with respect to its individual coordinates X^i, $i = 1,\ldots,n$. Likewise, we study only stochastic

[2] We refer to Scarsini (1989) and to Bielecki et al. (2008b) for relevant discussions.

structures subject to univariate marginal constraints. In more generality, one would study these two aspects of structural dependence with regard to groups of coordinates: $X^{\mathcal{I}_m} = (X^i,\ i \in \mathcal{I}_m)$, $m = 1,\ldots,M$, where $\{\mathcal{I}_m : m = 1,\ldots,M\}$ is a partition of the set $\{1,\ldots,n\}$.

The book is organized as follows. In Part One we provide a study of consistency for multivariate Markov families, multivariate Markov chains, multivariate conditional Markov chains, and multivariate special semimartingales. This study underlies the study of stochastic structures conducted in Part Two. Part Three is devoted to the investigation of issues related to consistency and structured dependence in the case of Archimedean survival processes and generalized multivariate Hawkes processes. In particular, we introduce and study generalized multivariate Hawkes processes, as well as the related concept of Hawkes structure. This concept turns out to be important and useful in numerous applications. Part Four illustrates the theory of structured dependence with numerous examples of the practical implementation of stochastic structures. Finally, we provide some relevant technical background in the Appendices.

PART ONE

CONSISTENCIES

2

Strong Markov Consistency of Multivariate Markov Families and Processes

In this chapter we introduce and study strong Markov consistency for multivariate Markov families and processes. We refer to Appendix B for the prerequisite information regarding Markov processes and families. In particular, we refer to Standing assumptions (B1) and (B2) formulated there as well as to Remark 8.9. So, we consider only conservative Markov processes admitting transition functions and conservative Markov families.

Markov consistency is one of the key concepts studied in this volume. As is well known, components of a multivariate Markov family or process may not be Markovian in any filtration. The strong Markov consistency of a multivariate Markov family or process, if satisfied, provides for invariance of the Markov property under coordinate projections, a property which is important in various practical applications.

We will examine strong Markov consistency in various contexts. In this chapter, we work in a rather general set-up. First, in Section 2.1, we study the so-called strong Markov consistency for multivariate Markov families and multivariate Markov processes taking values in an arbitrary metric space. This study is geared towards formulating a general framework within which strong Markov consistency can be conveniently analyzed. Then, in Section 2.2 we specify our study of the strong Markov consistency to the case of multivariate Feller–Markov families taking values in $\mathbb{R}^n$. The analysis is first carried out in the time-inhomogeneous case and then in the time-homogeneous case, where a more comprehensive study can be made.

2.1 Definition of Strong Markov Consistency and an Introduction to its Properties

This section introduces the first key concept that gave rise to this volume, specifically that of strong Markov consistency.

Historically, in Bielecki et al. (2008b), strong Markov consistency was introduced and studied for multivariate Markov processes. Here, we begin the study of strong Markov consistency for multivariate Markov families and then, as a natural follow-up, we discuss strong Markov consistency in the case of multivariate Markov processes.

2.1.1 Strong Markov Consistency of Multivariate Markov Families

For convenience, we work here with the canonical space $(\Omega, \mathcal{F})$ introduced in Appendix B. It needs to be stressed, though, that the concept of Markov consistency put forth in this section and later in the book does not require the canonical set-up.

A (canonical) multivariate Markov family, $\mathcal{MMF}$ for short, is defined in analogy to a (canonical) multivariate Markov process (see Definition B.15) as a Markov family corresponding to the canonical space $(\Omega, \mathcal{F})$ over $(\mathcal{X}, \boldsymbol{\mathcal{X}})$ with $\mathcal{X} = \times_{i=1}^{n} \mathcal{X}_i$ and $\boldsymbol{\mathcal{X}} = \bigotimes_{i=1}^{n} \boldsymbol{\mathcal{X}}_i$, where $(\mathcal{X}_i, \boldsymbol{\mathcal{X}}_i)$, $i = 1, \ldots, n$, are metric spaces endowed with Borel σ-algebras. In accordance with Remark B.19 we will consider only conservative multivariate Markov families.

Strong Markov consistency for a multivariate Markov family is defined as follows

Definition 2.1 A multivariate Markov family

$$\mathcal{MMF} = \{(\Omega, \mathcal{F}, \mathbb{F}_s, (X_t)_{t \geq s}, \mathbb{P}_{s,x}, P) \colon s \geq 0, x \in \mathcal{X}\}$$

is *strongly Markov consistent* with respect to the coordinate X^j (of X) if, for any $0 \leq s \leq t \leq u < \infty$, $x \in \mathcal{X}$, and $B_j \in \boldsymbol{\mathcal{X}}_j$,

$$\mathbb{P}_{s,x}(X_u^j \in B_j \mid \mathcal{F}_{s,t}) \quad \text{is} \quad \sigma(X_t^j)\text{-measurable} \tag{2.1}$$

or, equivalently, if

$$\mathbb{P}_{s,x}(X_u^j \in B_j \mid \mathcal{F}_{s,t}) = \mathbb{P}_{s,x}(X_u^j \in B_j \mid X_t^j), \tag{2.2}$$

which is also equivalent to

$$\mathbb{P}_{s,x}(X_u^j \in B_j \mid X_t) = \mathbb{P}_{s,x}(X_u^j \in B_j \mid X_t^j).$$

If an $\mathcal{MMF}$ is strongly Markov consistent with respect to the coordinate X^j for all $j \in \{1, \ldots, n\}$ then we say that $\mathcal{MMF}$ is strongly Markov consistent.

Remark 2.2 A weaker definition of Markov consistency reads as follows. A multivariate Markov family

$$\mathcal{MMF} = \{(\Omega, \mathcal{F}, \mathbb{F}_s, (X_t)_{t \geq s}, \mathbb{P}_{s,x}, P) : s \geq 0, x \in \mathcal{X}\}$$

is *weakly Markov consistent* with respect to the coordinate X^j (of X) if, for any $0 \leq s \leq t \leq u < \infty$, $x \in \mathcal{X}$, and $B_j \in \boldsymbol{\mathcal{X}}_j$,

$$\mathbb{P}_{s,x}(X_u^j \in B_j \mid \mathcal{F}_{s,t}^j) \quad \text{is} \quad \sigma(X_t^j)\text{-measurable} \tag{2.3}$$

or, equivalently, if

$$\mathbb{P}_{s,x}(X_u^j \in B_j \mid \mathcal{F}_{s,t}^j) = \mathbb{P}_{s,x}(X_u^j \in B_j \mid X_t^j). \tag{2.4}$$

If an $\mathcal{MMF}$ is weakly Markov consistent with respect to the coordinate X^j for all $j \in \{1, \ldots, n\}$ then we say that $\mathcal{MMF}$ is weakly Markov consistent.

It is clear that the strong Markov consistency of an $\mathcal{MMF}$ with respect to the coordinate X^j implies the weak Markov consistency of the $\mathcal{MMF}$ with respect to the coordinate X^j; we refer to Theorem 2.10 for more insight in this regard. The converse implication is not true in general.

We will not study weak Markov consistency in this chapter. A thorough study of weak Markov consistency will be made in the context of finite Markov chains in Chapter 3.

Remark 2.3 (i) In the case of a time-homogeneous multivariate Markov family

$$\mathcal{MMFH} = \{(\Omega, \mathcal{F}, \mathbb{F}, (X_t)_{t\geq 0}, \mathbb{P}_x, P) : x \in \mathcal{X}\},$$

the definition of the strong consistency of the family with respect to the coordinate X^j takes the following form: for any $0 \leq t \leq u < \infty$, $x \in \mathcal{X}$, and $B_j \in \boldsymbol{\mathcal{X}}_j$,

$$\mathbb{P}_x(X_u^j \in B_j \mid \mathcal{F}_t) \quad \text{is} \quad \sigma(X_t^j)\text{-measurable} \tag{2.5}$$

or, equivalently,

$$\mathbb{P}_x(X_u^j \in B_j \mid \mathcal{F}_t) = \mathbb{P}_x(X_u^j \in B_j \mid X_t) = \mathbb{P}_x(X_u^j \in B_j \mid X_t^j). \tag{2.6}$$

(ii) Note that if a multivariate Markov family $\mathcal{MMF}$ is strongly Markov consistent with respect to the coordinate X^j then $\mathbb{P}_{t,X_t}(X_u^j \in B_j)$ is $\sigma(X_t^j)$-measurable. Indeed, it is enough to combine (2.1) with

$$\mathbb{P}_{t,X_t}(X_u^j \in B_j) = \mathbb{P}_{s,x}(X_u^j \in B_j \mid \mathcal{F}_{s,t}),$$

which follows from condition 1 of Definition B.18.
(iii) In this book we are studying only the consistency of a multivariate family or process with respect to its individual coordinates. An analogous study of consistency can be carried out with respect to groups of coordinates.

Lemma 2.4 *It holds that* (2.1) *and* (2.2) *are, respectively, equivalent to the following:*

$$\mathbb{P}_{s,x}(A \mid \mathcal{F}_{s,t}) \quad \text{is} \quad \sigma(X_t^j)\text{-measurable} \tag{2.7}$$

and

$$\mathbb{P}_{s,x}(A \mid \mathcal{F}_{s,t}) = \mathbb{P}_{s,x}(A \mid X_t^j) \tag{2.8}$$

for any $A \in \mathcal{F}_{t,\infty}^j$.

Proof In order to see the equivalence between (2.1) and (2.7) one first notes that upon taking $A = \left\{X_u^j \in B_j\right\}$, (2.7) implies (2.1). To show the opposite implication, we start by observing that (2.1) can be equivalently written as follows: for every bounded measurable function $f : \mathcal{X}_j \to \mathbb{R}$ it holds that

$$\mathbb{E}_{s,x}(f(X_u^j) \mid \mathcal{F}_{s,t}) \quad \text{is} \quad \sigma(X_t^j)\text{-measurable}.$$

Using this, and applying conditioning and induction (see, for example, Theorem 7.2 in Kallenberg (2002)), one obtains that for every $m \geq 1$ and for every bounded measurable function $g : \underbrace{\mathcal{X}_j \times \cdots \times \mathcal{X}_j}_{m \text{ times}} \to \mathbb{R}$ it holds that

$$\mathbb{E}_{s,x}(g(X^j_{u_1}, \ldots, X^j_{u_m}) \mid \mathcal{F}_{s,t}) \quad \text{is} \quad \sigma(X^j_t)\text{-measurable}$$

for $t \leq u_1 < \cdots < u_m$; this is equivalent to (2.7).

Similar reasoning gives equivalence between (2.2) and (2.8). □

Proposition 2.5 *Let a Markov family $\mathcal{MMF}$ be strongly Markov consistent with respect to the coordinate X^j. Then, for any $s \geq 0$ and $A \in \mathcal{F}^j_{s,\infty}$, the function $\mathbb{P}_{s,\cdot}(A) \colon \mathcal{X} \to [0,1]$ does not depend on x_k for $k \neq j$.*

Proof The result follows from Lemma 2.4 upon taking $t = s$ in (2.7). □

Let us fix $j \in \{1, \ldots, n\}$. For $x = (x_1, \ldots, x_n) \in \mathcal{X}$, we set

$$x^{(j)} = (x_1, \ldots, x_{j-1}, x_{j+1}, \ldots, x_n),$$

and for $s \geq 0$ we define a measure $\mathbb{P}^{j,x^{(j)}}_{s,x_j}$ on $(\Omega, \mathcal{F}^j_{s,\infty})$ by

$$\mathbb{P}^{j,x^{(j)}}_{s,x_j}(A) := \mathbb{P}_{s,x}(A). \tag{2.9}$$

Proposition 2.5 implies

Corollary 2.6 *For a Markov family $\mathcal{MMF}$ which is strongly Markov consistent with respect to the coordinate X^j, the measure $\mathbb{P}^{j,x^{(j)}}_{s,x_j}$ does not depend on $x^{(j)}$.*

The above corollary leads to the following key definition.

Definition 2.7 Fix $j \in \{1, \ldots, n\}$. Let $\mathcal{MMF}$ be strongly Markov consistent with respect to the coordinate X^j. Then, for any $0 \leq s$ and $x_j \in \mathcal{X}^j$, we define a measure $\mathbb{P}^j_{s,x_j}$ on $(\Omega, \mathcal{F}^j_{s,\infty})$ as

$$\mathbb{P}^j_{s,x_j}(A) := \mathbb{P}^{j,x^{(j)}}_{s,x_j}(A). \tag{2.10}$$

Using this notation, we now conclude that if $\mathcal{MMF}$ is strongly Markov consistent with respect to the coordinate X^j then

$$\mathbb{E}^j_{s,x_j} f(X^j_t) := \mathbb{E}_{\mathbb{P}^j_{s,x_j}} f(X^j_t) = \mathbb{E}_{s,x} f(X^j_t), \tag{2.11}$$

for any positive Borel function $f : \mathcal{X}^j \to \mathbb{R}$ and for any $s \leq t$.

Remark 2.8 In the time-homogeneous case we will use the notation $\mathbb{P}^j_{x_j}$ for the probability measure on $(\Omega, \mathcal{F}^j_\infty)$ defined by

$$\mathbb{P}^j_{x_j}(A) := \mathbb{P}^{j,x^{(j)}}_{x_j}(A) = \mathbb{P}_y(A)$$

for every y such that $y_j = x_j$. Then, the counterpart of the equality (2.11) will read[1]

$$\mathbb{E}^j_{x_j} f(X^j_t) = \mathbb{E}_x f(X^j_t). \tag{2.12}$$

[1] By $\mathbb{E}_x$ we denote the expectation operator with respect to the probability measure $\mathbb{P}_x$.

We now have the following key result.

Theorem 2.9 *Let $\mathcal{MMF} = \{(\Omega, \mathcal{F}, \mathbb{F}_s, (X_t)_{t\geq s}, \mathbb{P}_{s,x}, P), s \geq 0, x \in \mathcal{X}\}$ be strongly Markov consistent with respect to the coordinate X^j. Fix $x_k \in \mathcal{X}_k$ for all $k \neq j$, and define the function P^j by*

$$P^j(s, x_j, t, B_j) := P(s, x, t, \mathcal{X}_1 \times \cdots \times \mathcal{X}_{j-1} \times B_j \times \mathcal{X}_{j+1} \times \cdots \times \mathcal{X}_n), \tag{2.13}$$

where $x = (x_1, \ldots, x_{j-1}, x_j, x_{j+1}, \ldots, x_n)$.

Then the collection of objects

$$\left\{(\Omega, \mathcal{F}, \mathbb{F}_s, (X_t^j)_{t\geq s}, \mathbb{P}_{s,x}, P^j) \colon s \geq 0, x_j \in \mathcal{X}_j\right\}$$

satisfies the following properties.

(i) *$X_t^j(\omega) = \omega_j(t)$ for $t \geq 0$ and $\omega \in \Omega$.*
(ii) *For any $s \geq 0$, $\left\{\mathbb{P}_{s,x}, x_j \in \mathcal{X}_j\right\}$ is a family of probability measures on $(\Omega, \mathcal{F}_{s,\infty})$.*
(iii) *P^j is a transition function.*
(iv) *$\mathbb{P}_{s,x}(X_s^j = x_j) = 1$ for any $s \geq 0$ and for any $x_j \in \mathcal{X}_j$.*
(v) *$\mathbb{P}_{s,x}(X_u^j \in B_j \mid \mathcal{F}_{s,t}) = \mathbb{P}_{t,X_t}(X_u^j \in B_j) = \mathbb{P}^j_{t,X_t^j}(X_u^j \in B_j)$, $\mathbb{P}_{s,x}$-a.s., for any $0 \leq s \leq t \leq u < \infty$, $x_j \in \mathcal{X}_j$, and $B_j \in \boldsymbol{\mathcal{X}}_j$.*

Proof Properties (i), (ii), (iv) and the first equality in property (v) follow directly from the fact that $\mathcal{MMF}$ is a Markov family. The second equality in property (v) follows from the fact that $\mathcal{MMF}$ is strongly Markov consistent with respect to the coordinate X^j, from (2.9), and from Corollary 2.6. Property (iii) follows from property (v), (B.8), Corollary 2.6, and Lemma B.20. □

One can prove the next result in a similar fashion, and thus we omit its proof.

Theorem 2.10 *Let $\mathcal{MMF}$ be strongly Markov consistent with respect to the coordinate X^j and fix $x_k \in \mathcal{X}_k$ for all $k \neq j$. Then, the collection of objects*

$$\left\{(\Omega, \mathcal{F}, \mathbb{F}_s^j, (X_t^j)_{t\geq s}, \mathbb{P}^j_{s,x_j}, P^j) : s \geq 0, x_j \in \mathcal{X}_j\right\},$$

where $\mathbb{P}^j_{s,x_j}$ is defined by (2.10), and where P^j is as in (2.13), satisfies the following properties:

(i) *$X_t^j(\omega) = \omega_j(t)$ for $t \geq 0$ and $\omega \in \Omega$.*
(ii) *For any $s \geq 0$, $\left\{\mathbb{P}^j_{s,x_j}, x_j \in \mathcal{X}_j\right\}$ is a family of probability measures on $(\Omega, \mathcal{F}^j_{s,\infty})$.*
(iii) *P^j is a transition function.*
(iv) *$\mathbb{P}^j_{s,x_j}(X_s^j = x_j) = 1$ for any $s \geq 0$, $x_j \in \mathcal{X}_j$.*
(v) *$\mathbb{P}^j_{s,x_j}(X_u^j \in B_j \mid \mathcal{F}^j_{s,t}) = \mathbb{P}_{t,X_t}(X_u^j \in B_j) = \mathbb{P}^j_{t,X_t^j}(X_u^j \in B_j)$, $\mathbb{P}_{s,x}$-a.s., for any $0 \leq s \leq t \leq u < \infty$, $x_j \in \mathcal{X}_j$, and $B_j \in \boldsymbol{\mathcal{X}}_j$.*

We note that the main difference between the results given in Theorem 2.9 and the results given in Theorem 2.10 is that in Theorem 2.9 the "large" filtration $\mathbb{F}_s$ is used, whereas in the Theorem 2.10 the "small" filtration $\mathbb{F}_s^j$ is used.

Remark 2.11 It is important to realize that the collections of objects

$$\left\{(\Omega,\mathcal{F},\mathbb{F}_s,(X_t^j)_{t\geq s},\mathbb{P}_{s,x},P^j) : s\geq 0, x_j\in\mathcal{X}_j\right\}$$

and

$$\left\{(\Omega,\mathcal{F},\mathbb{F}_s^j,(X_t^j)_{t\geq s},\mathbb{P}_{s,x_j}^j,P^j) : s\geq 0, x_j\in\mathcal{X}_j\right\}$$

are not canonical Markov families in the sense of Definition B.18, unless $n=1$. This is so because the process $(X_t^j)_{t\geq s}$ is not a canonical process on $(\Omega,\mathcal{F})$. Nevertheless, these collections feature all the relevant properties of Markov families, and therefore we will refer to them as *non-canonical families.*
In view of Remark B.17, one may produce canonical versions of the above two non-canonical families, but this is not really needed.

The above discussion justifies the following definition.

Definition 2.12 Fix $j\in\{1,\ldots,n\}$ and let $\mathcal{MMF}$ be strongly Markov consistent with respect to the coordinate X^j.

1. Fix $x_k\in\mathcal{X}_k$ for all $k\neq j$. We call the non-canonical Markov family

$$\mathcal{SMMF}^j=\left\{(\Omega,\mathcal{F},\mathbb{F}_s,(X_t^j)_{t\geq s},\mathbb{P}_{s,x},P^j) : s\geq 0, x_j\in\mathcal{X}_j\right\}$$

   the *strong jth coordinate* of $\mathcal{MMF}$ delineated by $(x_k,\ k\neq j)$.
2. We call the non-canonical Markov family

$$\mathcal{WMMF}^j=\left\{(\Omega,\mathcal{F},\mathbb{F}_s^j,(X_t^j)_{t\geq s},\mathbb{P}_{s,x_j}^j,P^j) : s\geq 0, x_j\in\mathcal{X}_j\right\}$$

   the *weak jth coordinate* of $\mathcal{MMF}$.

We may associate Markov processes with the non-canonical family $\mathcal{SMMF}^j$ (resp. $\mathcal{WMMF}^j$) in a way that is completely analogous to what is done in Definition B.24. Accordingly, we will call such Markov processes *non-canonical Markov processes associated with $\mathcal{SMMF}^j$ (resp. $\mathcal{WMMF}^j$).*

We end this section with a useful result which is a direct consequence of the above considerations, and thus its proof is skipped. Before we state the result we recall that the definition of the measure $\mathbb{Q}_s^{\gamma_s}$ is given in (B.11) and, for every probability measure γ_s^j on $(\mathcal{X}_j,\boldsymbol{\mathcal{X}}_j)$, we define a measure $\mathbb{P}_s^{j,\gamma_s^j}$ as follows:

$$\mathbb{P}_s^{j,\gamma_s^j}(A_j)=\int_{\mathcal{X}_j}\mathbb{P}_{s,x_j}^j(A_j)\gamma_s^j(dx_j),\quad A_j\in\mathcal{F}_{s,\infty}^j.$$

Having defined $\mathcal{SMMF}^j$ (resp. $\mathcal{WMMF}^j$) and using a probability measure γ_s on $(\mathcal{X},\boldsymbol{\mathcal{X}})$ (resp. γ_s^j on $(\mathcal{X}_j,\boldsymbol{\mathcal{X}}_j)$) we can consider the non-canonical Markov process associated with $\mathcal{SMMF}^j$ (resp. $\mathcal{WMMF}^j$).

Theorem 2.13 *Assume that $\mathcal{MMF}$ is strongly Markov consistent with respect to the coordinate X^j. Then, for every $s \in [0,\infty)$ and for every probability measure γ_s on $(\mathcal{X},\boldsymbol{\mathcal{X}})$ the collection of objects $(\Omega,\mathcal{F},\mathbb{F}_s,(X_t^j)_{t\geq s},\mathbb{Q}_s^{\gamma_s},P^j)$ is a non-canonical Markov process associated with $\mathcal{SMMF}^j$. Likewise, for every $s \in [0,\infty)$ and for every probability measure γ_s^j on $(\mathcal{X}_j,\boldsymbol{\mathcal{X}}_j)$, the collection of objects $(\Omega,\mathcal{F},\mathbb{F}_s^j,(X_t^j)_{t\geq s}, \mathbb{P}_s^{j,\gamma_s^j},P^j)$ is a non-canonical Markov processes associated with $\mathcal{WMMF}^j$.*

Remark 2.14 The specification of Definition 2.12 and of Theorem 2.13 to the time-homogeneous case is straightforward. In particular, if $\mathcal{MMFH}$ is strongly Markov consistent with respect to the coordinate X^j then we call the non-canonical Markov family

$$\mathcal{SMMFH}^j = \left\{(\Omega,\mathcal{F},\mathbb{F},(X_t^j)_{t\geq 0},\mathbb{P}_x,P^j),x_j \in \mathcal{X}_j\right\}$$

the *strong jth coordinate* of $\mathcal{MMFH}$ delineated by $(x_k,\ k \neq j)$, and we call the non-canonical Markov family

$$\mathcal{WMMFH}^j = \left\{(\Omega,\mathcal{F},\mathbb{F}^j,(X_t^j)_{t\geq 0},\ \mathbb{P}_{x_j}^j,P^j),\ s\geq 0,\ x_j \in \mathcal{X}_j\right\}$$

the *weak jth coordinate* of $\mathcal{MMFH}$. The P^j in the above non-canonical Markov families is the time-homogeneous counterpart of the P^j defined in (2.13).

2.1.2 Strong Markov Consistency of Multivariate Markov Processes

The strong Markov consistency of multivariate Markov processes is related to the strong Markov consistency of multivariate Markov families, of course. In fact, in order to study the strong Markov consistency of multivariate Markov processes it is enough to study the strong Markov consistency of the corresponding $\mathcal{MMF}$s. A brief relevant discussion is provided here.

For simplicity, we limit the discussion here to the case of processes starting at time $s = 0$.

Definition 2.15 We say that the process $\mathcal{MMP} = (\Omega,\mathcal{F},\mathbb{F}_0,(X_t)_{t\geq 0},\mathbb{Q},P)$ is *strongly Markov consistent with respect to the coordinate X^j* if, for any $0 \leq t \leq u$ and for any $B_j \in \boldsymbol{\mathcal{X}}_j$,

$$\mathbb{Q}(X_u^j \in B_j \mid \mathcal{F}_{0,t}) = \mathbb{Q}(X_u^j \in B_j \mid X_t^j), \quad \mathbb{Q}\text{-a.s.} \tag{2.14}$$

or, equivalently, if

$$\mathbb{Q}(X_u^j \in B_j \mid X_t) = \mathbb{Q}(X_u^j \in B_j \mid X_t^j), \quad \mathbb{Q}\text{-a.s.} \tag{2.15}$$

If the process $\mathcal{MMP}$ is strongly Markov consistent with respect to each coordinate X^j, $j = 1,\ldots,n$, then we say that it is strongly Markov consistent.

We now present a result that allows us to investigate the strong Markov consistency of an $\mathcal{MMP}$ by studying the strong Markov consistency of the $\mathcal{MMF}$ associated with it (see Definition B.26 for the notion of the associated family).

Theorem 2.16 *Let $\mathcal{MMP} = (\Omega, \mathcal{F}, \mathbb{F}_0, (X_t)_{t\geq 0}, \mathbb{Q}, P)$ be such that $\mathcal{MMF} = \{(\Omega, \mathcal{F}, \mathbb{F}_s, (X_t)_{t\geq s}, \mathbb{P}_{s,x}, P) : s \geq 0, x \in \mathcal{X}\}$ associated with it is strongly Markov consistent with respect to the coordinate X^j. Then, for $0 \leq t \leq u$ and $B_j \in \boldsymbol{\mathcal{X}}_j$,*

$$\mathbb{Q}(X_u^j \in B_j \mid \mathcal{F}_{0,t}) = \mathbb{P}^j_{t,X_t^j}(X_u^j \in B_j), \quad \mathbb{Q}\text{-a.s.} \tag{2.16}$$

where $\mathbb{P}^j_{t,x_j}$ is defined by (2.10). *In addition,*

$$\mathbb{Q}(X_u^j \in B_j \mid \mathcal{F}_{0,t}) = \mathbb{Q}(X_u^j \in B_j \mid X_t^j), \quad \mathbb{Q}\text{-a.s.} \tag{2.17}$$

Thus, $\mathcal{MMP} = (\Omega, \mathcal{F}, \mathbb{F}_0, (X_t)_{t\geq 0}, \mathbb{Q}, P)$ is strongly Markov consistent with respect to the coordinate X^j.

Proof In view of (2.9) and Corollary 2.6 we have

$$\mathbb{P}_{t,X_t}(X_u^j \in B_j) = \mathbb{P}^j_{t,X_t^j}(X_u^j \in B_j), \quad \mathbb{Q}\text{-a.s.} \tag{2.18}$$

Next, since P is the transition function of the $\mathcal{MMP}$ it holds that

$$\mathbb{Q}(X_u^j \in B_j \mid \mathcal{F}_{0,t}) = P(t, X_t, u, \widehat{B}_j), \quad \mathbb{Q}\text{-a.s.}, \tag{2.19}$$

where $\widehat{B}_j = \mathcal{X}_1 \times \cdots \times \mathcal{X}_{j-1} \times B_j \times \mathcal{X}_{j+1} \times \cdots \times \mathcal{X}_n$.

On the other hand, since P is the transition function of the $\mathcal{MMF}$ it follows from (B.8) that for any $y \in \mathcal{X}$ we have

$$P(t, y, u, \widehat{B}_j) = \mathbb{P}_{t,y}(X_u^j \in B_j),$$

so that, for any $x \in \mathcal{X}$,

$$P(t, X_t, u, \widehat{B}_j) = \mathbb{P}_{t,X_t}(X_u^j \in B_j), \quad \mathbb{P}_{0,x}\text{-a.s.},$$

and thus, since $\mathbb{Q}(\cdot) = \int_{\mathcal{X}} \mathbb{P}_{0,x}(\cdot)\mathbb{Q}(X_0 \in dx)$ (cf. (B.11) and (B.12) with $s = 0$, $\gamma_0(B) = \mathbb{Q}_0(X_0 \in B)$ and $\mathbb{Q}_0 = \mathbb{Q}$)

$$P(t, X_t, u, \widehat{B}_j) = \mathbb{P}_{t,X_t}(X_u^j \in B_j), \quad \mathbb{Q}\text{-a.s.} \tag{2.20}$$

Consequently, putting together (2.18)–(2.20) we complete the proof of (2.16).

Using (2.16) we obtain

$$\begin{aligned}\mathbb{Q}(X_u^j \in B \mid X_t^j) &= \mathbb{E}_{\mathbb{Q}}(\mathbb{Q}(X_u^j \in B \mid \mathcal{F}_{0,t}) \mid X_t^j) = \mathbb{E}_{\mathbb{Q}}\big(\mathbb{P}^j_{t,X_t^j}(X_u^j \in B) \mid X_t^j\big) \\ &= \mathbb{P}^j_{t,X_t^j}(X_u^j \in B) = \mathbb{Q}(X_u^j \in B \mid \mathcal{F}_{0,t}), \quad \mathbb{Q}\text{-a.s.},\end{aligned} \tag{2.21}$$

where $\mathbb{E}_{\mathbb{Q}}$ denotes expectation with respect to the measure $\mathbb{Q}$. This completes the proof of the theorem. □

We remark that the results presented in this section can be naturally adapted to any $\mathcal{MMP}_s$ for any $s > 0$.

2.2 Sufficient Conditions for Strong Markov Consistency

In all the applications of interest to us the multivariate Markov families and multivariate Markov processes take values either in $\mathbb{R}^n$, with $n > 1$, the case considered in this section, or in some finite set, the case considered in Chapter 3.

Accordingly, in this section we will consider the case $\mathcal{X}_j = \mathbb{R}$, $j = 1, \ldots, n$, and $\mathcal{X} = \times_{j=1}^{n} \mathcal{X}_j = \mathbb{R}^n$. First, in Section 2.2.1 we will study the sufficient conditions for strong Markov consistency for $\mathbb{R}^n$-valued Feller–Markov evolution families, which form an important class of Markov families. Then, in Section 2.2.2, we will study the sufficient conditions for strong Markov consistency for time-homogeneous Feller–Markov families.

We consider here only multivariate Markov families, since by Theorem 2.16 the investigation of strong Markov consistency for multivariate Markov processes boils down to investigation of the associated multivariate Markov families.

2.2.1 Sufficient Conditions for Strong Markov Consistency of $\mathbb{R}^n$-Feller–Markov Evolution Families

We refer to Appendix E for the background information about operator semigroups and their generators that is needed here.

In this section we fix $n > 1$ and consider multivariate $\mathbb{R}^n$-Feller–Markov evolution families. We will study their strong Markov consistency in terms of the symbols for the related infinitesimal generators. Accordingly, we take $(\Omega, \mathcal{F})$ to be the canonical space over $(\mathbb{R}^n, \mathcal{B}(\mathbb{R}^n))$.

Recall that we are considering only conservative Markov families.

Definition 2.17 Let $\mathcal{MMF} = \{(\Omega, \mathcal{F}, \mathbb{F}_s, (X_t)_{t \geq s}, \mathbb{P}_{s,x}, P) : s \geq 0, x \in \mathbb{R}^n\}$ be a Markov family. For all $0 \leq s \leq t$ define an operator $T_{s,t}$ on $C_0(\mathbb{R}^n)$ by

$$T_{s,t} f(x) = \mathbb{E}_{x,s} f(X_t) = \int_{\mathbb{R}^n} f(y) P(s,x,t,dy), \quad x \in \mathbb{R}^n. \tag{2.22}$$

If $T_{s,t}$, $0 \leq s \leq t$, defined in (2.22), is a $C_0(\mathbb{R}^n)$-Feller evolution system then $\mathcal{MMF}$ is called a *$\mathbb{R}^n$-Feller–Markov evolution family*. If, in addition, $C_c^\infty(\mathbb{R}^n) \subseteq \mathcal{D}(\mathcal{A})$, where $\mathcal{A} = (A_u,\ u \geq 0)$ is the generator system of $T_{s,t}$, $0 \leq s \leq t$, then the family $\mathcal{MMF}$ is called a *rich* $\mathbb{R}^n$-Feller–Markov evolution family.

We stated the above definition of *richness* in the context of a multivariate Markov evolution family $\mathcal{MMF}$. It needs to be said, though, that this definition applies to Markov evolution families that do not necessarily qualify as multivariate families. For example, an analogous definition applies to an $\mathbb{R}$-Feller–Markov evolution family, which we will encounter in what follows.

Remark 2.18 The generator system of $T_{s,t}$, $0 \leq s \leq t$, will be frequently referred to as the generator system of $\mathcal{MMF}$.

In view of Theorem I.6.3 in Gikhman and Skorokhod (2004) there exists a standard version, say $\widehat{\mathcal{MMF}} = \left\{(\Omega, \mathcal{F}, \widehat{\mathbb{F}}_s, (X_t)_{t\geq s}, \widehat{\mathbb{P}}_{s,x}, P) : s \geq 0, x \in \mathbb{R}^n\right\}$, of any $\mathbb{R}^n$-Feller–Markov evolution family $\mathcal{MMF}$, where the filtrations $\widehat{\mathbb{F}}_s$ are composed of appropriate completions (see Remark B.22) and right-continuous modifications of the canonical σ-fields $\mathcal{F}_{s,t}$, $s \leq t$. In particular, the trajectories $(X_t)_{t\geq s}$ are càdlàg functions $\widehat{\mathbb{P}}_{s,x}$-a.s. We note that the generator system $\widehat{A}_u$, $u \geq 0$, corresponding to the standard version of any $\mathbb{R}^n$-Feller–Markov evolution family is the same as the generator system A_u, $u \geq 0$, corresponding to the original family. Thus, the standard version of a rich $\mathcal{MMF}$ is also rich.

We will work with the standard version $\widehat{\mathcal{MMF}}$ of the rich $\mathcal{MMF}$ in the rest of this section. However, for simplicity, we will maintain the original notation

$$\mathcal{MMF} = \{(\Omega, \mathcal{F}, \mathbb{F}_s, (X_t)_{t\geq s}, \mathbb{P}_{s,x}, P), s \geq 0, x \in \mathbb{R}^n\}$$

instead of

$$\widehat{\mathcal{MMF}} = \left\{(\Omega, \mathcal{F}, \widehat{\mathbb{F}}_s, (X_t)_{t\geq s}, \widehat{\mathbb{P}}_{s,x}, P), s \geq 0, x \in \mathbb{R}^n\right\}.$$

Accordingly, we will continue to use the notation $T_{s,t}$ and A_s.

It was shown in Theorem 4.5 of Rüschendorf et al. (2016) that the restriction to $C_c^\infty(\mathbb{R}^n)$ of the generator system $\mathcal{A} = \{A_u,\ u \geq 0\}$ of a rich $\mathbb{R}^n$-Feller–Markov evolution family is given in terms of a system of time-dependent pseudo-differential operators (PDOs) with symbol q.[2] Specifically,

$$A_u f(x) = -(2\pi)^{-n/2} \int_{\mathbb{R}^n} e^{i\langle x,\xi\rangle} q(u,x,\xi) \hat{f}(\xi) d\xi, \quad u \geq 0,\ x \in \mathbb{R}^n \tag{2.23}$$

for every $f \in C_c^\infty(\mathbb{R}^n)$, where $i = \sqrt{-1}$ is the imaginary unit. The function $q(\cdot,\cdot,\cdot) : \mathbb{R}_+ \times \mathbb{R}^n \times \mathbb{R}^n \to \mathbb{C}$, which is also called the *symbol corresponding to* $\mathcal{A}$, satisfies properties G1–G3 in Appendix G.

Let $\mathcal{MMF}$ be a rich $\mathbb{R}^n$-Feller–Markov evolution family, and denote by q its symbol.

Let us fix $j \in \{1, \ldots, n\}$. We note that from Lemma 3.6.22 in Jacob (2001) it follows that

$$|\, q(u,x,\mathbf{e}_j\xi) \,| \leq c_{u,x}(1 + |\xi_j|^2), \quad u \geq 0,\ x, \xi \in \mathbb{R}^n,$$

for some positive constant $c_{u,x}$ and for

$$\mathbf{e}_j = (0, \ldots, 0, \underbrace{1}_{j\text{th position}}, 0, \ldots, 0).$$

Thus, for any $v \in C_c^\infty(\mathbb{R})$ and $u \geq 0$, we see that

$$\overline{A}_u^j v_j(x) = -(2\pi)^{-1/2} \int_{\mathbb{R}} e^{ix_j\xi_j} q(u,x,\mathbf{e}_j\xi_j) \hat{v}(\xi_j) d\xi_j, \quad x \in \mathbb{R}^n, \tag{2.24}$$

[2] We refer to Appendix G for basic concepts regarding PDOs and their symbols.

is a well-defined finite function, where v_j is the jth extension of v, i.e.

$$v_j(x_1,\ldots,x_j,\ldots,x_n) := v(x_j). \tag{2.25}$$

We will use the following conditions:

A1(j) For every $v \in C_c^\infty(\mathbb{R})$ the process M^{v_j} defined as

$$M_t^{v_j} = v_j(X_t) - \int_s^t \overline{A}_u^j v_j(X_u)du, \quad t \geq s, \tag{2.26}$$

is an $(\mathbb{F}_s, \mathbb{P}_{s,x})$-martingale for any $(s,x) \in [0,\infty) \times \mathbb{R}^n$.

A2(j) As a function of y, the symbol $q(t,y,\mathbf{e}_j z)$ depends on y_j only. That is, $q(t,y,\mathbf{e}_j z) = q(t,x,\mathbf{e}_j z)$ for all $x,y \in \mathbb{R}^n$ such that $x_j = y_j$.

A3(j) Given **A2**(j), define

$$q^j(t,y_j,z) := q(t,y,\mathbf{e}_j z), \quad t \geq 0,\ y \in \mathbb{R}^n,\ z \in \mathbb{R}, \tag{2.27}$$

where y_j is the jth coordinate of y.
Assume that q^j is a symbol of a PDO system, say $\mathcal{A}^j = (A_u^j)_{u \geq 0}$, i.e.,

$$A_t^j v(y_j) = -\frac{1}{\sqrt{2\pi}} \int_{\mathbb{R}} e^{iy_j\xi_j} \widehat{v}(\xi_j) q^j(t,y_j,\xi_j) d\xi_j, \quad y_j \in \mathbb{R}, \tag{2.28}$$

for $v \in C_c^\infty(\mathbb{R})$ and assume that the D-martingale problem for $\mathcal{A}^j$ is well posed.

The following theorem is the main result in this section.

Theorem 2.19 *Conditions **A1**(j)–**A3**(j) are sufficient for the family $\mathcal{MMF}$ to be strongly Markov consistent with respect to the coordinate X^j.*

Proof We let $j = 1$ without loss of generality, and let conditions **A1**(1)–**A3**(1) be satisfied.

Let us take $v \in C_c^\infty(\mathbb{R})$ and recall that $v_1(x) = v(x_1)$. Consequently, by condition **A1**(1) the process M^{v_1} is an $(\mathbb{F}_s, \mathbb{P}_{s,x})$-martingale for any $(s,x) \in [0,\infty) \times \mathbb{R}^n$.

Next, we observe that

$$\overline{A}_t^1 v_1(z) = -\frac{1}{\sqrt{2\pi}} \int_{\mathbb{R}} e^{iz\xi_1} \widehat{v}(\xi_1) q^1(t,z,\xi_1) d\xi_1 = A_t^1 v(z), \quad t \geq 0,\ z \in \mathbb{R}$$

where the first equality follows from (2.24) and (2.27), and the second follows from (2.28). Thus, we see that

$$M_t^{v_1} = v(X_t^1) - \int_s^t A_u^1 v(X_u^1) du, \quad t \geq s. \tag{2.29}$$

Since M^{v_1} is an $(\mathbb{F}_s, \mathbb{P}_{s,x})$-martingale for any $(s,x) \in [0,\infty) \times \mathbb{R}^n$, for each $(s,x) \in [0,\infty) \times \mathbb{R}^n$ the process X^1 solves the martingale problem for $(\mathcal{A}^1, \delta_{x_1})$ starting at s from x_1, with respect to $(\mathbb{F}_s, \mathbb{P}_{s,x})$. Note that X^1 is a càdlàg process and thus, using **A3**(1) and Theorem F.6, we conclude that $(\Omega, \mathcal{F}, \mathbb{F}_s, (X_t^1)_{t \geq s}, \mathbb{P}_{s,x})$ is a Markov process starting at time s from x_1, so that condition (2.2) is satisfied for X^1. This means that $\mathcal{MMF}$ is strongly Markov consistent with respect to the coordinate X^1. □

From Theorems 2.9 and 2.19 it follows for an $\mathcal{MMF}$ that for each j, under the conditions **A1**(j)-**A3**(j), the collection of objects

$$\left\{(\Omega,\mathcal{F},\mathbb{F}_s,(X^j_t)_{t\geq s},\mathbb{P}_{s,x},P^j): s\geq 0, x_j\in\mathbb{R}\right\}$$

is the strong jth coordinate of $\mathcal{MMF}$ delineated by $(x_k,\ k\neq j)$; see Definition 2.12. Analogously, from Theorems 2.10 and 2.19 it follows that for each j the collection of objects

$$\left\{(\Omega,\mathcal{F},\mathbb{F}^j_s,(X^j_t)_{t\geq s},\mathbb{P}^j_{s,x_j},P^j): s\geq 0, x_j\in\mathbb{R}\right\}$$

is the weak jth coordinate of $\mathcal{MMF}$; see Definition 2.12.

Propositions 2.21 and 2.22 below demonstrate that these collections of objects have other important properties. In order to prove these propositions we will need the following technical result,

Lemma 2.20 *Let $\rho\colon E\times\mathbb{R}^k\times\mathbb{R}^k\to\mathbb{C}^k$, where E is a closed subset of $\mathbb{R}^m$, be a measurable function such that:*

(i) $\xi\mapsto\rho(u,x,\xi)$ *is continuous and negative-definite for every* $(u,x)\in E\times\mathbb{R}^k$.
(ii) $(u,x)\mapsto\rho(u,x,\xi)$ *is continuous for every* $\xi\in\mathbb{R}^k$.

Then, for any $v\in C_c^\infty(\mathbb{R}^k)$,

$$(u,x)\mapsto -(2\pi)^{-k/2}\int_{\mathbb{R}^k} e^{i\langle x,\xi\rangle}\rho(u,x,\xi)\widehat{v}(\xi)d\xi \tag{2.30}$$

is continuous on $E\times\mathbb{R}^k$.

Proof In order to prove the continuity of the mapping given by (2.30) it suffices to prove its continuity on the compact subsets of $E\times\mathbb{R}^k$. Towards this end we fix K_1, a compact subset of E, and K_2, a compact subset of $\mathbb{R}^k$. From assumptions (i) and (ii) it follows that

$$|e^{ix\xi}\rho(u,x,\xi)\widehat{v}(\xi)|\leq c_{K_1,K_2}(1+|\xi|^2)|\widehat{v}(\xi)|,\quad \xi\in\mathbb{R}^k,$$

for all $(u,x)\in K_1\times K_2$, where $c_{K_1,K_2}=2\sup_{(u,x)\in K_1\times K_2,\|\xi\|\leq 1}\|q(u,x,\xi)\|$ (see Lemma 3.6.22 in Jacob (2001) and (2.128) in Jacob (2002)).

The fact that $v\in C_c^\infty(\mathbb{R}^k)$ implies that $\widehat{v}\in\mathcal{S}(\mathbb{R}^k)$ and hence that the function $\xi\mapsto(1+|\xi|^2)|\widehat{v}(\xi)|$ is in L^1. This and the continuity of ρ in (u,x) for every ξ allows us to use Lemma 2.3.22 in Jacob (2001) to conclude that

$$(u,x)\mapsto -(2\pi)^{-k/2}\int_{\mathbb{R}^k} e^{i\langle x,\xi\rangle}q(u,x,\xi)\widehat{v}(\xi)d\xi$$

is continuous on $K_1\times K_2$. This completes the proof. □

Let the family $\mathcal{MMF}$ be strongly consistent with respect to coordinate X^j. Corollary 2.6, Definition 2.7 and expression (2.11) combined with Theorem 2.9 allow us to define on $C_0(\mathbb{R})$ an evolution system $\mathcal{T}^j = (T^j_{s,t},\ 0 \le s \le t)$ by

$$T^j_{s,t} f(x_j) := \mathbb{E}_{s,x} f(X^j_t) = \mathbb{E}^j_{s,x_j} f(X^j_t) = \int_{\mathbb{R}} f(y_j) P^j(s,x_j,t,dy_j).$$

Clearly, this is an evolution system associated with $\mathcal{SMMF}^j$, the strong jth coordinate of $\mathcal{MMF}$ delineated by $(x_k,\ k \ne j)$. The corresponding generator system, say $\mathcal{B}^j = \{B^j_u, u \ge 0\}$, is given as

$$B^j_u f = \lim_{h \downarrow 0} \frac{T^j_{u,u+h} f - f}{h}, \tag{2.31}$$

where the limit is understood in the strong sense.

Proposition 2.21 *Fix $j \in \{1,\dots,n\}$, and fix $x_k \in \mathbb{R}$ for all $k \ne j$. Let conditions* ***A1****(j)–**A3**(j) be satisfied, and let q^j defined in* (2.27) *be continuous in (t,y_j). Suppose that the strong jth coordinate of $\mathcal{MMF}$ delineated by $(x_k,\ k \ne j)$, that is,*

$$\mathcal{SMMF}^j = \left\{(\Omega,\mathcal{F},\mathbb{F}_s,(X^j_t)_{t\ge s},\mathbb{P}_{s,x},P^j) : s \ge 0, x_j \in \mathbb{R}\right\},$$

is a rich $\mathbb{R}$-Feller–Markov evolution family with generator system $\mathcal{B}^j = (B^j_u,\ u \ge 0)$, whose symbol, say $\bar{q}^j(u,x_j,z)$, is continuous in (u,x_j). Then, the generator system of $\mathcal{SMMF}^j$ restricted to $C^\infty_c(\mathbb{R})$ is the PDO system $\mathcal{A}^j = (A^j_u,\ u \ge 0)$ appearing in the statement of assumption **A3**(j).

Proof Take $j = 1$ without loss of generality. Fix $s \ge 0$ and x_1. The collection

$$(\Omega,\mathcal{F},\mathbb{F}_s,(X^1_t)_{t\ge s},\mathbb{P}_{s,x},P^1)$$

is a non-canonical Markov process associated with $\mathcal{SMMF}^1$ with generator system $\mathcal{B}^1_s = (B^1_u,\ u \ge s)$.

Since $C^\infty_c(\mathbb{R}) \subset \bigcap_{u\ge s} \mathcal{D}(B^j_u)$, in view of the Dynkin formula (see equation (E.3)), for any $v \in C^\infty_c(\mathbb{R})$ the process $M^{1,v}$ given as

$$M^{1,v}_t = v(X^1_t) - \int_s^t B^1_u v(X^1_u)du, \quad t \ge s,$$

is a martingale with respect to $\mathbb{F}_s$ under $\mathbb{P}_{s,x}$. In view of (2.29), which is a consequence of **A1**(1) (see the proof of Theorem 2.19), we know that for any $v \in C^\infty_c(\mathbb{R})$ the process $N^{1,v}$ given as

$$N^{1,v}_t = v(X^1_t) - \int_s^t A^1_u v(X^1_u)du, \quad t \ge s,$$

is a martingale with respect to $\mathbb{F}_s$ under $\mathbb{P}_{s,x}$. Thus, the process

$$\int_s^t (A^1_u - B^1_u)v(X^1_u)du \quad t \ge s,$$

is a continuous $(\mathbb{F}_s, \mathbb{P}_{s,x})$-martingale of finite variation. Consequently,

$$(A_u^1 - B_u^1)v(X_u^j) = 0$$

for a.a. $u \geq s$ and $\mathbb{P}_{s,x}$ almost surely. Now using the representation of A_u^1 in terms of q^1, the representation of B_u^1 in terms of $\bar{q}^1$, and Lemma 2.20 for $k = 1$ applied to q^1 and $\bar{q}^1$, we conclude that

$$(u,z) \mapsto (A_u^1 - B_u^1)v(z)$$

is a continuous function. Taking into account càdlàg paths of X^j we conclude that

$$u \mapsto (A_u^1 - B_u^1)v(X_u^1)$$

is a càdlàg function on $[s,\infty)$-$\mathbb{P}_{s,x}$-a.s. Thus

$$0 = (A_s^1 - B_s^1)v(X_s^1) = (A_s^1 - B_s^1)v(x)$$

for every $(s,x) \in [0,\infty) \times \mathbb{R}^n$. □

Note that the evolution system $\mathcal{T}^j$ (resp. the generator system $\mathcal{B}^j$) is also the evolution system (resp. the generator system) for $\mathcal{WMMF}^j$, the weak jth coordinate of $\mathcal{MMF}$. Consequently, a result paralleling Proposition 2.21 holds for $\mathcal{WMMF}^j$. We state the result below, but skip its proof, as the proof is almost a carbon copy of the proof of Proposition 2.21.

Proposition 2.22 *Fix $j \in \{1,\dots,n\}$. Let conditions **A1(j)–A3(j)** be satisfied, and let q^j defined in* (2.27) *be continuous in (t, y_j). Suppose that the weak jth coordinate of $\mathcal{MMF}$, that is,*

$$\mathcal{WMMF}^j = \left\{(\Omega, \mathcal{F}, \mathbb{F}_s^j, (X_t^j)_{t\geq s}, \mathbb{P}_{s,x_j}^j, P^j) : s \geq 0, x_j \in \mathbb{R}\right\},$$

is a rich $\mathbb{R}$-Feller–Markov evolution family with generator system $\mathcal{B}^j = (B_u^j,\ u \geq 0)$, whose symbol, say $\bar{q}^j(u,x_j,z)$, is continuous in (u,x_j).

*Then, the generator system of $\mathcal{WMMF}^j$ restricted to $C_c^\infty(\mathbb{R})$ is the PDO system $\mathcal{A}^j = (A_u^j,\ u \geq 0)$ appearing in the statement of condition **A3(j)**.*

2.2.2 Strong Markov Consistency of Feller–Markov Families

We can conduct a more comprehensive study of strong Markov consistency in the case of time-homogeneous Feller–Markov families. In particular, we can prove in this case that, for each $j \in \{1,\dots,d\}$, the strong jth coordinate and the weak jth coordinate of an $\mathcal{MMFH}$ that is strongly Markovian consistent with respect to X^j are $\mathbb{R}$-Feller–Markov families. In the time-inhomogeneous case an analogous property was assumed rather than proved (cf. Propositions 2.21 and 2.22).

Definition 2.23 Let

$$\mathcal{MMFH} = \{(\Omega, \mathcal{F}, \mathbb{F}, (X_t)_{t\geq 0}, \mathbb{P}_x, P) : x \in \mathbb{R}^n\} \tag{2.32}$$

be a time-homogeneous Markov family. For all $t \geq 0$ we define an operator T_t on $C_0(\mathbb{R}^n)$ by

$$T_t u(x) = \mathbb{E}_x u(X_t) = \int_{\mathbb{R}^n} u(y)P(x,t,dy), \quad x \in \mathbb{R}^n. \tag{2.33}$$

If $(T_t,\ t \geq 0)$ is a $C_0(\mathbb{R}^n)$-Feller semigroup then $\mathcal{MMFH}$ is called a $\mathbb{R}^n$-*Feller–Markov family*. If, in addition, $C_c^\infty(\mathbb{R}^n) \subseteq D(A)$, where A is the generator of $(T_t,\ t \geq 0)$, then the family $\mathcal{MMFH}$ is called a *nice* $\mathbb{R}^n$-Feller–Markov family.

Remark 2.24 We stated the above definition of *niceness* in the context of a time-homogeneous multivariate Markov family $\mathcal{MMFH}$. It needs to be said, though, that this definition applies to Markov families that do not necessarily qualify as multivariate families. For example, an analogous definition applies to a time-homogeneous $\mathbb{R}$-Feller–Markov family $\mathcal{MFH}$.

We fix $j \in \{1,\ldots,n\}$ throughout the rest of this section. Recall that in case of a time-homogeneous family the definition of strong Markov consistency of the family $\mathcal{MMFH}$ with respect to the coordinate X^j takes the form (2.5) (or, equivalently, (2.6)).

Recall that if a Markov family $\mathcal{MMFH}$ is strongly Markov consistent with respect to the coordinate X^j then the strong jth coordinate of $\mathcal{MMFH}$ delineated by $(x_k,\ k \neq j)$ and the weak jth coordinate of $\mathcal{MMF}$ are given, respectively, as

$$\mathcal{SMMFH}^j = \left\{(\Omega,\mathcal{F},\mathbb{F},(X_t^j)_{t\geq 0},\mathbb{P}_x,P^j) : x_j \in \mathbb{R}\right\}$$

and

$$\mathcal{WMMFH}^j = \left\{(\Omega,\mathcal{F},\mathbb{F}^j,(X_t^j)_{t\geq 0},\mathbb{P}_{x_j}^j,P^j) : x_j \in \mathbb{R}\right\}$$

(see Remark 2.14).

Definition 2.25 Let $\mathcal{MFH}^j := \left\{(\Omega^j,\mathcal{G},\mathbb{G}^j,(Y_t^j)_{t\geq 0},\mathbb{Q}_y^j,Q^j), y \in \mathbb{R}\right\}$ be a canonical time-homogeneous nice $\mathbb{R}$-Feller–Markov family. A Markov family $\mathcal{MMFH}$ satisfies the *strong Markov consistency property with respect to X^j relative to $\mathcal{MFH}^j$* if $\mathcal{MMFH}$ is strongly Markov consistent with respect to the coordinate X^j and if the transition function P^j coincides with Q^j. We will denote this property as

$$\mathcal{MMFH} \curvearrowright \mathcal{MFH}^j. \tag{2.34}$$

Remark 2.26 The transition function of a time-homogeneous Markov family determines the corresponding generators via (2.33) and the time-homogeneous versions of (E.1) and (E.2), and vice versa. Thus, in fact, Definition 2.25 is about the structural properties of the Markov families that this definition is concerned with.

In this section we will study the strong Markov consistency property with respect to X^j of a Markov family $\mathcal{MMFH}$, as well as the strong Markov consistency property with respect to X^j relative to an $\mathcal{MFH}^j$.

Strong Markov Consistency Property of $\mathcal{MMFH}$ with respect to X^j

Consider the nice $\mathbb{R}^n$-Feller–Markov family given in Definition 2.23. Sometimes, for conciseness of presentation, we will refer to this $\mathcal{MMFH}$ as to X.

According to the results of Courrège (1965-1966) (see also Appendix G), the generator A of T_t, $t \geq 0$, acting on $u \in C_c^\infty(\mathbb{R}^n)$ has a representation

$$Au(x) = -q(x,D)u(x) = -(2\pi)^{-n/2} \int_{\mathbb{R}^n} e^{i\langle x,\xi\rangle} q(x,\xi)\widehat{u}(\xi)d\xi, \tag{2.35}$$

where $q(x,\xi)$ is the symbol of the PDO $-q(x,D)$. Note, in particular, that here the symbol q does not depend on the time variable, in contrast with the symbol considered in the time-inhomogeneous case (see (2.23)).

According to our convention the function $q(x,\xi)$ is called the symbol of A. It is known (see Appendix G, Courrège's theorem) that the symbol $q(x,\xi)$ has the following generic form[3]

$$\begin{aligned} q(x,\xi) &= -i\langle b(x),\xi\rangle + \langle \xi, a(x)\xi\rangle \\ &\quad + \int_{\mathbb{R}^n\setminus\{0\}} \left(1 - e^{i\langle y,\xi\rangle} + \frac{i\langle y,\xi\rangle}{1+|y|^2}\right)\mu(x,dy), \quad x,\xi \in \mathbb{R}^n, \end{aligned} \tag{2.36}$$

where a,b are Borel measurable functions: $b(x) \in \mathbb{R}^n$, $a(x) \in L(\mathbb{R}^n,\mathbb{R}^n)$ is a symmetric positive semi-definite matrix, and $\mu(x,dy)$ is a Lévy kernel.

Moreover, if q is continuous (in all variables) then q maps $C_c^\infty(\mathbb{R}^n)$ into $C(\mathbb{R}^n)$ (Theorem 4.5.7, p.337 in Jacob (2001)). In Theorem 3.10 in Schnurr (2009) a probabilistic interpretation of (b,a,μ) is given, namely, (B,C,ν) are the semimartingale characteristics of X, where

$$B_t := \int_0^t b(X_u)du, \quad C_t := 2\int_0^t a(X_u)du,$$

$$\nu(du,dy) := \mu(X_u,dy)du. \tag{2.37}$$

From now on we will occasionally use the following

Convention We will also refer to q as the symbol of $\mathcal{MMFH}$ and as the symbol of X.

Remark 2.27 We stress that not every symbol of a PDO is a symbol of a generator of a nice $\mathbb{R}^n$-Feller–Markov family. If a symbol q is, however, a symbol of a generator of a nice $\mathbb{R}^n$-Feller–Markov family then we say that it generates this family. At the end of this section we shall state additional conditions on a symbol of a PDO such

[3] Since we are dealing here with Markovian, as opposed to sub-Markovian, processes, the killing rate c is zero, and that is why it is not seen in formula (2.36).

that it generates a nice $\mathbb{R}^n$-Feller–Markov family. Specifically, see Propositions 2.38, 2.39, 2.41, and 2.45–2.47.

It is important to observe that the form of the symbol of A does not have a unique representation. In fact, the function $y \mapsto (1+|y|^2)^{-1}$ in (2.36) can be replaced by another weight function $w : \mathbb{R}^n \mapsto \mathbb{R}$ satisfying appropriate conditions. The new form of the symbol q of A with respect to the weight function w is given by

$$q(x,\xi) = -i\langle b^w(x),\xi\rangle + \langle \xi, a(x)\xi\rangle + \int_{\mathbb{R}^n\setminus\{0\}} \left(1 - e^{i\langle y,\xi\rangle} + i\langle y,\xi\rangle w(y)\right) \mu(x,dy). \tag{2.38}$$

The function b^w can be expressed in terms of b as follows:

$$b^w(x) = b(x) + \int_{\mathbb{R}^n\setminus\{0\}} y \left(w(y) - \frac{1}{1+|y|^2}\right) \mu(x,dy),$$

provided that the last integral is finite. In particular, letting $w \equiv 0$ yields

$$q(x,\xi) = -i\langle b^0(x),\xi\rangle + \langle \xi, a(x)\xi\rangle + \int_{\mathbb{R}^n\setminus\{0\}} \left(1 - e^{i\langle y,\xi\rangle}\right) \mu(x,dy). \tag{2.39}$$

In view of the above discussion we now give

Definition 2.28 We call the right-hand side of (2.36) the type I representation of the symbol q, and we call the right-hand side of (2.39) the type II representation of the symbol q.

However, it needs to be stressed that the weight function $w \equiv 0$ is not always admissible, i.e. not every symbol q admits a type II representation. This representation is well defined under the assumption that, for every $x \in \mathbb{R}^n$, the measure $\mu(x,\cdot)$ satisfies

$$\int_{\mathbb{R}^n\setminus\{0\}} \frac{|y|}{1+|y|^2} \mu(x,dy) < \infty.$$

The type I representation of q is generic. So, we will mainly work with the type I representation. However, whenever possible and convenient we will work with the type II representation.

In the remainder of the section we consider a multivariate Markov family

$$\mathcal{MMFH} = \{(\Omega, \mathcal{F}, \mathbb{F}, (X_t)_{t\geq 0}, \mathbb{P}_x, P) : x \in \mathbb{R}^n\},$$

abbreviated as X, with generator A associated with the symbol q, and we will investigate the strong Markov consistency of X with respect to X^j. We shall need the following conditions:

C1(j) For all $w \in C_c^\infty(\mathbb{R})$ we have that the process N^w given as

$$N_t^w = w(X_t^j) - \int_0^t (\bar{A}^j w)(X_u)du, \quad t \geq 0, \tag{2.40}$$

is an $(\mathbb{F},\mathbb{P}_x)$-martingale for any $x \in \mathbb{R}^n$, where

$$(\bar{A}^j w)(x) := -(2\pi)^{-1/2} \int_{\mathbb{R}} e^{ix_j\xi_j} \widehat{w}(\xi_j) q(x, \mathbf{e}_j\xi_j) d\xi_j, \quad x \in \mathbb{R}^n. \tag{2.41}$$

C2(j) The function $q(x,\mathbf{e}_j\xi_j)$ as a function of x depends only on x_j; specifically,

$$q(x,\mathbf{e}_j\xi_j) = q(y,\mathbf{e}_j\xi_j) \quad \text{for all } x,y \in \mathbb{R}^n \text{ such that } y_j = x_j.$$

Assuming that condition **C2**(j) holds, we define

$$\tilde{q}_j(x_j,\xi_j) := q(x,\mathbf{e}_j\xi_j), \tag{2.42}$$

and we postulate

C3(j) There exists an operator $(A^j,\mathcal{D}(A^j))$ such that $A^j f(z) = -\tilde{q}_j(z,D)f(z)$ for every $f \in C_c^\infty(\mathbb{R}) \subseteq \mathcal{D}(A^j)$ and $(A^j,\mathcal{D}(A^j))$ is the infinitesimal generator of a nice $\mathbb{R}$-Feller–Markov family.

C4(j) For each $f \in \mathcal{D}(A^j)$ there exists a sequence $f^n \in C_c^\infty(\mathbb{R})$ such that $f^n(y) \to f(y)$ for each $y \in \mathbb{R}$, as $n \to \infty$, and

$$\lim_{n\to\infty} \mathbb{E}_x\left(\int_0^\infty e^{-\lambda s} |A^j(f^n(X_s^j) - f(X_s^j))| ds \right) = 0,$$

for each $x \in \mathbb{R}^n$.

Remark 2.29 (i) Conditions **C1**(j) and **C2**(j) correspond to conditions **A1**(j) and **A2**(j), respectively.
(ii) We stress that condition **C2**(j) is the key condition underlying the property of strong Markov consistency discussed here.
(iii) Conditions **C1**(j), **C3**(j), and **C4**(j) will be discussed later in this section.

The following theorem is the key result in this section.

Theorem 2.30 *Let X be an $\mathcal{MMFH}$ with generator A associated with symbol q. Assume that conditions **C1(j)–C4(j)** are satisfied. Then we have the following.*

(i) *The family $\mathcal{MMFH}$ is strongly Markov consistent with respect to X^j.*

(ii) *Fix $x_k \in \mathbb{R}$ for all $k \neq j$. The strong jth coordinate of $\mathcal{MMFH}$ delineated by $(x_k,\ k \neq j)$, that is (see Remark 2.14)*

$$\mathcal{SMMFH}^j = \left\{ (\Omega,\mathcal{F},\mathbb{F},(X_t^j)_{t\ge0},\mathbb{P}_x,P^j) : x_j \in \mathbb{R} \right\},$$

is a (non-canonical) nice $\mathbb{R}$-Feller–Markov family with generator (see (2.31)*) $B^j = A^j$.*

(iii) *The weak jth coordinate of $\mathcal{MMFH}$, that is (see Remark 2.14)*

$$\mathcal{WMMFH}^j = \left\{ (\Omega,\mathcal{F},\mathbb{F}^j,(X_t^j)_{t\ge0},\mathbb{P}_{x_j}^j,P^j) : x_j \in \mathbb{R} \right\},$$

is a (non-canonical) nice $\mathbb{R}$-Feller–Markov family with generator $B^j = A^j$.

Proof We will prove only parts (i) and (ii). The proof of part (iii) is a direct consequence of the proofs of parts (i) and (ii).

We shall first verify that (see (2.41))

$$\bar{A}^j w(x) = A^j w(x_j), \quad x = (x_1, \ldots, x_j, \ldots, x_n) \in \mathbb{R}^n, \ w \in C_c^\infty(\mathbb{R}). \tag{2.43}$$

Indeed, for any $x \in \mathbb{R}^n$, $w \in C_c^\infty(\mathbb{R})$, in view of condition **C3**(j) we have

$$A^j w(x_j) = -(2\pi)^{-1/2} \int_{\mathbb{R}} e^{ix_j\xi_j} \widehat{w}(\xi_j) \widetilde{q}_j(x_j, \xi_j) d\xi_j$$

and, by conditions **C1**(j) and **C2**(j),

$$\begin{aligned}\bar{A}^j w(x) &= -(2\pi)^{-1/2} \int_{\mathbb{R}} e^{ix_j\xi_j} \widehat{w}(\xi_j) q(x, \mathbf{e}_j\xi_j) d\xi_j \\ &= -(2\pi)^{-1/2} \int_{\mathbb{R}} e^{ix_j\xi_j} \widehat{w}(\xi_j) \widetilde{q}_j(x_j, \xi_j) d\xi_j \\ &= A^j w(x_j),\end{aligned}$$

which proves (2.43). Hence, the process M^w given as

$$M_t^w = w(X_t^j) - \int_0^t A^j w(X_u^j) du, \quad t \geq 0, \tag{2.44}$$

is an $(\mathbb{F}, \mathbb{P}_x)$-martingale for any $w \in C_c^\infty(\mathbb{R})$ and any $x \in \mathbb{R}^n$.

Consequently, for any $x \in \mathbb{R}^n$ the process X^j is a solution, under $\mathbb{P}_x$, to the (time-homogeneous) martingale problem for (A^j, δ_{x_j}) relative to the full filtration of process X, that is, with respect to $\mathbb{F}$.

Now we follow the reasoning from Theorem 4.4.1 in Ethier and Kurtz (1986). Since X^j is a solution to the martingale problem for (A^j, δ_x) with respect to $\mathbb{F}^X$, using Lemma 4.3.2 in Ethier and Kurtz (1986) we see that, for every $f \in C_c^\infty(\mathbb{R})$,

$$f(X_t^j) = \mathbb{E}_x\Big(\int_0^\infty e^{-\lambda s} (\lambda f(X_{t+s}^j) - A^j f(X_{t+s}^j)) ds \Big| \mathcal{F}_t^X \Big). \tag{2.45}$$

Let us note that the right-hand side of (2.45) is well defined and finite for every $f \in \mathcal{D}(A^j)$.

Making use of **C4**(j), and passing to the limit, we conclude that (2.45) holds for every $f \in \mathcal{D}(A^j)$.

Let us take an arbitrary $h \in C_0^\infty(\mathbb{R})$. Fix $\lambda > 0$ and let $f = \lambda^{-1}(\lambda - A^j)^{-1} h$. Using condition **C3**(j) we have $f \in \mathcal{D}(A^j)$ since $\mathcal{D}(A^j) = (\lambda - A^j)^{-1}(C_0(\mathbb{R}))$ (see Lemma 1.27 in Böttcher et al. (2013)). So, applying (2.45) to $f = \lambda^{-1}(\lambda - A^j)^{-1} h$ we obtain

$$\begin{aligned}(I - \lambda^{-1} A^j)^{-1} h(X_t^j) &= \mathbb{E}_x\Big(\int_0^\infty e^{-\lambda s} \lambda h(X_{t+s}^j) ds \Big| \mathcal{F}_t^X \Big) \\ &= \mathbb{E}_x\Big(\int_0^\infty e^{-s} h(X_{t+s/\lambda}^j) ds \Big| \mathcal{F}_t^X \Big).\end{aligned}$$

Hence, using the above equalities with t replaced by $t + s_1/\lambda$ we get

$$(I - \lambda^{-1} A^j)^{-1} h(X_{t+s_1/\lambda}^j) = \mathbb{E}_x\Big(\int_0^\infty e^{-s_2} h(X_{t+s_1/\lambda+s_2/\lambda}^j) ds_2 \Big| \mathcal{F}_{t+s_1/\lambda}^X \Big),$$

which leads to

$$
\begin{aligned}
&(I-\lambda^{-1}A^j)^{-2}h(X_t^j)\\
&=\mathbb{E}_x\Big(\int_0^\infty e^{-s_1}\mathbb{E}_x\Big(\int_0^\infty e^{-s_2}h(X^j_{t+s_1/\lambda+s_2/\lambda})ds_2\Big|\mathcal{F}^X_{t+s_1/\lambda}\Big)ds_1\Big|\mathcal{F}^X_t\Big)\\
&=\mathbb{E}_x\Big(\int_0^\infty\int_0^\infty e^{-(s_1+s_2)}h(X^j_{t+(s_1+s_2)/\lambda})ds_2ds_1\Big|\mathcal{F}^X_t\Big)\\
&=\mathbb{E}_x\Big(\int_0^\infty \frac{se^{-s}}{\Gamma(1)}h(X^j_{t+s/\lambda})ds\Big|\mathcal{F}^X_t\Big).
\end{aligned}
$$

We may iterate this procedure since $(\lambda I-A^j)^{-1}h\in\mathcal{D}(A^j)$. This yields

$$
(I-\lambda^{-1}A^j)^{-k}h(X_t^j)=\mathbb{E}_x\Big(\int_0^\infty \frac{s^{k-1}}{\Gamma(k)}e^{-s}h(X^j_{t+s/\lambda})ds\Big|\mathcal{F}^X_t\Big). \tag{2.46}
$$

Now we consider $h\in C_c^\infty(\mathbb{R})$. Note that we have

$$
h(X^j_{t+s/\lambda})=M^h_{t+s/\lambda}-M^h_{t+u}+h(X^j_{t+u})-\int_u^{s/\lambda}A^jh(X^j_{t+v})dv,
$$

where M^h is as in (2.44) but with h in place of w. This, (2.46) and $\int_0^\infty(s^{k-1}/\Gamma(k))e^{-s}ds=1$ yields

$$
\begin{aligned}
&(I-\lambda^{-1}A^j)^{-[nu]}h(X_t^j)\\
&=\mathbb{E}_x\Big(h(X^j_{t+u})\Big|\mathcal{F}^X_t\Big)+\mathbb{E}_x\Big(\int_0^\infty \frac{s^{[nu]-1}}{\Gamma([nu])}e^{-s}\Big(\int_u^{s/n}A^jh(X^j_{t+v})dv\Big)ds\Big|\mathcal{F}^X_t\Big),
\end{aligned}
$$

where $[z]$ denotes the integer part of z. Hence, since

$$
\int_0^\infty \frac{s^{[nu]-1}}{\Gamma([nu])}e^{-s}\Big|\int_u^{s/n}A^jh(X^j_{t+v})dv\Big|ds\le\|A^jh\|_\infty\int_0^\infty\frac{s^{[nu]-1}}{\Gamma([nu])}e^{-s}\Big|\frac{s}{n}-u\Big|ds
$$

and

$$
\int_0^\infty\frac{s^{[nu]-1}}{\Gamma([nu])}e^{-s}\Big|\frac{s}{n}-u\Big|ds=\mathbb{E}\Big|\frac{\sum_{i=1}^{[nu]}\gamma_i}{n}-u\Big|,
$$

where $(\gamma_i)_{i\ge1}$ are independent and exponentially distributed with parameter 1, we see, using the strong law of large numbers, that for every $h\in C_c^\infty(\mathbb{R})$,

$$
\lim_{n\to\infty}(I-n^{-1}A^j)^{-[nu]}h(X_t^j)=\mathbb{E}_x(h(X^j_{t+u})|\mathcal{F}^X_t). \tag{2.47}
$$

Because A^j generates a strongly continuous contraction semigroup, say $(T_t^j;t\ge0)$, on $C_0(\mathbb{R})$ and since, for $f\in C_0(\mathbb{R})$,

$$
T_u^jf=\lim_{n\to\infty}(I-n^{-1}A^j)^{-[nu]}f
$$

uniformly on bounded intervals, see Corollary 1.6.8 in Ethier and Kurtz (1986), we conclude from (2.47) that

$$
T_u^jh(X_t^j)=\mathbb{E}_x(h(X^j_{t+u})|\mathcal{F}^X_t) \tag{2.48}
$$

for each $h \in C_c^\infty(\mathbb{R})$. Since $C_c^\infty(\mathbb{R})$ is dense in $C_0(\mathbb{R})$, formula (2.48) holds for every function in $C_0(\mathbb{R})$. Consequently, we deduce that $(\Omega, \mathcal{F}, \mathbb{F}, (X_t^j)_{t \geq 0}, \mathbb{P}_x)$ is a (non-canonical) Markov process starting at time $t = 0$ from x_j, the jth coordinate of x, and with corresponding semigroup T_t^j, $t \geq 0$. In particular, $\mathbb{P}_x(X_u^j \in B | \mathcal{F}_t) = \mathbb{P}_x(X_u^j \in B | X_t^j)$ is a measurable function of X_t^j for any Borel subset B, which means that condition (2.6) is satisfied. Thus, since x is arbitrary, our $\mathcal{MMF}$ is strongly Markov consistent with respect to X^j. This finishes the proof of the first part of the theorem.

In view of (the time-homogeneous version of) Theorem 2.10 and Remark 2.14, the collection of objects $\mathcal{SMMF}^j = \{(\Omega, \mathcal{F}, \mathbb{F}, (X_t^j)_{t \geq 0}, \mathbb{P}_x, P^j), x_j \in \mathbb{R}\}$ is the strong jth coordinate non-canonical Markov family of $\mathcal{MMF}$. By condition **C3**(j) we have $C_c^\infty(\mathbb{R}) \subset \mathcal{D}(A^j)$, so we can conclude that $\mathcal{SMMF}^j$ is a non-canonical nice $\mathbb{R}$-Feller–Markov family with generator $B^j = A^j$. □

Strong Markov Consistency Property of $\mathcal{MMFH}$ with respect to X^j and Relative to $\mathcal{MFH}^j$

Recall that we fix $j \in \{1, \ldots, n\}$.

Before we state the second main result of the section (Theorem 2.33) we first prove the following two auxiliary results.

Proposition 2.31 *Consider a nice $\mathbb{R}^n$-Feller–Markov family*

$$\mathcal{MMFH} = \{(\Omega, \mathcal{F}, \mathbb{F}, (X_t)_{t \geq 0}, \mathbb{P}_x, P) : x \in \mathbb{R}^n\}$$

with corresponding symbol q such that $x \mapsto q(x, \xi)$ is continuous for every $\xi \in \mathbb{R}^k$ and

$$|q(x, \xi)| \leq c(1 + |\xi|^2), \quad x, \xi \in \mathbb{R}^n,$$

*for some constant $c > 0$. Assume that $\mathcal{MMFH}$ is strongly Markov consistent with respect to X^j and that condition **C1**(j) is satisfied. Additionally, assume that the weak jth coordinate of $\mathcal{MMFH}$ is a nice $\mathbb{R}$-Feller–Markov family with generator A^j admitting symbol q_j. Then*

$$q(x, \mathbf{e}_j \xi_j) = q_j(x_j, \xi_j), \quad x = (x_1, \ldots, x_j, \ldots, x_n) \in \mathbb{R}^n, \ \xi_j \in \mathbb{R}. \tag{2.49}$$

Proof Let us fix $x \in \mathbb{R}^n$. For any $w \in C_c^\infty(\mathbb{R})$, we have

$$\begin{aligned}
\lim_{t \to 0+} \frac{\mathbb{E}_x \left(w(X_t^j) - w(x_j) \right)}{t} &= \lim_{t \to 0+} \frac{\mathbb{E}_x \int_0^t \left(\bar{A}^j w \right)(X_u) du}{t} \\
&= \mathbb{E}_x \lim_{t \to 0+} \frac{\int_0^t \left(\bar{A}^j w \right)(X_u) du}{t} \\
&= \left(\bar{A}^j w \right)(x) = -(2\pi)^{-1/2} \int_{\mathbb{R}} e^{i x_j \xi_j} \widehat{w}(\xi_j) q(x, \mathbf{e}_j \xi_j) d\xi_j.
\end{aligned} \tag{2.50}$$

The first equality follows from the martingale property given by condition **C1**(j). The second follows from Lemma 2.34 below. In fact, in view of this lemma we obtain that the random variable $t^{-1}\int_0^t (\bar{A}^j w)(X_u)du$ is bounded, so we may change the order of the limit and the expectation. The third equality is a consequence of (2.41), of the continuity of q, of Lemma 2.20 with $E=\mathbb{R}^{n-1}$, $k=1$, and $\rho(x^{(j)},x_j,\xi_j)=q(x,\mathbf{e}_j\xi_j)$, where for $x=(x_1,\ldots,x_n)\in\mathcal{X}$, we set

$$x^{(j)}=(x_1,\ldots,x_{j-1},x_{j+1},\ldots,x_n),$$

of the right-continuity of X^j, and of the fundamental theorem of calculus.

Now, since $\mathcal{MMFH}$ is strongly Markov consistent with respect to X^j and the weak jth coordinate of $\mathcal{MMFH}$ is a nice $\mathbb{R}$-Feller–Markov family with generator given by the symbol q_j, we have

$$\begin{aligned}\lim_{t\to 0^+}\frac{\mathbb{E}_x(w_j(X_t)-w_j(x))}{t} &= \lim_{t\to 0^+}\frac{\mathbb{E}^j_{x_j}(w(X^j_t)-w(x_j))}{t}\\ &= A^j w(x_j)=-(2\pi)^{-1/2}\int_{\mathbb{R}} e^{ix_j\xi_j}\widehat{w}(\xi_j)q_j(x_j,\xi_j)d\xi_j,\end{aligned} \tag{2.51}$$

where the first equality follows from (2.12), with $\mathbb{E}^j_{x_j}$ denoting the expectation under $\mathbb{P}^j_{x_j}$, and the second and the third equalities follow from the assumption that the weak jth coordinate of $\mathcal{MMFH}$ is a nice $\mathbb{R}$-Feller–Markov family with generator A^j and with symbol q_j.

Therefore by (2.50) and (2.51) we have

$$\int_{\mathbb{R}} e^{ix_j\xi_j}\widehat{w}(\xi_j)q(x,\mathbf{e}_j\xi_j)d\xi_j=\int_{\mathbb{R}} e^{ix_j\xi_j}\widehat{w}(\xi_j)q_j(x_j,\xi_j)d\xi_j \tag{2.52}$$

for all $w\in C_c^\infty(\mathbb{R})$. Because $C_c^\infty(\mathbb{R})$ is dense in $\mathcal{S}(\mathbb{R})$ and the Fourier transform is a bijective mapping on $\mathcal{S}(\mathbb{R})$ with continuous inverse, we conclude that the image of $C_c^\infty(\mathbb{R})$ under Fourier transformation is dense in $\mathcal{S}(\mathbb{R})$. Therefore the image of $C_c^\infty(\mathbb{R})$ under Fourier transformation is dense in $C_0(\mathbb{R})$. Thus (2.52) implies that

$$q(x,\mathbf{e}_j\xi_j)=q_j(x_j,\xi_j)$$

for all $x\in\mathbb{R}^n$ and $\xi_j\in\mathbb{R}$. □

The next proposition will be used in the proof of Theorem 2.33, but it is of independent interest as well.

Proposition 2.32 *Let* $\left\{(\Omega,\mathcal{F},\mathbb{F}^j,(X^j_t)_{t\ge 0},\bar{\mathbb{P}}_{x_j},\bar{P}): x_j\in\mathbb{R}\right\}$ *and* $\left\{(\Omega,\mathcal{F},\mathbb{F}^j,(X^j_t)_{t\ge 0},\widehat{\mathbb{P}}_{x_j},\widehat{P}): x\in\mathbb{R}\right\}$ *be nice* $\mathbb{R}$*-Feller–Markov families with symbols* $\bar{q}$ *and* $\widehat{q}$*, respectively. Then*

$$\bar{q}=\widehat{q}\Longleftrightarrow \bar{P}=\widehat{P}.$$

Proof The result follows from Corollary 1.21 in Schnurr (2009), Proposition 4.1.6 in Ethier and Kurtz (1986), and the fact that $C_c^\infty(\mathbb{R}^n)$ is a separating class. □

Finally, assuming that conditions **C1**(j)–**C4**(j) are satisfied, we will deliver sufficient and necessary conditions under which a Markov family $\mathcal{MMFH}$ satisfies the strong Markov consistency property with respect to X^j and relative to $\mathcal{MFH}^j$, that is the conditions under which $\mathcal{MMFH}$ is strongly Markov consistent with respect to the coordinate X^j and under which the transition function P^j coincides with Q^j.

Theorem 2.33 *Let q be the symbol of $\mathcal{MMFH}$, the nice $\mathbb{R}^n$-Feller–Markov family given in Definition 2.23, and let ρ_j be the symbol of $\mathcal{MFH}^j$, the nice $\mathbb{R}$-Feller–Markov family given in Definition 2.25. Assume that conditions **C1(j)–C4(j)** are satisfied. Then, $\mathcal{MMFH}$ satisfies the strong Markov consistency property with respect to X^j relative to $\mathcal{MFH}^j$ if and only if*

$$q(x, \mathbf{e}_j \xi_j) = \rho_j(x_j, \xi_j) \quad \text{for all } x \in \mathbb{R}^n \text{ and } \xi_j \in \mathbb{R}. \tag{2.53}$$

In particular the symbol of $\mathcal{WMMF}^j$, and the symbol of $\mathcal{SMMF}^j$, are both equal to ρ_j, and thus the generator of $\mathcal{SMMF}^j$ and the generator of $\mathcal{WMMF}^j$, restricted to $C_c^\infty(\mathbb{R})$, are both equal to the PDO associated with ρ_j, i.e. $-\rho_j(x, D)$.

Proof First observe that, in view of Theorem 2.30 and Proposition 2.31, conditions **C1**(j)–**C4**(j) imply that $\mathcal{MMFH}$ satisfies the strong Markov consistency property with respect to X^j and that

$$q(x, \mathbf{e}_j \xi_j) = q_j(x_j, \xi_j), \quad x \in \mathbb{R}^n \text{ and } \xi_j \in \mathbb{R}, \tag{2.54}$$

where q_j is the symbol of the $\mathcal{WMMF}^j$ family.

Thus, if (2.53) holds then $\rho_j = q_j$. Hence, in view of Proposition 2.32, we have

$$P^j = Q^j. \tag{2.55}$$

So $\mathcal{MMFH} \curvearrowright \mathcal{MFH}^j$.

Conversely, if (2.55) holds, then, again in view of Proposition 2.32, we have

$$q_j(x_j, \xi_j) = \rho_j(x_j, \xi_j) \quad \text{for all } x_j, \xi_j \in \mathbb{R}, \tag{2.56}$$

which, in combination with (2.54) implies that (2.53) is satisfied. □

Discussion of Conditions C1(j), C3(j), and C4(j)

We start with the following technical result.

Lemma 2.34 *Consider a nice $\mathbb{R}^n$-Feller–Markov family*

$$\mathcal{MMFH} = \{(\Omega, \mathcal{F}, \mathbb{F}, (X_t)_{t \geq 0}, \mathbb{P}_x, P) : x \in \mathbb{R}^n\}$$

with generator A and corresponding symbol q, and let $w \in C_c^\infty(\mathbb{R})$. Next, fix $v \in C_c^\infty(\mathbb{R}^{n-1})$ such that $v(0) = 1$, $||v||_\infty \leq 1$. Finally, define a function u^k as follows:

$$u^k(x) = w(x_j) v\left(\frac{1}{k}(x_1, \ldots, x_{j-1}, x_{j+1}, \ldots, x_n)\right), \quad x \in \mathbb{R}^n. \tag{2.57}$$

Then $(u^k)_{k \geq 1}$ is a uniformly bounded sequence such that

$$\lim_{k \to \infty} u^k(x) = w(x_j), \quad x = (x_1, \ldots, x_{j-1}, x_j, x_{j+1}, \ldots, x_n) \in \mathbb{R}^n, \tag{2.58}$$

$$\lim_{k\to\infty} Au^k(x) = -(2\pi)^{-1/2}\int_{\mathbb{R}} e^{ix_j\xi_j}\widehat{w}(\xi_j)q(x,\mathbf{e}_j\xi_j)d\xi_j, \quad x\in\mathbb{R}^n. \tag{2.59}$$

Moreover, if $x\mapsto q(x,\xi)$ *is continuous for every* $\xi\in\mathbb{R}^k$ *and is such that*

$$|q(x,\xi)| \le c(1+|\xi|^2), \quad x,\xi\in\mathbb{R}^n, \tag{2.60}$$

for a positive constant c *then the sequence* $(Au^k)_{k\ge1}$ *is uniformly bounded.*

Proof Without loss of generality we can take $j=1$. Note that

$$u^k(x) = w(x_1)v^k(\bar{x}),$$

where $\bar{x}=(x_2,\dots,x_n)$ and

$$v^k(\bar{x}) = v\left(\frac{1}{k}\bar{x}\right).$$

Clearly, $(v^k)_{k\ge1}$ is a sequence of functions of the class $C_c^\infty(\mathbb{R}^{n-1})$ that are uniformly bounded by 1, and it converges pointwise in $\mathbb{R}^{n-1}$ to 1. So (2.58) holds.

Using (2.35) and the fact that $\widehat{v^k}(\bar{\xi}) = k^{n-1}\widehat{v}(k\bar{\xi})$ we see that

$$\begin{aligned}
Au^k(x) &= -(2\pi)^{-n/2}\int_{\mathbb{R}^n} e^{i\langle x,\xi\rangle}q(x,\xi)\widehat{u^k}(\xi)d\xi\\
&= -(2\pi)^{-n/2}\int_{\mathbb{R}^n} e^{i\langle x,\xi\rangle}q(x,\xi)\widehat{w}(\xi_1)k^{n-1}\widehat{v}(k\bar{\xi})d\xi_1 d\bar{\xi}\\
&= -(2\pi)^{-n/2}\int_{\mathbb{R}^n} e^{ix_1\xi_1+ik^{-1}\langle\bar{x},\bar{\xi}\rangle}q\left((x_1,\bar{x}),\left(\xi_1,\frac{\bar{\xi}}{k}\right)\right)\widehat{w}(\xi_1)\widehat{v}(\bar{\xi})d\xi_1 d\bar{\xi}.
\end{aligned}$$

We now observe that

$$\begin{aligned}
\left|q\left((x_1,\bar{x}),\left(\xi_1,\frac{\bar{\xi}}{k}\right)\right)\widehat{w}(\xi_1)\widehat{v}(\bar{\xi})\right| &\le c(x)\left(1+|\xi_1|^2+\frac{1}{k^2}|\bar{\xi}|^2\right)|\widehat{wv}(\xi_1,\bar{\xi})|\\
&\le c(x)(1+|\xi_1|^2+|\bar{\xi}|^2)|\widehat{wv}(\xi_1,\bar{\xi})|,
\end{aligned}$$

where the first inequality is implied by

$$|q(x,\xi)| \le c(x)(1+\|\xi\|^2), \tag{2.61}$$

which in turn follows from the fact that q is a symbol of a Feller semigroup (see Lemma 3.6.22 and Theorem 4.5.6 in Jacob (2001)). In addition, since $wv\in C_c^\infty(\mathbb{R}^n)$ it follows that $\widehat{wv}\in\mathcal{S}(\mathbb{R}^n)$, which in turn implies that $\xi\mapsto(1+\|\xi\|^2)\widehat{vw}(\xi)$ is in $L^1(\mathbb{R}^n)$. Taking these facts into account we have

$$|Au^k(x)| \le -(2\pi)^{-n/2}c(x)\int_{\mathbb{R}^n}(1+\|\xi\|^2)\widehat{vw}(\xi)d\xi < \infty. \tag{2.62}$$

Thus invoking the dominated convergence theorem we see that

$$\lim_{k\to\infty} Au^k(x)$$

$$= -(2\pi)^{-n/2}\int_{\mathbb{R}^n}\lim_{k\to\infty} e^{ix_1\xi_1+ik^{-1}\langle\bar{x},\bar{\xi}\rangle} q\left((x_1,\bar{x}),\left(\xi_1,\frac{\bar{\xi}}{k}\right)\right)\widehat{w}(\xi_1)\widehat{v}(\bar{\xi})d\bar{\xi}d\xi_1$$

$$= -(2\pi)^{-1/2}\int_{\mathbb{R}} e^{ix_1\xi_1}\widehat{w}(\xi_1)q(x,e_1\xi_1)d\xi_1,$$

where the last equality holds since

$$(2\pi)^{-(n-1)/2}\int_{\mathbb{R}^{n-1}} e^{i\langle 0,\bar{\xi}\rangle}\widehat{v}(\bar{\xi})d\bar{\xi} = v(0) = 1.$$

This demonstrates (2.59).

Proceeding as above, but now using assumption (2.60) instead of (2.61), we obtain (2.62) with a constant c which does not depend on x. So

$$\sup_{x\in\mathbb{R}^n}|Au^k(x)| \le -(2\pi)^{-n/2}c\int_{\mathbb{R}^n}(1+\|\xi\|^2)\widehat{vw}(\xi)d\xi = K < \infty. \qquad \square$$

Now, we will give sufficient conditions for condition **C1**(j) to hold.

Proposition 2.35 *Consider a nice $\mathbb{R}^n$-Feller–Markov family*

$$\mathcal{MMFH} = \{(\Omega,\mathcal{F},\mathbb{F},(X_t)_{t\ge 0},\mathbb{P}_x,P) : x\in\mathbb{R}^n\}$$

with symbol q. Assume that q is continuous in x and satisfies

$$|q(x,\xi)| \le c(1+|\xi|^2), \qquad x,\xi\in\mathbb{R}^n, \tag{2.63}$$

*for a positive constant c. Then condition **C1**(j) holds for every $j=1,\ldots,n$.*

Proof It is enough to demonstrate the result for $j=1$.

In order to prove that condition **C1**(1) holds we need to show that for each $w\in C_c^\infty(\mathbb{R})$ the process N^w given by (2.40) and (2.41) is an $(\mathbb{F},\mathbb{P}_x)$-martingale for any $x\in\mathbb{R}^n$.

Using Lemma (2.34), we first note that for fixed $w\in C_c^\infty(\mathbb{R})$ the sequence $(u^k)_{k\ge 1}$ defined in (2.57) is uniformly bounded and converges pointwise.

Since for a nice $\mathbb{R}^n$-Feller–Markov family with symbol q it holds that $u^k\in C_c^\infty(\mathbb{R}^n)\subseteq\mathcal{D}(A)$, the process N^k given as

$$N_t^k = u^k(X_t) - \int_0^t Au^k(X_u)du, \quad t\ge 0,$$

is an $(\mathbb{F},\mathbb{P}_x)$-martingale for any $x\in\mathbb{R}^n$. By (2.58) $u^k(X_t)\to w(X_t^1)$ $\mathbb{P}_x$-a.s., and by (2.59) $Au^k(X_t)\to A^1w(X_t)$ $\mathbb{P}_x$-a.s. as $k\to\infty$. Moreover, $(Au^k)_{k\ge 1}$ is uniformly bounded, so $N_t^k\to N_t^w$ $\mathbb{P}_x$-a.s. and in $L^1(\Omega,\mathcal{F},\mathbb{P}_x)$. Therefore N^w is an $(\mathbb{F},\mathbb{P}_x)$-martingale since N^k, $k\ge 1$, are $(\mathbb{F},\mathbb{P}_x)$-martingales.

The proof is complete. $\square$

Remark 2.36 By a result of Schilling (1998, Lemma 2.1) the condition (2.63) is equivalent to

$$\|b\|_\infty + \|a\|_\infty + \left\| \int_{\mathbb{R}^n \setminus 0} \frac{|y|^2}{1+|y|^2} \mu(\cdot, dy) \right\|_\infty < \infty.$$

We now proceed to give sufficient conditions for the condition **C3**(j) to hold . This will be done in Proposition 2.41, but we need some preparation for this.

Towards this end, we let $\psi : \mathbb{R}^n \to \mathbb{R}$ be a continuous negative-definite function such that for some positive constants r and c we have

$$\psi(\xi) \geq c\|\xi\|^r, \quad \text{for } \|\xi\| \geq 1.$$

Next, we define

$$\lambda(\xi) := (1 + \psi(\xi))^{1/2}.$$

Finally, we let M be the smallest integer such that $M > (n/r \vee 2) + n$ and set $k = 2M + 1 - n$.

Now, following Hoh (1998), we consider a symbol $q : \mathbb{R}^n \times \mathbb{R}^n \to \mathbb{C}$ and impose the following additional conditions on q:

H0(n) The function q is real-valued and continuous in both variables.

H1(n) The map $x \mapsto q(x, \xi)$ is k times continuously differentiable and

$$\left\| \partial_x^\beta q(x, \xi) \right\| \leq c\lambda^2(\xi), \quad \beta \in \mathbb{N}_0^n, \ \|\beta\| \leq k.$$

H2(n) For some strictly positive function $\gamma : \mathbb{R}^n \to \mathbb{R}$,

$$q(x, \xi) \geq \gamma(x)\lambda^2(\xi), \quad \text{for } \|\xi\| \geq 1, \ x \in \mathbb{R}^n.$$

H3(n)

$$\sup_{x \in \mathbb{R}^n} |q(x, \xi)| \underset{\xi \to 0}{\longrightarrow} 0.$$

Recall that the PDO associated with q is denoted as $-q(x, D)$ and defined for $v \in C_c^\infty(\mathbb{R}^n)$ by

$$-q(y, D)v(y) := -(2\pi)^{n/2} \int_{\mathbb{R}^n} e^{i\langle y, \xi \rangle} q(y, \xi) \widehat{v}(\xi) d\xi. \tag{2.64}$$

The following lemma is, essentially, a restatement of Theorem 5.24 in Hoh (1998), and it formulates the conditions under which q generates an $\mathcal{MMFH}$.

Lemma 2.37 *Assume that conditions **H0**(n)–**H3**(n) are satisfied. Let $(A, \mathcal{D}(A))$ be the PDO associated with q, so that $Af(x) = -q(x, D)f(x)$ for every $f \in C_c^\infty(\mathbb{R}^n) \subseteq \mathcal{D}(A)$. Then the time-homogeneous D-martingale problem for A (see Definition F.7)* *is well posed. In particular, for any $x \in \mathbb{R}^n$, the D-martingale problem for (A, δ_x) admits a unique solution, say $\widehat{\mathbb{P}}_x$. Moreover, the semigroup defined by*

$$T_t f(x) = \mathbb{E}_{\widehat{\mathbb{P}}_x} f(\widehat{Z}_t), \quad t \geq 0,\, x \in \mathbb{R}^n,\, f \in C_0(\mathbb{R}^n),$$

where $\widehat{Z}$ is a canonical process, is a $C_0(\mathbb{R}^n)$-Feller semigroup with infinitesimal generator $(A, \mathcal{D}(A))$. Moreover $\left\{(\widehat{\Omega}, \widehat{\mathcal{F}}, \widehat{\mathbb{F}}, (Z_t)_{t\geq 0}, \widehat{\mathbb{P}}_x, P) : x \in \mathbb{R}^n\right\}$ is a nice $\mathbb{R}^n$-Feller–Markov process.

Given the above lemma we can now prove the following important result.

Proposition 2.38 *Assume that a symbol q satisfies conditions **H0(n)–H3(n)** and **C2(j)**. Let $-\tilde{q}_j(x, D)$ be defined as in (2.64) with $\tilde{q}_j(x_j, \xi_j)$ given by (2.42). Then there exists a nice $\mathbb{R}$-Feller–Markov family with infinitesimal generator $(A^j, \mathcal{D}(A^j))$ such that $A^j f(x) = -\tilde{q}_j(x, D) f(x)$ for every $f \in C_c^\infty(\mathbb{R}) \subseteq \mathcal{D}(A^j)$. In particular, condition **C3(j)** holds.*

Proof Since q satisfies conditions **H0**(n)–**H3**(n) and **C2**(j), it is easy to verify that $\tilde{q}^j$ given in (2.42) satisfies conditions **H0**(1)–**H3**(1). Therefore, the result follows from Lemma 2.37. □

Given Proposition 2.38 we can now provide the respective analysis of the strong Markov consistency property. For this, let $\mathcal{MMFH}$ be a nice $\mathbb{R}^n$-Feller–Markov family with symbol q satisfying conditions **H0**(n)–**H3**(n), **C2**(j), and **C4**(j). Then

(i) $\mathcal{MMFH}$ is strongly Markov consistent with respect to X^j.
(ii) The strong jth coordinate $\mathcal{SMMF}^j$ of X delineated by $(x_k, k \neq j)$ is a nice $\mathbb{R}$-Feller–Markov family with symbol $\tilde{q}_j$.
(iii) The weak jth coordinate $\mathcal{WMMF}^j$ of X is a nice $\mathbb{R}$-Feller–Markov family with symbol $\tilde{q}_j$.

In fact, Lemma 3.6.22 in Jacob (2001) implies that, for a continuous negative-definite function ψ, we have

$$|\psi(\xi)| \leq c_\psi(1 + |\xi|^2).$$

Hence, and from condition **H1**(n), we conclude that (2.63) holds. So, by Proposition 2.35, condition **C1**(j) holds. By Proposition 2.38 we conclude that $\tilde{q}_j$ satisfies condition **C3**(j). Consequently, implications (i)–(iii) above follow from Theorem 2.30.

We now present another set of assumptions imposed on q, under which we can obtain a similar result to that in the case where Hoh conditions apply. In the next proposition, which is a counterpart of Proposition 2.38, we adopt, and adapt, conditions of Stroock (1975).

Proposition 2.39 *Let q be a function given as follows:*

$$q(x,\xi) = -i\langle b(x),\xi\rangle + \langle \xi, a(x)\xi\rangle + \int_{\mathbb{R}^n\setminus\{0\}} \left(1 - e^{i\langle y,\xi\rangle} + \frac{i\langle y,\xi\rangle}{1+|y|^2}\right)\mu(x,dy), \quad x,\xi \in \mathbb{R}^n. \tag{2.65}$$

Assume that the following conditions hold.

S1(n) *a is bounded, continuous, and positive definite.*

S2(n) *b is bounded and continuous.*

S3(n) *For all $C \in \mathcal{B}(\mathbb{R}^n \setminus \{0\})$, the mapping*

$$x \mapsto \int_C \frac{y}{1+|y|^2}\,\mu(x,dy) \tag{2.66}$$

is bounded and continuous.

S4(n) *The mapping*

$$x \mapsto \int_{\mathbb{R}^n\setminus\{0\}} \left(1 - e^{i\langle y,\xi\rangle} + \frac{i\langle y,\xi\rangle}{1+|y|^2}\right)\mu(x,dy) \tag{2.67}$$

is bounded and continuous.

*Additionally, assume that condition **H3**(n) holds. Then the function q is a symbol satisfying* (2.63) *and there exists a nice $\mathbb{R}^n$-Feller–Markov family with infinitesimal generator $(A,\mathcal{D}(A))$ such that $Af(x) = -q(x,D)f(x)$ for every $f \in C_c^\infty(\mathbb{R}^n) \subseteq \mathcal{D}(A)$.*

Proof Since the function q is given by (2.65), it is, by Theorem 3.7.8 in Jacob (1998), continuous and negative definite in ξ. From conditions **S1**(n)–**S4**(n) we obtain that q is continuous, so it is locally bounded and measurable. Thus q is a symbol. Since q is a symbol, then -$q(x,D)$ has the Waldenfels representation (see (G.2)). So, by results of Stroock (1975), the conditions **S1**(n)–**S3**(n) imply that the D-martingale problem for q is well posed. Moreover (2.63) is satisfied (see Remark 2.36). Thus the result follows from Theorem 5.23 in Hoh (1998). □

Remark 2.40 Let us note that if q is a function given by (2.65) with $\mu \equiv 0$ and conditions **S1**(n)–**S2**(n) hold then condition **H3**(n) holds. So, since conditions **S3**(n) – **S4**(n) clearly are satisfied, Proposition 2.39 holds in this case if only one assumes that conditions **S1**(n)–**S2**(n) are satisfied.

The following result, being a counterpart of Proposition 2.38, is a direct consequence of the above considerations.

Proposition 2.41 *Let q be a function given by* (2.65)*. Assume that conditions **S1**(n)–**S4**(n), **H3**(n), and **C2**(j) are satisfied for q. Let $-\tilde{q}_j(x,D)$ be defined as in* (2.64) *with $\tilde{q}_j(x_j,\xi_j)$ given by* (2.42)*. Then there exists a nice $\mathbb{R}$-Feller–Markov family with infinitesimal generator $(A^j,\mathcal{D}(A^j))$ such that $A^jf(x) = -\tilde{q}_j(x,D)f(x)$ for every $f \in C_c^\infty(\mathbb{R}) \subseteq \mathcal{D}(A^j)$. In particular condition **C3**(j) holds.*

Proof By (2.42) it is clear that $\tilde{q}_j$ has the following generic representation in terms of the triple (b_j, a_j, μ_j):

$$\tilde{q}_j(x_j,\xi_j) = -ib_j(x),\xi_j + \xi_j^2 a_j(x) + \int_{\mathbb{R}\setminus\{0\}} \left(1 - e^{iy_j\xi_j} + \frac{iy_j\xi_j}{1+|y_j|^2}\right)\mu^j(x,dy_j). \tag{2.68}$$

In view of our assumptions the triple (b_j, a_j, μ_j) satisfies conditions **S1**(1)–**S4**(1). Clearly $\tilde{q}_j$ satisfies condition **H3**(1). The assertion now follows from Proposition 2.39. □

Next, we discuss condition **C4**(j). In particular, we will show that condition **C4**(j) holds if the generator A^j admits $C_c^\infty(\mathbb{R})$ as its core.

We begin with the following straightforward, albeit important, result.

Proposition 2.42 *Let $\widetilde{A}$ be the strong generator of a semigroup on $C_0(\mathbb{R})$. Suppose that $C_c^\infty(\mathbb{R})$ is a core of $\widetilde{A}$. Then for each $f \in \mathcal{D}(\widetilde{A})$ there exists a sequence $f^k \in C_c^\infty(\mathbb{R})$ such that $f^k(y) \to f(y)$ for each $y \in \mathbb{R}$ as $k \to \infty$ and*

$$\lim_{k\to\infty} \mathbb{E}_{\mathbb{P}}\Big(\int_0^\infty e^{-\lambda s}|\widetilde{A}(f^k(Z_s) - f(Z_s))|ds\Big) = 0$$

for an arbitrary probability measure $\mathbb{P}$ and an arbitrary real-valued càdlàg process Z, both defined on some $(\Omega, \mathcal{F})$.

Proof By the definition of a core of $\widetilde{A}$, for every $f \in \mathcal{D}(\widetilde{A})$ there exists a sequence $(f^k)_{k\ge 1} \subset C_c^\infty(\mathbb{R})$ such that

$$\lim_{k\to\infty}\left(\left\|f - f^k\right\|_\infty + \left\|\widetilde{A}f - \widetilde{A}f^k\right\|_\infty\right) = 0.$$

This, combined with the inequality

$$\mathbb{E}_{\mathbb{P}}\Big(\int_0^\infty e^{-\lambda s}|\widetilde{A}(f^k(Z_s) - f(Z_s))|ds\Big) \le \int_0^\infty e^{-\lambda u}\left\|\widetilde{A}f - \widetilde{A}f^k\right\|_\infty du$$

proves the result. □

Consequently, by taking $Z = X^j$ and $\mathbb{P} = \mathbb{P}_x$, $x \in \mathbb{R}^n$, we obtain

Corollary 2.43 *Suppose that $C_c^\infty(\mathbb{R})$ is a core of the generator A^j given in condition C3(j). Then condition **C4**(j) holds.*

Remark 2.44 (i) If A^j given in condition **C3**(j) admits a symbol $\tilde{q}_j$ that does not depend on the variable x_j then the set $C_c^\infty(\mathbb{R})$ is a core of A^j (see Corollary 2.10 in Böttcher et al. (2013)).

(ii) Suppose that A^j admits a symbol $\widetilde{q}_j$ of the form

$$\widetilde{q}_j(x_j,\xi_j) = (b' + \beta x_j)\xi_j + \alpha x_j \xi_j^2 + \int_{\mathbb{R}_+} (e^{iy\xi_j} - 1 - y\xi_j \mathbb{1}_{|y|\le 1})(m(dy) + x_j\mu(dy)),$$

where $b' = b + \int_{\mathbb{R}_+}(|y| \wedge 1)m(dy)$, $\alpha, b \in \mathbb{R}_+$, $\beta \in \mathbb{R}$, and m and μ are nonnegative Borel measures satisfying

$$\int_{\mathbb{R}_+} (|y| \wedge 1)m(dy) + \int_{\mathbb{R}_+} (|y|^2 \wedge 1)\mu(dy) < \infty.$$

The symbol $\tilde{q}_i$ corresponds to a one-dimensional affine Markov family. It follows from Proposition 8.2 in Duffie et al. (2003) that for a PDO A_j associated with symbol $\tilde{q}_j$, the set of functions $C_c^\infty(\mathbb{R})$ is a core of $(A^j, \mathcal{D}(A^j))$.

We end this section by providing some additional insight into the conditions **C1**(j), **C3**(j), and **C4**(j). The following results will be used in Chapter 6 to construct examples of strong Markov structures.

Proposition 2.45 *Let $\tilde{q}_j$ be given as*

$$\tilde{q}_j(x_j, \xi_j) = -i\xi_j d_j(x_j) + c_j(x_j)\xi_j^2, \quad x_j \in \mathbb{R},$$

where d_j and $c_j \geq 0$ are real-valued bounded functions which satisfy at least one of the following three conditions:

(i) *d_j is Lipschitz and c_j is in $C^2(\mathbb{R})$ with bounded second derivatives.*
(ii) *$c_j > \delta > 0$; d_j and c_j are bounded and α-Hölder continuous with $\alpha \in (0,1]$.*
(iii) *d_j and $\sqrt{c}_j$ are functions in $C_b^2(\mathbb{R})$.*

*Then there exists a nice $\mathbb{R}$-Feller–Markov family with infinitesimal generator $(A^j, \mathcal{D}(A^j))$ such that $A^j f(x) = -\tilde{q}_j(x, D)f(x)$ for every $f \in C_c^\infty(\mathbb{R}) \subseteq \mathcal{D}(A^j)$. In particular, condition **C3(j)** holds. Moreover, $C_c^\infty(\mathbb{R})$ is a core of A^j, so that condition **C4(j)** holds for the generator A^j.*

Proof Using the Waldenfels representation of $-\tilde{q}_j(x_j, D)$ (see (G.2)) we have

$$-\tilde{q}_j(x_j, D)f(x_j) = d_j(x_j)\frac{\partial}{\partial x_j}f(x_j) + c_j(x_j)\frac{\partial^2}{\partial x_j^2}f(x_j).$$

Thus, condition (i) implies the assertion, by Theorem 8.2.5 in Ethier and Kurtz (1986). Condition (ii) implies the assertion, by Theorem 8.1.6 in that monograph. Finally, condition (iii) implies the assertion, by Proposition 8.2.4 in the same monograph. □

Proposition 2.46 *Let q be given as*

$$q(x, \xi) = -i\langle b(x), \xi\rangle + \langle \xi, a(x)\xi\rangle,$$

where the functions $b : \mathbb{R}^n \to \mathbb{R}^n$ and $a : \mathbb{R}^n \to L(\mathbb{R}^n, \mathbb{R}^n)$ satisfy at least one of the following three conditions:

(i) *b is Lipschitz and all elements of the matrix a are $C^2(\mathbb{R}^n)$ functions with bounded second-order derivatives.*

(ii) *a and b are bounded and α-Holder continuous with $\alpha \in (0,1]$ and a is elliptic, i.e.*

$$\inf_{x\in\mathbb{R}^n} \inf_{|\xi|=1} \langle \xi, a(x)\xi \rangle > 0.$$

(iii) *b and σ are $C_b^2(\mathbb{R}^n)$ functions, where σ satisfies $a(x) = \sigma(x)\sigma^\top(x)$.*

*Then there exists a nice $\mathbb{R}^n$-Feller–Markov family with infinitesimal generator $(A, \mathcal{D}(A))$ such that $Af(x) = -q(x,D)f(x)$ for every $f \in C_c^\infty(\mathbb{R}^n) \subseteq \mathcal{D}(A)$. Moreover, $C_c^\infty(\mathbb{R}^n)$ is a core of A and condition **C1(j)** holds.*

Proof The proof of the first assertion, using the Waldenfels representation, is analogous to the proof of Proposition 2.45. It remains to prove the second assertion. For this, we observe that, given our assumptions, it is clear that (2.63) holds and hence, in view of Proposition 2.35, condition **C1**(j) is satisfied. □

We close our discussion of conditions **C1**(j), **C3**(j), and **C4**(j) with the following result.

Proposition 2.47 *Suppose that the assumptions of Proposition 2.46 hold, and assume that condition **C2(j)** is satisfied. Let $-\tilde{q}_j(x,D)$ be defined as in (2.64) with $\tilde{q}_j(x_j,\xi_j)$ given by (2.42). Then there exists a nice $\mathbb{R}$-Feller–Markov family with infinitesimal generator $(A^j, \mathcal{D}(A^j))$ such that $A^j f(x) = -\tilde{q}_j(x,D)f(x)$ for every $f \in C_c^\infty(\mathbb{R}) \subseteq \mathcal{D}(A^j)$. In particular, conditions **C3(j)** and **C4(j)** hold.*

Proof The assertion follows from Proposition 2.46 for $n = 1$ and Corollary 2.43. □

3

Consistency of Finite Multivariate Markov Chains

In this chapter we study strong Markov consistency and weak Markov consistency for finite time-inhomogeneous multivariate Markov chains. We refer to Appendix C for some auxiliary technical framework regarding finite Markov chains. As in the case of general Markov families and processes discussed in Chapter 2, we only consider conservative Markov chains.

Let $(\Omega, \mathcal{F}, \mathbb{P})$ be the underlying (not necessarily canonical) complete probability space. On this space we consider a multivariate Markov chain $X = (X_t = (X_t^1, \ldots, X_t^n), t \geq 0)$ with values in a finite product space, say $\mathcal{X} = \times_{i=1}^n \mathcal{X}_i$, where $\mathcal{X}_i$ is a finite set with cardinality $\text{card}(\mathcal{X}_i) = m_i$. The process X starts at time $t = 0$ with some initial distribution. Thus, we are dealing here with the special case of the Markov process introduced in Definition B.1, for $s = 0$ and a finite state space $\mathcal{X} = \times_{i=1}^n \mathcal{X}_i$. As mentioned above, we will study here the strong and weak Markov consistency of X. However, the techniques employed in this chapter are different from those of Chapter 2.

In order to somewhat simplify the notation and the ensuing formulae, in most of the chapter we consider a bivariate processes X only, that is, we take $n = 2$; furthermore, we take $\Lambda(t) = [\lambda_{xy}(t)]_{x,y \in \mathcal{X}}$, $t \geq 0$, as a generic symbol for the infinitesimal generator function of X (see Remark E.4). Thus, $\Lambda(t)$ is an $m \times m$ matrix, where $m = m_1 m_2$. We stress that the restriction to the bivariate case is for notational convenience only.

The reader will get the flavor of the formulae pertaining to the case of general n in Chapter 4, in which we study the consistency of multivariate conditional Markov chains (CMCs). Several results presented there are generalizations of results derived in the present chapter. However, the present chapter contains more in-depth discussion and some results, which may not be generalized to the case of CMCs.

In what follows we consider various functions of $t \geq 0$. We take the liberty of using the notations f, $f(\cdot)$, $f(t)$, or $f(t)$, $t \geq 0$, to denote any such function. So, for example, we may write $\Lambda = [\lambda_{xy}]_{x,y \in \mathcal{X}}$ rather than $\Lambda(t) = [\lambda_{xy}(t)]_{x,y \in \mathcal{X}}$, $t \geq 0$.

3.1 Definition and Characterization of Strong Markov Consistency

We begin with the specification of Definition 2.15 to the present case of Markov chains.

Definition 3.1 1. Let us fix $i \in \{1,\dots,n\}$. We say that the Markov chain $X = (X^1,\dots,X^n)$ satisfies the *strong Markov consistency condition with respect to the coordinate* X^i if, for every $B \subset \mathcal{X}_i$ and all $t,s \geq 0$,

$$\mathbb{P}\left(X^i_{t+s} \in B | \mathcal{F}^X_t\right) = \mathbb{P}\left(X^i_{t+s} \in B | X^i_t\right), \quad \mathbb{P}\text{-a.s.}, \tag{3.1}$$

or, equivalently,

$$\mathbb{P}\left(X^i_{t+s} \in B | X_t\right) = \mathbb{P}\left(X^i_{t+s} \in B | X^i_t\right), \quad \mathbb{P}\text{-a.s.}, \tag{3.2}$$

so that X^i is a Markov chain in the filtration of X, i.e. in the filtration $\mathbb{F}^X$.

2. If X satisfies the strong Markov consistency condition with respect to X^i for each $i \in \{1,\dots,n\}$ then we say that X satisfies the strong Markov consistency condition.

Remark 3.2 It will be shown in Theorem 3.3 that the strong Markov consistency property formulated in Definition 3.1 can be characterized in terms of the structural properties of X and of X^i, $i = 1,\dots,n$. Specifically, it is characterized in terms of the infinitesimal generator functions of these processes.

3.1.1 Necessary and Sufficient Conditions for Strong Markov Consistency

In this section we shall discuss the necessary and sufficient conditions for the strong Markov consistency of our finite Markov chain X. According to our convention in this chapter, we take $n = 2$.

We recall that we take $\Lambda = [\lambda_{xy}]_{x,y\in\mathcal{X}}$ as a generic symbol for the infinitesimal generator function of X.

Theorem 3.3 *The process X satisfies the strong Markov consistency condition with respect to the coordinate X^1 if and only if, for any $x_1, y_1 \in \mathcal{X}^1$, $x_1 \neq y_1$,*

$$\mathbb{1}_{\{X^1_t = x_1\}} \sum_{y_2 \in \mathcal{X}_2} \lambda_{(x_1 X^2_t)(y_1 y_2)}(t) = \mathbb{1}_{\{X^1_t = x_1\}} \lambda^1_{x_1 y_1}(t), \quad dt \otimes d\mathbb{P}\text{-a.e.}, \tag{3.3}$$

for some locally integrable functions $\lambda^1_{x_1 y_1}(\cdot)$. The infinitesimal generator function of X^1 is $\Lambda^1(t) = [\lambda^1_{x_1 y_1}(t)]_{x_1,y_1\in\mathcal{X}_1}$, $t \geq 0$, with $\lambda^1_{x_1 x_1}(\cdot)$ given by

$$\lambda^1_{x_1 x_1}(t) = -\sum_{y_1 \in \mathcal{X}_1, y_1 \neq x_1} \lambda^1_{x_1 y_1}(t), \quad x_1 \in \mathcal{X}_1.$$

The process X satisfies the strong Markov consistency condition with respect to the coordinate X^2 if and only if, for any $x_2, y_2 \in \mathcal{X}^2$, $x_2 \neq y_2$, we have

$$\mathbb{1}_{\{X^2_t = x_2\}} \sum_{y_1 \in \mathcal{X}_1} \lambda_{(X^1_t x^2)(y_1 y_2)}(t) = \mathbb{1}_{\{X^2_t = x_2\}} \lambda^2_{x_2 y_2}(t), \quad dt \otimes d\mathbb{P}\text{-a.s.}, \tag{3.4}$$

for some locally integrable functions $\lambda^2_{x_2y_2}(\cdot)$. *The infinitesimal generator function of* X^2 *is* $\Lambda^2(t) = [\lambda^2_{x_2y_2}(t)]_{x_2,y_2\in\mathcal{X}_2}$, $t \geq 0$, *with* $\lambda^2_{x_2x_2}(\cdot)$ *given by*

$$\lambda^2_{x_2x_2}(t) = -\sum_{y_2\in\mathcal{X}_2, y_2\neq x_2} \lambda^2_{x_2y_2}(t), \quad x_2 \in \mathcal{X}_2.$$

Proof We will give only the proof regarding the coordinate X^1 of X. For the coordinate X^2 the proof is analogous.

Assume that (3.3) holds. Since X is a Markov chain, by (C.11), (C.12), and (3.3), we see that for each $x_1, y_1 \in \mathcal{X}_1$ the $\mathbb{F}^X$-compensator of $N^1_{x_1y_1}$ has the form

$$\begin{aligned} \nu^1_{x_1y_1}(dt) &= \sum_{x_2,y_2\in\mathcal{X}_2} \lambda_{(x_1x_2)(y_1y_2)}(t)\mathbb{1}_{\{(X^1_t,X^2_t)=(x_1,x_2)\}}dt \\ &= \mathbb{1}_{\{X^1_t=x_1\}}\sum_{y_2\in\mathcal{X}_2} \lambda_{(x_1X^2_t)(y_1y_2)}(t)dt = \mathbb{1}_{\{X^1_t=x_1\}}\lambda^1_{x_1y_1}(t)dt \end{aligned}$$

for some locally integrable deterministic function $\lambda^1_{x_1y_1}$. Then, by martingale characterization (see Proposition C.2), X^1 is a Markov chain with respect to $\mathbb{F}^X$ with generator $\Lambda^1(t) = [\lambda^1_{x_1y_1}(t)]_{x_1,y_1\in\mathcal{X}_1}$.

Conversely, assume that X^1 is a Markov chain with respect to the filtration $\mathbb{F}^X$ with generator $\Lambda^1(t) = [\lambda^1_{x_1y_1}(t)]_{x_1,y_1\in\mathcal{X}_1}$. Then (3.3) follows from martingale characterization, (C.11), and (C.12). Indeed, we have

$$\begin{aligned} \mathbb{1}_{\{X^1_t=x_1\}}\lambda^1_{x_1y_1}(t)dt = \nu^1_{x_1y_1}(dt) &= \sum_{x_2,y_2\in\mathcal{X}_2} \lambda_{(x_1x_2)(y_1y_2)}(t)\mathbb{1}_{\{(X^1_t,X^2_t)=(x_1,x_2)\}}dt \\ &= \mathbb{1}_{\{X^1_t=x_1\}}\sum_{y_2\in\mathcal{X}_2} \lambda_{(x_1X^2_t)(y_1y_2)}(t)dt. \end{aligned}$$

□

Let us recall Condition (M), first introduced in Bielecki et al. (2008b):

Condition(M) *The infinitesimal generator function* Λ *satisfies:*

M(1) *For every* $x_2, \bar{x}_2 \in \mathcal{X}_2$, $x_1, y_1 \in \mathcal{X}_1$, $x_1 \neq y_1$,

$$\sum_{y_2\in\mathcal{X}_2} \lambda_{(x_1x_2)(y_1y_2)}(t) = \sum_{y_2\in\mathcal{X}_2} \lambda_{(x_1\bar{x}_2)(y_1y_2)}(t), \quad t \geq 0,$$

and

M(2) *For every* $x_1, \bar{x}_1 \in \mathcal{X}_1$, $x_2, y_2 \in \mathcal{X}_2$, $x_2 \neq y_2$,

$$\sum_{y_1\in\mathcal{X}_1} \lambda_{(x_1x_2)(y_1y_2)}(t) = \sum_{y_1\in\mathcal{X}_1} \lambda_{(\bar{x}_1x_2)(y_1y_2)}(t), \quad t \geq 0.$$

It is clear that the conditions **M**(1) and **M**(2) imply conditions (3.3) and (3.4), respectively. Thus, we have the following:

Corollary 3.4 *Assume that Condition (M) is satisfied. Then the Markov chain $X = (X^1, X^2)$ is strongly Markov consistent.*

Remark 3.5 (i) It is clear that conditions $\mathbf{M}(1)$ and $\mathbf{M}(2)$ can be weakened, to the effect that rather than taking all $t \geq 0$ one takes almost all $t \geq 0$, and Corollary 3.4 will still be true.
(ii) Conditions $\mathbf{M}(1)$ and $\mathbf{M}(2)$ are sufficient for strong Markov consistency to hold for X with respect to both its coordinates. It turns out, however, that in general conditions $\mathbf{M}(1)$ and $\mathbf{M}(2)$ are too strong; in particular, they are not necessary for strong Markov consistency to hold for X with respect to its coordinates. In this regard we refer to the discussion in Example 4.9 below, where it is demonstrated, in the context of conditional Markov chains, that conditions such as $\mathbf{M}(1)$ and $\mathbf{M}(2)$ are not necessary, in general, for strong Markov consistency to hold.
(iii) Even though conditions $\mathbf{M}(1)$ and $\mathbf{M}(2)$ are stronger than needed to establish strong Markov consistency, they are very convenient for that purpose. In particular, they can be conveniently used to construct a strong Markov structure (see Section 7.1).

In the next section we shall provide operator interpretation of conditions $\mathbf{M}(1)$ and $\mathbf{M}(2)$.

3.1.2 Operator Interpretation of Condition (M)

In Proposition 3.7 below we present an operator interpretation of Condition (M). Towards this end, for $i = 1,2$ we first introduce an extension operator $C^{i,*}$ as follows: for any function f^i on $\mathcal{X}_i$, the function $C^{i,*} f^i$ is defined on $\mathcal{X}$ by

$$(C^{i,*} f^i)(x) = f^i(x_i), \quad x = (x_1, x_2) \in \mathcal{X}. \tag{3.5}$$

Remark 3.6 We may, and we will, identify operators $C^{i,*}$, $i = 1,2$, with their matrix representations and real functions on $\mathcal{X}$ or $\mathcal{X}_i$ with their vector representations. Whenever needed and convenient, we also identify matrices $\Gamma = [\gamma_{v,w}]_{v,w \in \mathcal{V}}$ with linear operators acting on functions $f \colon \mathcal{V} \mapsto \mathbb{R}$ defined by

$$(\Gamma f)(v) := \sum_{w \in \mathcal{V}} \gamma_{v,w} f(w), \quad v \in \mathcal{V},$$

where $\mathcal{V}$ denotes a generic finite set.

We now have the following:

Proposition 3.7 *For $i = 1,2$, condition $\mathbf{M}(i)$ is equivalent to condition $\mathbf{N}(i)$:*

N(*i*) *There exist generator functions $\Lambda^i = [\lambda^i_{x_i y_i}]_{x_i, y_i \in \mathcal{X}_i}$ such that for any real-valued function g on $\mathcal{X}_i$ we have*

$$C^{i,*}(\Lambda^i(t) g) = \Lambda(t)(C^{i,*} g), \quad t \geq 0, \tag{3.6}$$

or, in the matrix representation of $C^{i,*}$, Λ and Λ^i,

$$C^{i,*}\Lambda^i(t) = \Lambda(t)C^{i,*}, \quad t \geq 0. \tag{3.7}$$

Proof We prove the result only for $i = 1$. The proof for $i = 2$ is analogous. We note that (3.7) is equivalent to the equality

$$(C^{1,*}\Lambda^1(t)g)(x_1,x_2) = (\Lambda(t)C^{1,*}g)(x_1,x_2), \quad (x_1,x_2) \in \mathcal{X}_1 \times \mathcal{X}_2, \tag{3.8}$$

for an arbitrary function g on $\mathcal{X}_1$ and $t \geq 0$. By definition, the right-hand side of (3.8) satisfies

$$\begin{aligned}(\Lambda(t)C^{1,*}g)(x_1,x_2) &= \sum_{(y_1,y_2)\in\mathcal{X}} \lambda_{(x_1x_2)(y_1y_2)}(t)(C^{1,*}g)(y_1,y_2)\\ &= \sum_{(y_1,y_2)\in\mathcal{X}} \lambda_{(x_1x_2)(y_1y_2)}(t)g(y_1) = \sum_{y_1\in\mathcal{X}_1}\left(\sum_{y_2\in\mathcal{X}_2} \lambda_{(x_1x_2)(y_1y_2)}(t)\right)g(y_1),\end{aligned}$$

and the left-hand side of (3.8) is given by

$$(C^{1,*}\Lambda^1(t)g)(x_1,x_2) = \sum_{y_1\in\mathcal{X}_1} \lambda^1_{x_1y_1}(t)g(y_1).$$

Since g is arbitrary, we obtain that condition **N(1)** is equivalent to the existence of a matrix function $\Lambda^1 = [\lambda^1_{x_1y_1}]_{x_1,y_1\in\mathcal{X}_1}$ such that for each $t \geq 0$ we have

$$\lambda^1_{x_1y_1}(t) = \sum_{y_2\in\mathcal{X}_2} \lambda_{(x_1x_2)(y_1y_2)}(t), \quad x_1,y_1 \in \mathcal{X}_1,\ x_2 \in \mathcal{X}_2. \tag{3.9}$$

Hence, we see that condition $\mathbf{N}(1)$ is equivalent to condition $\mathbf{M}(1)$. Finally, note also that, in view of (3.9) and the fact that Λ is the infinitesimal generator function of a Markov chain, it is straightforward to verify that the matrix function Λ^1 is a valid generator function. □

The sufficient conditions (3.7) for strong Markov consistency can be interpreted in the context of the martingale characterization of Markov chains (see Appendix C), or just recalling Dynkin's formula (see e.g. Rogers and Williams (2000)), as follows:

Let C^i, $i = 1,2$, be the projection from $\mathcal{X}_1 \times \mathcal{X}_2$ onto the ith coordinate, that is $C^i(x) = x_i$ for any $x = (x_1,x_2)$. Since X is a Markov chain, for $i \in \{1,2\}$, $0 \leq s \leq t$, and for any function f^i on $\mathcal{X}_i$ we have

$$C^{i,*}f^i(X_t) = C^{i,*}f^i(X_s) + \int_s^t (\Lambda(u)(C^{i,*}f^i))(X_u)du + M_t^{C^{i,*}f^i} - M_s^{C^{i,*}f^i}, \tag{3.10}$$

where $M^{C^{i,*}f^i}$ is a martingale with respect to $\mathbb{F}^X$. Thus,

$$f^i(C^iX_t) = f^i(C^iX_s) + \int_s^t (\Lambda(u)(C^{i,*}f^i))(X_u)du + M_t^{C^{i,*}f^i} - M_s^{C^{i,*}f^i}. \tag{3.11}$$

If condition $\mathbf{N}(i)$ holds then we may rewrite (3.11) as

$$f^i(X^i_t) = f^i(X^i_s) + \int_s^t (\Lambda^i(u)f^i)(X^i_u)du + M_t^{C^{i,*}f^i} - M_s^{C^{i,*}f^i}, \tag{3.12}$$

which shows that X^i is a Markov chain with respect to $\mathbb{F}^X$ with matrix generator function Λ^i.

3.2 Definition and Characterization of Weak Markov Consistency

The concept of weak Markov consistency for a multivariate Markov chain X is defined as follows.

Definition 3.8 1. Let us fix $i \in \{1,\ldots,n\}$. We say that the process X satisfies the *weak Markov consistency condition with respect to the coordinate* X^i if, for every $B \subset \mathcal{X}_i$ and all $t,s \geq 0$,

$$\mathbb{P}\left(X^i_{t+s} \in B \middle| \mathcal{F}^{X^i}_t\right) = \mathbb{P}\left(X^i_{t+s} \in B|X^i_t\right), \quad \mathbb{P}\text{-a.s.}, \tag{3.13}$$

so that the coordinate X^i of X is a Markov process in its own filtration $\mathbb{F}^{X^i}$.

2. If X satisfies the weak Markov consistency condition with respect to X^i for each $i \in \{1,\ldots,n\}$ then we say that X satisfies the weak Markov consistency condition.

Remark 3.9 It will be shown in Theorem 3.12 that the weak Markov consistency property given in Definition 3.8 can be characterized in terms of the structural properties of X and of X^i, $i = 1,\ldots,n$. Specifically, it is characterized in terms of the infinitesimal generator functions of these processes.

Obviously, strong Markov consistency implies weak Markov consistency but not vice versa, as will be seen in Example 7.7 in Section 7.3. This observation leads to

Definition 3.10 1. Let us fix $i \in \{1,\ldots,n\}$. We say that the process X satisfies the *strictly weak Markov consistency condition with respect to the coordinate* X^i if X is weakly Markov consistent, but not strongly Markov consistent, with respect to X^i.

2. If X satisfies the strictly weak Markov consistency condition with respect to X^i for each $i \in \{1,\ldots,n\}$ then we say that X satisfies the strictly weak Markov consistency condition.

In Chapter 7 we will give an example of a strictly weak Markov consistent Markov chain.

Remark 3.11 In many respects, the concept of weak Markov consistency is more important in practical applications than the concept of strong Markov consistency. For example, in the context of credit risk, strong Markov consistency prohibits the so-called default contagion except for the extreme case of joint defaults. In contrast, weak Markov consistency allows us to model not only default contagion but, more generally, the contagion of credit migrations. We refer to Example 2.4 in Bielecki et al. (2014b) for some insight into these issues.

3.2.1 Necessary and Sufficient Conditions for Weak Markov Consistency

We take $n=2$ and study the weak Markov consistency of X with respect to X^1. A completely analogous discussion can be carried out with regard to X^2.

Theorem 3.12 *The process X satisfies the weak Markov consistency property with respect to coordinate X^1 if and only if, for any $x_1, y_1 \in \mathcal{X}_1$, $x_1 \neq y_1$, we have*

$$\mathbb{1}_{\{X_t^1=x_1\}} \sum_{x_2,y_2\in\mathcal{X}_2} \lambda_{(x_1x_2)(y_1y_2)}(t)\mathbb{E}_{\mathbb{P}}\left(\mathbb{1}_{\{X_t^2=x_2\}}|\mathcal{F}_t^{X^1}\right) \\ = \mathbb{1}_{\{X_t^1=x_1\}}\lambda^1_{x_1y_1}(t), \quad dt\otimes d\mathbb{P}\text{-a.e.} \tag{3.14}$$

for some locally integrable functions $\lambda^1_{x_1y_1}(\cdot)$. The infinitesimal generator function of X^1 is $\Lambda^1 = [\lambda^1_{x_1y_1}]_{x_1,y_1\in\mathcal{X}_1}$, with $\lambda^1_{x_1x_1}(\cdot)$ given by

$$\lambda^1_{x_1x_1}(t) = -\sum_{y_1\in\mathcal{X}_1, y_1\neq x_1} \lambda^1_{x_1y_1}(t), \quad x_1\in\mathcal{X}_1,\ t\geq 0.$$

Proof Assume that (3.14) holds. Since X is a Markov chain, using Lemma C.4 and (3.14) we see that for each $x_1, y_1 \in \mathcal{X}_1$ the $\mathbb{B}^{X^1}$-compensator of $N^1_{x_1,y_1}$ has the form

$$\widehat{\nu}^1_{x_1y_1}(dt) = \mathbb{1}_{\{X_t^1=x_1\}}\lambda^1_{x_1y_1}(t)dt$$

for some locally integrable deterministic function $\lambda^1_{x_1y_1}(t)$, $t\geq 0$. In particular, note that (3.14) implies that $\lambda^1_{x_1y_1}$ is nonnegative for $x_1\neq y_1$. Then, by Proposition C.2, X^1 is a Markov chain with generator $\Lambda^1(t) = [\lambda^1_{x_1y_1}(t)]_{x_1,y_1\in\mathcal{X}_1}$, $t\geq 0$.

Conversely, assume that X^1 is a Markov chain with respect to its natural filtration with generator $\Lambda^1(t) = [\lambda^1_{x_1y_1}(t)]_{x_1,y_1\in\mathcal{X}_1}$, $t\geq 0$. Then (3.14) follows from Lemma C.4 and Proposition C.2. □

Remark 3.13 Note that (3.14) implies that, for all $x_1, y_1\in\mathcal{X}_1$ such that $x_1\neq y_1$, we have

$$\mathbb{1}_{\{X_t^1=x_1\}} \sum_{x_2,y_2\in\mathcal{X}_2} \lambda_{(x_1x_2)(y_1y_2)}(t)\mathbb{E}_{\mathbb{P}}\left(\mathbb{1}_{\{X_t^2=x_2\}}\middle|X_t^1=x_1\right) \\ = \mathbb{1}_{\{X_t^1=x_1\}}\lambda^1_{x_1y_1}(t), \quad dt\otimes d\mathbb{P}\text{-a.e.}, \tag{3.15}$$

and thus

$$\sum_{x_2,y_2\in\mathcal{X}_2} \lambda_{(x_1x_2)(y_1y_2)}(t)\mathbb{E}_{\mathbb{P}}\left(\mathbb{1}_{\{X_t^2=x_2\}}\middle|X_t^1=x_1\right) = \lambda^1_{x_1y_1}(t) \tag{3.16}$$

for almost all $t\geq 0$. Thus, condition (3.16) is necessary for the weak Markov consistency of X with respect to X^1.

A result analogous to Theorem 3.12, but with respect to coordinate X^2, reads: *the process X satisfies the weak Markov consistency property with respect to coordinate*

X^2 if and only if, for any $x_2, y_2 \in \mathcal{X}_2$, $x_2 \neq y_2$ we have

$$\begin{aligned} \mathbb{1}_{\{X_t^2 = x_2\}} \sum_{x_1, y_1 \in \mathcal{X}_1} \lambda_{(x_1 x_2)(y_1 y_2)}(t) \mathbb{E}_{\mathbb{P}} \left(\mathbb{1}_{\{X_t^1 = x_1\}} \middle| \mathcal{F}_t^{X^2} \right) \\ = \mathbb{1}_{\{X_t^2 = x_2\}} \lambda^2_{x_2 y_2}(t), \quad dt \otimes d\mathbb{P}\text{-a.e.} \end{aligned} \tag{3.17}$$

for some locally integrable functions $\lambda^2_{x_2 y_2}$. Then the generator of X^2 is $\Lambda^2(t) = [\lambda^2_{x_2 y_2}(t)]_{x_2, y_2 \in \mathcal{X}_2}$, $t \geq 0$, with $\lambda^2_{x_2 x_2}$ given by

$$\lambda^2_{x_2 x_2}(t) = - \sum_{y_2 \in \mathcal{X}_2, y_2 \neq x_2} \lambda^2_{x_2 y_2}(t), \quad x_2 \in \mathcal{X}_2,\ t \geq 0.$$

As already mentioned, on the one hand conditions $\mathbf{M}(1)$ and $\mathbf{M}(2)$ imply conditions (3.3) and (3.4), respectively. On the other hand, it is clear that conditions (3.3) and (3.4) imply (3.16) and (3.17), respectively. This agrees with the fact that strong Markov consistency implies weak Markov consistency.

Remark 3.14 Ball and Yeo (1993) considered the so-called lumpability problem, which can be described as follows. For a given Markov chain X with state space $\mathcal{X}$ and a function $h : \mathcal{X} \to \mathcal{Y}$, we define a lumped process Y by $Y(t) = h(X(t))$, $t \geq 0$. If the process Y is a Markov chain for an arbitrary initial distribution of X then we say that Y is strongly lumpable with respect to the function h. If Y is a Markov chain for at least one initial distribution of X then Y is said to be weakly lumpable with respect to h. The term "lumping" is motivated by the fact that we may have $h(x) = h(\widetilde{x})$ for two different states x and $\widetilde{x}$ of X, and thus these two states are lumped via the function h into one state $h(x)$ of Y. It is clear that the lumpability problem of a multivariate Markov chain X with respect to a coordinate projection function is a Markov consistency problem. It is important to note, though, that in the study of lumpability problems the issue of filtration for which Y is a Markov chain was not addressed explicitly. Inspection of the proofs in Ball and Yeo (1993) indicates that the Markov property of Y in the filtration of $\mathbb{F}^Y$ is proved there. Consequently, the lumpability property with respect to the coordinate projection functions is related to the weak Markov consistency property. It needs to be observed, however, that Ball and Yeo (1993) obtained their results under strong assumptions of the irreducibility and positive recurrence of X, which we do not impose in our study of strong and weak Markov consistency.

3.2.2 Operator Interpretation of Necessary Conditions for Weak Markov Consistency

For $i = 1, 2$ and $t \geq 0$ we define an operator Q^i_t, acting on any function f defined on $\mathcal{X} = \mathcal{X}_1 \times \mathcal{X}_2$, by

$$(Q^i_t f)(x_i) = \mathbb{E}_{\mathbb{P}}(f(X_t) | X^i_t = x_i), \quad x_i \in \mathcal{X}_i. \tag{3.18}$$

We recall that $C^{i,*}$ is an extension operator, defined in (3.5).

The next result we prove will be important in Section 7.2 in the context of weak Markov structures.

Theorem 3.15 *Fix $i \in \{1,2\}$. The condition*

$$Q_t^i \Lambda(t) C^{i,*} = \Lambda^i(t) \quad \text{for almost all } t \geq 0, \tag{3.19}$$

where $\Lambda^i(t) = [\lambda^i_{x_i y_i}(t)]_{x_i, y_i \in \mathcal{X}_i}$, with functions $\lambda^i_{x_i y_i}$ as in (3.14) *for $i = 1$ and as in* (3.17) *for $i = 2$, is necessary for weak Markov consistency with respect to X^i.*

Proof We give the proof for $i = 1$. It is enough to observe that (3.16) is equivalent to (3.19). Indeed, first note that (3.19) is equivalent to the equality

$$(Q_t^1 \Lambda(t) C^{1,*} g)(x_1) = \sum_{y_1 \in \mathcal{X}_1} \lambda^1_{x_1 y_1}(t) g(y_1) \tag{3.20}$$

being satisfied for an arbitrary function g on $\mathcal{X}_1$ and $x_1 \in \mathcal{X}_1$ and for almost all $t \geq 0$. Now, using (3.18) we rewrite the left hand side of (3.20) as

$$\begin{aligned}
&(Q_t^1 \Lambda(t) C^{1,*} g)(x_1) \\
&= \mathbb{E}\left(\sum_{(z_1, x_2) \in \mathcal{X}} \mathbb{1}_{\{X_t^1 = z_1, X_t^2 = x_2\}} \sum_{(y_1, y_2) \in \mathcal{X}} \lambda_{(z_1 x_2)(y_1 y_2)}(t) g(y_1) \middle| X_t^1 = x_1 \right) \\
&= \sum_{x_2 \in \mathcal{X}_2} \left(\mathbb{E}\left(\mathbb{1}_{\{X_t^2 = x_2\}} \middle| X_t^1 = x_1 \right) \sum_{(y_1, y_2) \in \mathcal{X}} \lambda_{(x_1 x_2)(y_1 y_2)}(t) g(y_1) \right) \\
&= \sum_{y_1 \in \mathcal{X}_1} \left(\sum_{x_2 \in \mathcal{X}_2} \sum_{y_2 \in \mathcal{X}_2} \mathbb{E}\left(\mathbb{1}_{\{X_t^2 = x_2\}} \middle| X_t^1 = x_1 \right) \lambda_{(x_1 x_2)(y_1 y_2)}(t) \right) g(y_1).
\end{aligned}$$

Since g is arbitrary, (3.20) is equivalent to

$$\lambda^1_{x_1 y_1}(t) = \sum_{x_2 \in \mathcal{X}_2} \sum_{y_2 \in \mathcal{X}_2} \mathbb{E}\left(\mathbb{1}_{\{X_t^2 = x_2\}} \middle| X_t^1 = x_1 \right) \lambda_{(x_1 x_2)(y_1 y_2)}(t),$$

for almost all $t \geq 0$, which is exactly (3.16). □

We end this section by relating the sufficient algebraic conditions for strong Markov consistency to the necessary algebraic conditions for weak Markov consistency.

Proposition 3.16 *Condition* (3.7) *implies condition* (3.19).

Proof It can be easily verified that we have $Q_t^i C^{i,*} = I$ (where I is the respective identity operator) for $i = 1,2$. Thus (3.6) gives

$$Q_t^i \Lambda(t) C^{i,*} = Q_t^i C^{i,*} \Lambda^i(t) = \Lambda^i(t), \quad t \geq 0, i = 1,2. \tag{3.21}$$

□

Remark 3.17 Another possible proof of Proposition 3.16 is the following. Condition (3.7) for $i = 1,2$ is sufficient for the strong Markov consistency of X, which implies the weak Markov consistency of X, for which (3.19) for $i = 1,2$ is a necessary condition.

3.3 When does Weak Markov Consistency Imply Strong Markov Consistency?

It is well known that if a process X is a $\mathbb{P}$-Markov chain with respect to a filtration $\mathbb{H}$, and if it is adapted with respect to a filtration $\hat{\mathbb{H}} \subset \mathbb{H}$, then X is a $\mathbb{P}$-Markov chain with respect to $\hat{\mathbb{H}}$. However, the converse is not true in general. Nevertheless, if X is a $\mathbb{P}$-Markov chain with respect to $\hat{\mathbb{H}}$, and $\hat{\mathbb{H}}$ is $\mathbb{P}$-*immersed* in $\mathbb{H}$ (see Appendix A), then we can deduce from the martingale characterization of Markov chains that X is also a $\mathbb{P}$-Markov chain with respect to $\mathbb{H}$.

Thus, if $\mathbb{F}^{X^i}$ is $\mathbb{P}$-immersed in $\mathbb{F}^X$ then the weak Markov consistency of X with respect to X^i will imply the strong Markov consistency of X with respect to X^i. In the following theorem we demonstrate that in fact this property is equivalent to $\mathbb{P}$-immersion between $\mathbb{F}^{X^i}$ and $\mathbb{F}^X$, given that weak Markov consistency holds.

Theorem 3.18 *Let $i = 1,2$ and assume that X satisfies the weak Markov consistency condition with respect to X^i. Then X satisfies the strong Markov consistency condition with respect to X^i if and only if $\mathbb{F}^{X^i}$ is $\mathbb{P}$-immersed in $\mathbb{F}^X$.*

Proof "$\Longrightarrow$" We give a proof in the case $i = 1$.

Since X^1 is a Markov chain with respect to its own filtration, by Proposition C.2, for any $x_1, y_1 \in \mathcal{X}_1$, $x_1 \neq y_1$, the process $M^1_{x_1y_1}$ given as

$$M^1_{x_1y_1}(t) = N^1_{x_1y_1}((0,t]) - \widehat{\nu}^1_{x_1y_1}((0,t]), \quad t \geq 0, \tag{3.22}$$

is an $\mathbb{F}^{X^1}$-martingale, where $N^1_{x_1y_1}$ and $\widehat{\nu}^1_{x_1y_1}$ are given by (C.9) and (C.11) in Section C.2, respectively.

By Proposition 5.9.1.1 in Jeanblanc et al. (2009), in order to show that $\mathbb{F}^{X^i}$ is $\mathbb{P}$-*immersed* in $\mathbb{F}^X$ it is sufficient to show that every $\mathbb{F}^{X^1}$-square-integrable martingale Z is also an $\mathbb{F}^X$-square-integrable martingale under $\mathbb{P}$.

Now, let Z be an $\mathbb{F}^{X^1}$-square-integrable martingale under $\mathbb{P}$. Taking the family $(M^1_{x_1y_1})_{x_1,y_1 \in \mathcal{X}_1 : x_1 \neq y_1}$ given by (3.22) and using the martingale representation theorem (see Theorem IV.21.15 in Rogers and Williams (2000)) we have,

$$Z_t = Z_0 + \sum_{x_1 \neq y_1} \int_{(0,t]} g(s,x_1,y_1,\omega)dM^1_{x_1y_1}(s), \quad t \geq 0, \tag{3.23}$$

for some function $g : (0,\infty) \times \mathcal{X}_1 \times \mathcal{X}_1 \times \Omega \to \mathbb{R}$ such that, for every x_1, y_1, the mapping $(t,\omega) \mapsto g(t,x_1,y_1,\omega)$ is $\mathbb{F}^{X^1}$-predictable and $g(t,x_1,x_1,\omega) = 0$, $\mathbb{P}$-a.s. The $\mathbb{F}^{X^1}$-oblique bracket of $M^1_{x_1y_1}$ (i.e. the $\mathbb{F}^{X^1}$-compensator of $(M^1_{x_1,y_1})^2$) is equal to $(\widehat{\nu}^1_{x_1y_1}((0,t]))_{t\geq 0}$. Indeed, first note that using integration by parts it follows that

$$(M^1_{x_1,y_1}(t))^2 - [M^1_{x_1,y_1}](t) = (M^1_{x_1,y_1}(0))^2 + 2\int_0^t M^1_{x_1,y_1}(u-)dM^1_{x_1,y_1}(u), \quad t \geq 0$$

is an $\mathbb{F}^{X^1}$-local martingale, where, as usual, by $[M]$ we denote the square bracket (see Appendix A) of M. By convention, $M^1_{x_1,y_1}(u-)$ denotes the left-hand limit of $M^1_{x_1,y_1}$ at u. Moreover, we have

$$[M^1_{x_1,y_1}](t) = N^1_{x_1,y_1}(t), \quad t \geq 0,$$

and thus

$$[M^1_{x_1,y_1}](t) - \widehat{\nu}^1_{x_1 y_1}((0,t]), \quad t \geq 0,$$

is an $\mathbb{F}^{X^1}$-local martingale. These relations imply that

$$(M^1_{x_1,y_1}(t))^2 - \widehat{\nu}^1_{x_1 y_1}((0,t]) = (M^1_{x_1,y_1}(t))^2 - [M^1_{x_1,y_1}]_t \\ + [M^1_{x_1,y_1}]_t - \widehat{\nu}^1_{x_1 y_1}((0,t]), \quad t \geq 0,$$

is an $\mathbb{F}^{X^1}$-local martingale since it is a sum of two $\mathbb{F}^{X^1}$-local martingales. Since the process $(\widehat{\nu}^1_{x_1 y_1}((0,t]))_{t \geq 0}$ is $\mathbb{F}^{X^1}$-predictable it is the $\mathbb{F}^{X^1}$-oblique bracket of $M^1_{x_1,y_1}$.

Consequently, since Z is square integrable, g satisfies the integrability condition

$$\mathbb{E}\left(\sum_{x_1 \neq y_1} \int_{(0,t]} |g(s,x_1,y_1)|^2 \widehat{\nu}^1_{x_1 y_1}(ds) \right) < \infty, \quad t \geq 0. \tag{3.24}$$

From the strong Markov consistency of X with respect to X^1 we infer that X^1 is a Markov chain with respect to $\mathbb{F}^X$ and, therefore, for every $x_1 \neq y_1$, $M^1_{x_1 y_1}$ is an $\mathbb{F}^X$-martingale. Thus, the $\mathbb{F}^X$-oblique bracket of $M^1_{x_1 y_1}$ is equal to $(\widehat{\nu}^1_{x_1 y_1}((0,t]))_{t \geq 0}$.

Moreover, for every x_1, y_1 the mapping $(t, \omega) \to g(t, x_1, y_1, \omega)$ is $\mathbb{F}^X$-predictable. Hence, using (3.23) and (3.24) we deduce that Z is an $\mathbb{F}^X$-square-integrable martingale.

"$\Longleftarrow$" Assume that $\mathbb{F}^{X^1}$ is immersed in $\mathbb{F}^X$. The weak Markov consistency of X with respect to X^1 implies that, for any $x_1, y_1 \in \mathcal{X}_1$, $x_1 \neq y_1$, the process $M^1_{x_1 y_1}$ is an $\mathbb{F}^{X^1}$-martingale. Using our immersion hypothesis we conclude that, for any $x_1, y_1 \in \mathcal{X}_1$, $x_1 \neq y_1$, the process $M^1_{x_1 y_1}$ is an $\mathbb{F}^X$-martingale. Thus, Proposition C.2 implies that X^1 is a Markov process with respect to $\mathbb{F}^X$. □

Corollary 3.19 *Fix $i \in \{1,2\}$. Assume that X satisfies the weak Markov consistency condition with respect to X^i. Then X satisfies the strictly weak Markov consistency condition with respect to X^i if and only if $\mathbb{F}^{X^i}$ is not $\mathbb{P}$-immersed in $\mathbb{F}^X$.*

4

Consistency of Finite Multivariate Conditional Markov Chains

Conditional Markov chains (CMCs) form an important class of stochastic processes, and thus the study of the related consistency problems is important. Appendix D provides all the background material on CMCs needed here.

Finite conditional Markov chains generalize classical finite Markov chains. Thus, in many ways, the study of Markov consistency for finite multivariate conditional Markov chains done in this chapter is a generalization of the study done in Chapter 3. Nevertheless, the presence of both chapters in this book is justified. For one, the mathematics employed in Chapter 3 is more straightforward than the mathematics used here; thus making the ideas presented in Chapter 3 easier to discuss, and consequently rendering that chapter to serve as natural preparatory material for the present chapter. In particular, the results derived here are nicely illustrated by their counterparts in the simpler set-up of Chapter 3. Also, given that simpler set-up, we were able to conduct a more comprehensive study in Chapter 3; it is still a challenge to carry some of the analysis done in Chapter 3 to the case of conditional finite Markov chains.

4.1 Strong Markov Consistency of Conditional Markov Chains

Let $X = (X^1, \ldots, X^n)$ be a multivariate $(\mathbb{F}, \mathbb{F}^X)$-CMC with values in $\mathcal{X} := \times_{k=1}^n \mathcal{X}_k$, where $\mathcal{X}_k$ is a finite subset of some vector space, $k = 1, \ldots, n$. We recall that we consider CMCs on the finite time interval $[0, T]$.

Definition 4.1 1. Let us fix $k \in \{1, \ldots, n\}$. We say that the process X satisfies the strong Markov consistency property with respect to $(X^k, \mathbb{F})$ if, for every $x_1^k, \ldots, x_m^k \in \mathcal{X}_k$ and for all $0 \leq t \leq t_1 \leq \cdots \leq t_m \leq T$, it holds that

$$\begin{aligned} &\mathbb{P}\left(X_{t_m}^k = x_m^k, \ldots, X_{t_1}^k = x_1^k \,\middle|\, \mathcal{F}_t \vee \mathcal{F}_t^X\right) \\ &= \mathbb{P}\left(X_{t_m}^k = x_m^k, \ldots, X_{t_1}^k = x_1^k \,\middle|\, \mathcal{F}_t \vee \sigma(X_t^k)\right), \end{aligned} \tag{4.1}$$

or, equivalently, if X^k is an $(\mathbb{F}, \mathbb{F}^X)$-CMC.

2. If X satisfies the strong Markov consistency property with respect to $(X^k,\mathbb{F})$ for all $k \in \{1,\ldots,n\}$ then we say that X satisfies the strong Markov consistency property with respect to $\mathbb{F}$.

The above definition of strong Markov consistency can be naturally extended to the case of a process $X = (X^1,\ldots,X^n)$ which is a multivariate $(\mathbb{F},\mathbb{G})$-CMC, with $\mathbb{F}^X \subseteq \mathbb{G}$. Here, essentially without any loss of generality but with a gain of transparency, we limit ourselves to the study of strong consistency for multivariate $(\mathbb{F},\mathbb{F}^X)$-CMCs.

Remark 4.2 We say that the collection $\{X^i,\ i \neq k\}$ Granger-causes X^k if $E(f(X^k_s)|X_t) \neq E(f(X^k_s)|X^k_t)$ for some $0 \leq t \leq s$ (see Granger (1969)). There is a relation between the strong Markov consistency of X with respect to $(X^k,\mathbb{F})$ and the concept of Granger's causality: suppose that the process X satisfies the strong Markov consistency property with respect to $(X^k,\mathbb{F})$. If the reference filtration $\mathbb{F}$ is trivial then the collection $\{X^i,\ i \neq k\}$ does not Granger-cause X^k. By extension, we may say that, in the case when the reference filtration $\mathbb{F}$ is not trivial, "conditionally on $\mathbb{F}$", the collection $\{X^i,\ i \neq k\}$ does not Granger-cause X^k.

The next definition extends the previous one by requiring that the laws of the marginal processes $X^k,\ k = 1,\ldots,n$, are predetermined. This definition will be a gateway to the concept of strong CMC structures in Chapter 8.

Definition 4.3 Let $\mathcal{Y} = \{Y^1,\ldots,Y^n\}$ be a collection of processes such that each Y^k is an $(\mathbb{F},\mathbb{F}^{Y^k})$-CMC with values in $\mathcal{X}_k$.

1. Let us fix $k \in \{1,\ldots,n\}$. We say that a process X satisfies the strong Markov consistency property with respect to $(X^k,\mathbb{F},Y^k)$ if both the following conditions hold

   (i) The process X satisfies the strong Markov consistency property with respect to $(X^k,\mathbb{F})$.

   (ii) The conditional law of X^k given $\mathcal{F}_T$ coincides with the conditional law of Y^k given $\mathcal{F}_T$, i.e. if, for arbitrary $0 = t_0 \leq t_1 \leq \cdots \leq t_n \leq t \leq T$ and $(x^k_1,\ldots,x^k_n) \in \mathcal{X}^n_k$, we have

$$\mathbb{P}(X^k_{t_m} = x^k_m, \ldots X^k_{t_1} = x^k_1 \mid \mathcal{F}_T) = \mathbb{P}(Y^k_{t_1} = x^k_m, \ldots Y^k_{t_1} = x^k_1 \mid \mathcal{F}_T). \tag{4.2}$$

2. If X satisfies the strong Markov consistency property with respect to $(X^k,\mathbb{F},Y^k)$ for every $k \in \{1,\ldots,n\}$ then we say that X satisfies the strong Markov consistency property with respect to $(\mathbb{F},\mathcal{Y})$.

It will be clear from Theorems 4.6 and 4.7, respectively, that properties introduced in Definitions 4.1 and 4.3, respectively, can be characterized in terms of the structural properties, specifically, in terms of the $\mathbb{F}$-intensities of all the processes referred to in these definitions.

4.1.1 Necessary and Sufficient Conditions for Strong Markov Consistency of a Multivariate $(\mathbb{F},\mathbb{F}^X)$-CDMC

We will study the necessary and sufficient conditions for strong Markov consistency for process X that is an $(\mathbb{F},\mathbb{F}^X)$-CDMC. We refer to Definition D.11 for the general notion of $(\mathbb{F},\mathbb{G})$-CDMC and Definition D.3 for the notion of $\mathbb{F}$-intensity.

Accordingly, in the rest of this section we will impose

Assumption 4.4

(i) X is an $(\mathbb{F},\mathbb{F}^X)$-CDMC admitting an $\mathbb{F}$-intensity Λ:

$$\Lambda_t = [\lambda_t^{xy}]_{x,y\in\mathcal{X}} = [\lambda_t^{(x^1,\dots,x^n)(y^1,\dots,y^n)}]_{x^k,y^k\in\mathcal{X}_k,k=1,\dots,n}, \quad t\in[0,T].$$

(ii) $\mathbb{P}(X_0 = x_0|\mathcal{F}_T) = \mathbb{P}(X_0 = x_0|\mathcal{F}_0)$ for every $x_0 \in \mathcal{X}$.

Postulate (ii) above has a very clear interpretation and motivation behind it. Indeed, it is natural to postulate that the future information contained in the reference filtration has no additional conditioning impact on the distribution of the initial state X_0 beyond the impact already delivered by the information contained in $\mathcal{F}_0$.

In view of Proposition D.42, we see that Assumption 4.4 is satisfied for the $(\mathbb{F},\mathbb{G})$-CMC processes constructed in Section D.4, with $\mathbb{G} = \mathbb{F}^X$.

Our next goal is to provide a condition characterizing the strong Markov consistency of a process X satisfying Assumption 4.4. Towards this end we first introduce the following condition (which is a counterpart of conditions (3.3) and (3.4)):

Condition (SM-*k*) There exist $\mathbb{F}$-adapted processes $\lambda^{k;x^k y^k}$, $x^k, y^k \in \mathcal{X}_k$, $x^k \neq y^k$, defined on the time interval $[0,T]$, such that

$$\begin{aligned}
&\mathbb{1}_{\{X_t^k=x^k\}} \sum_{\substack{y^m\in\mathcal{X}_m,\\ m=1,\dots,n,m\neq k}} \lambda_t^{(X_t^1,\dots,X_t^{k-1},x^k,X_t^{k+1},\dots,X_t^n)(y^1,\dots,y^k,\dots,y^n)}\\
&\quad = \mathbb{1}_{\{X_t^k=x^k\}} \lambda_t^{k;x^k y^k}, \quad dt\otimes d\mathbb{P}\text{-}a.e.
\end{aligned} \tag{4.3}$$

It is worth noting that whether Condition (SM-*k*) holds depends on the initial distribution of process X (see Example 4.11).

Recall that the $\mathbb{F}$-intensity of X is not unique in general. Nevertheless, we have the following result, which is a direct consequence of Proposition D.8; thus we omit its proof.

Proposition 4.5 *Let $\widehat{\Lambda}$ be an $\mathbb{F}$-intensity of X. Then Condition (SM-k) holds for $\widehat{\Lambda}$ if and only if it holds for Λ.*

The next theorem provides sufficient and necessary conditions for the strong Markov consistency property of process X with respect to $(X^k,\mathbb{F})$. This theorem elevates the results of Theorem 3.3 to the universe of conditional Markov chains studied in this chapter.

Theorem 4.6 *Let us fix $k \in \{1,\dots,n\}$. A process X is strongly Markov consistent with respect to $(X^k,\mathbb{F})$ if and only if Condition* (SM-k) *is satisfied. Moreover, in this case, process X^k admits an $\mathbb{F}$–intensity $\Lambda^k = [\lambda^{k;x^k y^k}]_{x^k,y^k\in\mathcal{X}_k}$, with $\lambda^{k;x^k y^k}$ as in Condition* (SM-k), *and with $\lambda^{k;x^k x^k}$ given by*

$$\lambda_t^{k;x^k x^k} = -\sum_{y^k\in\mathcal{X}_k, y^k\neq x^k} \lambda_t^{k;x^k y^k}, \quad t\in[0,T].$$

Proof For simplicity of notation, but without loss of generality, we give the proof for $k=1$ and $n=2$, so that $S=\mathcal{X}_1\times\mathcal{X}_2$, $X=(X^1,X^2)$. In this case, (4.3) takes the form

$$\mathbb{1}_{\{X_t^1=x^1\}} \sum_{y^2\in\mathcal{X}_2} \lambda_t^{(x^1,X_t^2)(y^1,y^2)} = \mathbb{1}_{\{X_t^1=x^1\}}\lambda_t^{1;x^1 y^1} \quad dt\otimes d\mathbb{P}\text{-}a.e. \tag{4.4}$$

for all $x^1,y^1\in\mathcal{X}_1, x^1\neq y^1$.

Step 1: Recall that (see Appendix D)

$$H_t^{(x^1,x^2)(y^1,y^2)} = \sum_{0<u\leq t} \mathbb{1}_{\{X_{u-}^1=x^1,X_{u-}^2=x^2,X_u^1=y^1,X_u^2=y^2\}}, \quad t\in[0,T],$$

$$H_t^{(x^1,x^2)} = \mathbb{1}_{\{X_u^1=x^1,X_u^2=x^2\}}, \quad t\in[0,T].$$

For $x^1,y^1\in\mathcal{X}_1$, $x^1\neq y^1$, and for $x^2\in\mathcal{X}_2$ we define auxiliary processes $H^{1;x^1y^1}$, $H^{1;x^1}$, $H^{2;x^2}$ by

$$H_t^{1;x^1y^1} := \sum_{0<u\leq t} \mathbb{1}_{\{X_{u-}^1=x^1,X_u^1=y^1\}} = \sum_{x^2,y^2\in\mathcal{X}_2} H_t^{(x^1,x^2)(y^1,y^2)}, \quad t\in[0,T],$$

and

$$H_t^{1;x^1} := \mathbb{1}_{\{X_t^1=x^1\}}, \quad H_t^{2;x^2} := \mathbb{1}_{\{X_t^2=x^2\}}, \quad t\in[0,T].$$

Next, we consider a process $K^{(x^1,x^2)(y^1,y^2)}$ given by

$$\begin{aligned}
K_t^{(x^1,x^2)(y^1,y^2)} &= H_t^{(x^1,x^2)(y^1,y^2)} - \int_0^t H_u^{(x^1,x^2)}\lambda_u^{(x^1,x^2)(y^1,y^2)}\,du \\
&= H_t^{(x^1,x^2)(y^1,y^2)} - \int_0^t H_u^{1;x^1}H_u^{2;x^2}\lambda_u^{(x^1,x^2)(y^1,y^2)}\,du, \quad t\in[0,T].
\end{aligned}$$

In view of Theorem D.9, the process $K^{(x^1,x^2)(y^1,y^2)}$ is an $\mathbb{F}\vee\mathbb{F}^X$-local martingale. Since, in view of Assumption 4.4, X is also an $(\mathbb{F},\mathbb{F}^X)$-DMC, Theorem D.22 implies that $K^{(x^1,x^2)(y^1,y^2)}$ is also an $\widehat{\mathbb{F}}^X$-local martingale, where $\widehat{\mathbb{F}}^X := (\mathcal{F}_T\vee\mathcal{F}_t^X)_{t\in[0,T]}$. Consequently, the process $K^{x^1y^1}$ given by

$$K_t^{x^1y^1} = \sum_{x^2,y^2\in\mathcal{X}_2} K_t^{(x^1,x^2)(y^1,y^2)}$$
$$= H_t^{1;x^1y^1} - \int_0^t \sum_{x^2,y^2\in\mathcal{X}_2} H_u^{1;x^1} H_u^{2;x^2} \lambda_u^{(x^1,x^2)(y^1,y^2)} du, \quad t \in [0,T], \tag{4.5}$$

is an $\mathbb{F} \vee \mathbb{F}^X$ - local martingale as well as an $\widehat{\mathbb{F}}^X$ - local martingale.

Step 2: Now we prove sufficiency. Assume that (4.4) holds. Then, using (4.5) we obtain that for $t \in [0,T]$,

$$K_t^{x^1y^1} = H_t^{1;x^1y^1} - \int_0^t H_u^{1;x^1} \sum_{x^2\in\mathcal{X}_2} \left[H_u^{2;x^2} \left(\sum_{y^2\in\mathcal{X}_2} \lambda_u^{(x^1,x^2)(y^1,y^2)} \right) \right] du$$
$$= H_t^{1;x^1y^1} - \int_0^t H_u^{1;x^1} \left[\sum_{y^2\in\mathcal{X}_2} \lambda_u^{(x^1,X_u^2)(y^1,y^2)} \right] du$$
$$= H_t^{1;x^1y^1} - \int_0^t H_u^{1;x^1} \lambda_u^{1;x^1y^1} du.$$

Since $K^{x^1y^1}$ is an $\widehat{\mathbb{F}}^X$ - local martingale, by Theorem D.22 the process X^1 is an $(\mathbb{F},\mathbb{F}^X)$-DMC with intensity process Λ^1. Since X is an $(\mathbb{F},\mathbb{F}^X)$–DMC, the filtration $\mathbb{F}$ is immersed in $\mathbb{F} \vee \mathbb{F}^X$ (see Corollary D.18). Consequently, applying Proposition D.26 we conclude that X^1 is an $(\mathbb{F},\mathbb{F}^X)$-CMC.

Step 3: Conversely, assume that X^1 is an $(\mathbb{F},\mathbb{F}^X)$-CMC with an $\mathbb{F}$-intensity Λ^1. So, the process $\widehat{K}^{x^1y^1}$ given by

$$\widehat{K}_t^{x^1y^1} = H_t^{1;x^1y^1} - \int_0^t H_u^{1;x^1} \lambda_u^{1;x^1y^1} du, \quad t \in [0,T],$$

is an $\mathbb{F} \vee \mathbb{F}^X$ - local martingale. Since the process $K^{x^1y^1}$ defined in (4.5) is an $\mathbb{F} \vee \mathbb{F}^X$ - local martingale, the difference $\widehat{K}^{x^1y^1} - K^{x^1y^1}$, which satisfies

$$\widehat{K}_t^{x^1y^1} - K_t^{x^1y^1} = \int_0^t H_u^{1;x^1} \left(\sum_{x^2,y^2\in\mathcal{X}_2} H_u^{2;x^2} \lambda_u^{(x^1,x^2)(y^1,y^2)} - \lambda_u^{1;x^1y^1} \right) du$$
$$= \int_0^t H_u^{1;x^1} \left[\sum_{y^2\in\mathcal{X}_2} \lambda_u^{(x^1,X_u^2)(y^1,y^2)} - \lambda_u^{1;x^1y^1} \right] du, \quad t \in [0,T],$$

is a continuous $\mathbb{F} \vee \mathbb{F}^X$ - local martingale of finite variation and therefore is equal to the null process. This implies (4.4). The proof of the theorem is complete. □

The next theorem gives sufficient and necessary conditions for the strong Markov consistency property of X with respect to $(\mathbb{F},\mathcal{Y})$. This theorem will be used to prove Proposition 4.12, which will be important in the study of strong CMC structures in Chapter 8.

Theorem 4.7 *Let $\mathcal{Y} = \{Y^1, \ldots, Y^n\}$ be a collection of processes such that each Y^k is an $(\mathbb{F}, \mathbb{F}^{Y^k})$-CDMC, with values in $\mathcal{X}_k$ and with an $\mathbb{F}$-intensity $\Psi_t^k = [\psi_t^{k;x^k y^k}]_{x^k, y^k \in \mathcal{X}_k}$, $t \in [0, T]$. Let a process X satisfy Assumption 4.4. Then, X satisfies the strong Markov consistency property with respect to $(\mathbb{F}, \mathcal{Y})$ if and only if, for all $k = 1, \ldots, n$, the following hold:*

(i) *Condition (SM-k) is satisfied with*

$$\lambda^{k;x^k y^k} = \psi^{k;x^k y^k}, \quad x^k, y^k \in \mathcal{X}_k, \quad x^k \neq y^k.$$

(ii) *The law of X_0^k given $\mathcal{F}_T$ coincides with the law of Y_0^k given $\mathcal{F}_T$.*

Proof First, we prove sufficiency. Assuming (i), we conclude from Theorem 4.6 that, for each $k = 1, \ldots, n$, the process X is strongly Markov consistent with respect to $(X^k, \mathbb{F})$ and that X^k admits an $\mathbb{F}$-intensity Ψ^k. This, combined with (ii), implies, in view of Lemma D.31, that X satisfies the strong Markov consistency property with respect to $(\mathbb{F}, \mathcal{Y})$.

Now we prove necessity. Since X satisfies the strong Markov consistency property with respect to $(\mathbb{F}, \mathcal{Y})$, clearly the law of X_0^k given $\mathcal{F}_T$ coincides with the law of Y_0^k given $\mathcal{F}_T$ for all $k = 1, \ldots, n$. In addition, in view of Theorem 4.6 and Lemma D.31, we conclude that (4.3) is satisfied for $\Lambda^k = \Psi^k$, for all $k = 1, \ldots, n$. □

4.1.2 Algebraic Conditions for Strong Markov Consistency

The necessary and sufficient condition for strong Markov consistency stated in Theorem 4.6 may not be easily verifiable. Here, we provide an algebraic sufficient condition for strong Markov consistency, which typically is easily verified. This condition, for a fixed $k \in \{1, \ldots, n\}$, reads as follows:

Condition (ASM-k) *A version of the $\mathbb{F}$-intensity process of X, say Λ, satisfies, for all x^k, $y^k \in \mathcal{X}_k$, $x^k \neq y^k$, and x, $\bar{x} \in \mathcal{X}$ such that $x^k = \bar{x}^k$, the following condition:*

$$\sum_{\substack{y^m \in \mathcal{X}_m, \\ m=1,\ldots,n, m \neq k}} \lambda_t^{(x^1,\ldots,x^k,\ldots,x^n)(y^1,\ldots,y^k,\ldots,y^n)} \tag{4.6}$$
$$= \sum_{\substack{y^m \in \mathcal{X}_m, \\ m=1,\ldots,n, m \neq k}} \lambda_t^{(\bar{x}^1,\ldots,\bar{x}^{k-1},x^k,\bar{x}^{k+1},\ldots,\bar{x}^n)(y^1,\ldots,y^k,\ldots,y^n)} \quad dt \otimes d\mathbb{P}\text{-a.e.}$$

We note that Condition (ASM-k) generalizes Condition $\mathbf{M}(k)$ of Section 3.1 (stated there for $k = 1$ and $k = 2$).

The importance of Condition (ASM-k) stems from the fact that it is easily verifiable and from the following result.

Proposition 4.8 *Let a process X satisfy Assumption 4.4, and let us fix $k \in \{1, \ldots, n\}$. Then, Condition (ASM-k) is sufficient for the strong Markov consistency of X with* *respect to* $(X^k, \mathbb{F})$ *and for* $\Lambda^k = [\lambda^{k;x^k y^k}]_{x^k, y^k \in \mathcal{X}_k}$ *to be an* $\mathbb{F}$*-intensity process of* X^k*, where* $\lambda^{k;x^k y^k}$ *is given by*

$$\lambda^{k;x^k y^k} = \sum_{\substack{y^m \in \mathcal{X}_m, \\ m=1,\dots,n, m \neq k}} \lambda^{(x^1,\dots,x^k,\dots,x^n)(y^1,\dots,y^k,\dots,y^n)} \tag{4.7}$$

for $x^k \neq y^k$*, and*

$$\lambda^{k;x^k x^k} = - \sum_{y^k \in \mathcal{X}_k, y^k \neq x^k} \lambda^{k;x^k y^k}.$$

In particular, Condition (ASM-k) implies Condition (SM-k).

Proof Condition (ASM-k) implies that, for any $x^k, y^k \in \mathcal{X}_k$, $x^k \neq y^k$, the sum

$$\sum_{\substack{y^m \in \mathcal{X}_m, \\ m=1,\dots,n, m \neq k}} \lambda_t^{(x^1,\dots,x^k,\dots,x^n)(y^1,\dots,y^k,\dots,y^n)} \tag{4.8}$$

does not depend on $x^1, \dots, x^{k-1}, x^{k+1}, \dots, x^n$. Thus, (4.3) holds with $\lambda^{k;x^k y^k}$ given by (4.7). Consequently, the result follows by application of Theorem 4.6. □

Example 4.9 below shows that, in general, Condition (ASM-k) is not necessary for the strong Markov consistency of X with respect to $(X^k, \mathbb{F})$. Thus, Condition (SM-k) is, in general, weaker than Condition (ASM-k). In fact, as will be seen in Example 4.9, Condition (ASM-k) implies the strong Markov consistency of X with respect to $(X^k, \mathbb{F})$ regardless of the initial distribution of the process X, whereas, as already observed, whether Condition (SM-k) holds does depend on the initial distribution of X.

Example 4.9 Consider a bivariate process $X = (X^1, X^2)$ taking values in the state space $S = \{(0,0), (0,1), (1,0), (1,1)\}$ and such that it is an $(\mathbb{F}, \mathbb{F}^X)$-CDMC. Suppose that X admits an $\mathbb{F}$-intensity Λ of the form [1]

$$\Lambda_t = [\lambda_t^{xy}]_{x,y \in \mathcal{X}} = \begin{array}{c} \\ (0,0) \\ (0,1) \\ (1,0) \\ (1,1) \end{array} \begin{array}{c} \begin{array}{cccc} (0,0) & (0,1) & (1,0) & (1,1) \end{array} \\ \begin{pmatrix} -a_t & 0 & 0 & a_t \\ 0 & 0 & 0 & 0 \\ 0 & 0 & 0 & 0 \\ b_t & 0 & 0 & -b_t \end{pmatrix} \end{array}, \tag{4.9}$$

where a and b are strictly positive processes.

[1] It is shown in Proposition D.42 that one can always construct an $(\mathbb{F}, \mathbb{F}^X)$-CDMC with a given $\mathbb{F}$-intensity Λ.

Let the $\mathcal{F}_T$-conditional distribution of X_0 be given as

$$\begin{aligned}\mathbb{P}(X_0=(0,1)|\mathcal{F}_T)=\mathbb{P}(X_0=(1,0)|\mathcal{F}_T)=0,\\ \mathbb{P}(X_0=(0,0)|\mathcal{F}_T)=m_0,\quad \mathbb{P}(X_0=(1,1)|\mathcal{F}_T)=m_1,\end{aligned} \tag{4.10}$$

where m_0, m_1 are $\mathcal{F}_0$-measurable random variables. Thus, process X satisfies Assumption 4.4.

Now let us investigate Condition (SM-1) with respect to X. One can verify that the c-transition field of X (see Definition D.15) has the following form:

$$\begin{array}{c} \\ (0,0)\\ (0,1)\\ (1,0)\\ (1,1)\end{array}\begin{array}{c}\begin{array}{cccc}(0,0) & (0,1) & (1,0) & (1,1)\end{array}\\ \begin{pmatrix} P^1_{00}(s,t) & 0 & 0 & P^1_{01}(s,t)\\ 0 & 1 & 0 & 0\\ 0 & 0 & 1 & 0\\ P^1_{10}(s,t) & 0 & 0 & P^1_{11}(s,t)\end{pmatrix}\end{array}.$$

Thus, in view of Proposition D.17, we conclude that, for any $t\in[0,T]$,

$$\begin{aligned}\mathbb{P}(X_t=(0,1)|\mathcal{F}_T)=\mathbb{P}(X_t=(1,0)|\mathcal{F}_T)=0,\\ \mathbb{P}(X_t=(0,0)|\mathcal{F}_T)=m_0P^1_{00}(0,t)+m_1P^1_{10}(0,t).\end{aligned} \tag{4.11}$$

Consequently, Condition (SM-1) (i.e. (4.3) for $k=1$) is satisfied here. Indeed, taking $x^1=0,\ y^1=1$ and invoking (4.10), we obtain that

$$\begin{aligned}&\mathbb{1}_{\{X^1_t=0\}}(\lambda_t^{(0,X^2_t)(1,0)}+\lambda_t^{(0,X^2_t)(1,1)})\\ &\quad=\mathbb{1}_{\{X^1_t=0,X^2_t=0\}}(\lambda_t^{(0,0)(1,0)}+\lambda_t^{(0,0)(1,1)})\\ &\qquad+\mathbb{1}_{\{X^1_t=0,X^2_t=1\}}(\lambda_t^{(0,1)(1,0)}+\lambda_t^{(0,1)(1,1)})\\ &\quad=\mathbb{1}_{\{X^1_t=0,X^2_t=0\}}a_t+\mathbb{1}_{\{X^1_t=0,X^2_t=1\}}0=(\mathbb{1}_{\{X^1_t=0,X^2_t=0\}}+\mathbb{1}_{\{X^1_t=0,X^2_t=1\}})a_t\\ &\quad=\mathbb{1}_{\{X^1_t=0\}}a_t,\end{aligned}$$

where the third equality follows from the fact that

$$\mathbb{1}_{\{X^1_t=0,X^2_t=1\}}=0\quad dt\otimes d\mathbb{P}, \tag{4.12}$$

which is a consequence of (4.11). Analogously, for $x^1=1,\ y^1=0$ it holds that

$$\begin{aligned}&\mathbb{1}_{\{X^1_t=1\}}(\lambda_t^{(1,X^2_t)(0,0)}+\lambda_t^{(1,X^2_t)(0,1)})\\ &\quad=\mathbb{1}_{\{X^1_t=1,X^2_t=0\}}(\lambda_t^{(1,0)(0,0)}+\lambda_t^{(1,0)(0,1)})\\ &\qquad+\mathbb{1}_{\{X^1_t=1,X^2_t=1\}}(\lambda_t^{(1,1)(0,0)}+\lambda_t^{(1,1)(0,1)})\\ &\quad=\mathbb{1}_{\{X^1_t=1,X^2_t=0\}}0+\mathbb{1}_{\{X^1_t=1,X^2_t=1\}}b_t=(\mathbb{1}_{\{X^1_t=1,X^2_t=0\}}+\mathbb{1}_{\{X^1_t=1,X^2_t=1\}})b_t\\ &\quad=\mathbb{1}_{\{X^1_t=1\}}b_t,\end{aligned}$$

where we have used the fact that

$$\mathbb{1}_{\{X_t^1=1,X_t^2=0\}} = 0 \quad dt \otimes d\mathbb{P}. \tag{4.13}$$

Thus, Condition (SM-1) holds here for $\lambda_t^{1;01} = a_t$, $\lambda_t^{1;11} = b_t$. Similarly, one can show that Condition (SM-2) is fulfilled for $\lambda_t^{2;01} = a_t$, $\lambda_t^{2;11} = b_t$. Thus, X is strongly Markov consistent with respect to $\mathbb{F}$. However, Condition (ASM-1) is not satisfied here (regardless of the initial distribution of X) since, for every $t \in [0,T]$, we have

$$\lambda_t^{(0,0)(1,0)} + \lambda_t^{(0,0)(1,1)} = a_t \neq 0 = \lambda_t^{(0,1)(1,0)} + \lambda_t^{(0,1)(1,1)}.$$

Remark 4.10 As already observed, in general the strong Markov consistency of X with respect to $\mathbb{F}$ depends on the initial distribution of X. Consequently, we may have two processes, X which is $(\mathbb{F},\mathbb{F}^X)$-CMC and Y which is $(\mathbb{F},\mathbb{F}^Y)$-CMC, with the same $\mathbb{F}$-intensity, such that one of them is strongly Markov consistent with respect to $\mathbb{F}$ and the other one is not. In fact, let Y be an $(\mathbb{F},\mathbb{F}^Y)$-CMC taking values in a finite state space $S = \{(0,0),(0,1),(1,0),(1,1)\}$, endowed with the same $\mathbb{F}$-intensity as in Example 4.9 and with a conditional initial distribution such that either $\mathbb{P}(\mathbb{P}(Y_0 = (0,1)|\mathcal{F}_T) > 0) > 0$ or $\mathbb{P}(\mathbb{P}(Y_0 = (1,0)|\mathcal{F}_T) > 0) > 0$. For the process Y equality (4.12) is not satisfied and thus Condition (SM-1) does not hold. Consequently, process Y is not strongly Markov consistent with respect to $\mathbb{F}$.

In the next example we will show that an $(\mathbb{F},\mathbb{F}^X)$-CMC may have an intensity for which Condition (ASM-k) does not hold, and it may admit another version of intensity, in the sense of Definition D.7, for which Condition (ASM-k) is fulfilled.

Example 4.11 Let us take X as in Example 4.9. In that example we proved that Conditions (ASM-1) and (ASM-2) are not satisfied by the $\mathbb{F}$-intensity Λ given in (4.9). However, there exists another version of the $\mathbb{F}$-intensity of X, say Γ, for which Conditions (ASM-1) and (ASM-2) are satisfied. Indeed, let us consider a process Γ defined by

$$\Gamma_t = [\gamma_t^{xy}]_{x,y \in \mathcal{X}} = \begin{array}{c} (0,0) \\ (0,1) \\ (1,0) \\ (1,1) \end{array} \overset{\begin{array}{cccc} (0,0) & (0,1) & (1,0) & (1,1) \end{array}}{\begin{pmatrix} -a_t & 0 & 0 & a_t \\ b_t & -a_t - b_t & 0 & a_t \\ b_t & 0 & -a_t - b_t & a_t \\ b_t & 0 & 0 & -b_t \end{pmatrix}}.$$

By Proposition D.8(ii) the process Γ is an $\mathbb{F}$-intensity of X since, in view of (4.12) and (4.13), it holds that

$$\int_0^t (\Gamma_u - \Lambda_u)^\top H_u du = 0, \quad t \in [0,T].$$

Finally, we see that Conditions (ASM-1) and (ASM-2) are satisfied for Γ, because

$$\begin{aligned}
\gamma_t^{(0,0)(1,0)} + \gamma_t^{(0,0)(1,1)} &= a_t = \gamma_t^{(0,1)(1,0)} + \gamma_t^{(0,1)(1,1)}, \\
\gamma_t^{(1,1)(0,0)} + \gamma_t^{(1,1)(0,1)} &= b_t = \gamma_t^{(1,0)(0,0)} + \gamma_t^{(1,0)(0,1)}, \\
\gamma_t^{(0,0)(0,1)} + \gamma_t^{(0,0)(1,1)} &= a_t = \gamma_t^{(1,0)(0,1)} + \gamma_t^{(1,0)(1,1)}, \\
\gamma_t^{(1,1)(0,0)} + \gamma_t^{(1,1)(1,0)} &= b_t = \gamma_t^{(0,1)(0,0)} + \gamma_t^{(0,1)(1,0)}.
\end{aligned}$$

We stress that the result presented in this example very much depends on the conditional initial distribution of the process X.

We end this section with an important result that will play a key role in Chapter 8.

Proposition 4.12 *Let $\mathcal{Y} = \{Y^1, \ldots, Y^n\}$ be a collection of processes such that each Y^k is an $(\mathbb{F}, \mathbb{F}^{Y^k})$-CDMC with values in $\mathcal{X}_k$ and with an $\mathbb{F}$-intensity $\Psi_t^k = [\psi_t^{k;x^k y^k}]_{x^k, y^k \in \mathcal{X}_k}$, $t \in [0,T]$. Let the process X satisfy Assumption 4.4. Assume also that*

(i) *There exists a version of the $\mathbb{F}$-intensity Λ, also denoted by Λ, which satisfies the following condition:*
for each $k = 1, \ldots, n$, $x^k, y^k \in \mathcal{X}_k$, $x^k \neq y^k$, $t \in [0,T]$,

$$\psi_t^{k;x^k y^k} = \sum_{\substack{y^m \in \mathcal{X}_m, \\ m=1,\ldots,n, m \neq k}} \lambda_t^{(x^1,\ldots,x^k,\ldots,x^n)(y^1,\ldots,y^k,\ldots,y^n)}. \tag{4.14}$$

(ii) *The conditional law of X_0^k given $\mathcal{F}_T$ coincides with the conditional law of Y_0^k given $\mathcal{F}_T$ for all $k = 1, \ldots, n$.*

Then X satisfies the strong Markov consistency property with respect to $(\mathbb{F}, \mathcal{Y})$.

Proof We observe that for an $\mathbb{F}$-intensity Λ satisfying (i), Condition (ASM-k) holds for every $k = 1, \ldots, n$. Thus by Proposition 4.8 it follows that (4.3) holds with $\lambda^{k;x^k y^k} = \psi^{k;x^k y^k}$ for all $x^k, y^k \in \mathcal{X}_k$, $x^k \neq y^k$. From (ii) and Theorem 4.7 we conclude that X is strongly Markov consistent with respect to $(\mathbb{F}, \mathcal{Y})$. □

4.2 Weak Markovian Consistency of Conditional Markov Chains

We now proceed with a study of the weak Markov consistency of conditional Markov chains. As in Section 4.1 we consider a process $X = (X^1, \ldots, X^n)$, which is a multivariate $(\mathbb{F}, \mathbb{F}^X)$-CMC with values in $\mathcal{X} := \times_{k=1}^n \mathcal{X}_k$ (recall that $\mathcal{X}_k$ is a finite set, $k = 1, \ldots, n$), and admitting an $\mathbb{F}$-intensity Λ.

As will be seen below, the definitions and results regarding weak Markov consistency are to some extent parallel to those regarding strong Markov consistency.

But, as always, "the devil is in the details", so the reader is asked to be patient with the presentation that follows.

Definition 4.13 1. Let us fix $k \in \{1,\dots,n\}$. We say that the process X satisfies the weak Markov consistency property with respect to $(X^k, \mathbb{F})$ if, for every $x_1^k,\dots,x_m^k \in \mathcal{X}_k$ and for all $0 \le t \le t_1 \le \dots \le t_m \le T$, it holds that

$$\begin{aligned}\mathbb{P}&\left(X_{t_m}^k = x_m^k,\dots,X_{t_1}^k = x_1^k \,\middle|\, \mathcal{F}_t \vee \mathcal{F}_t^{X^k}\right)\\&= \mathbb{P}\left(X_{t_m}^k = x_m^k,\dots,X_{t_1}^k = x_1^k \,\middle|\, \mathcal{F}_t \vee \sigma(X_t^k)\right),\end{aligned} \tag{4.15}$$

or, equivalently, if X^k is an $(\mathbb{F},\mathbb{F}^{X^k})$-CMC.

2. If X satisfies the weak Markov consistency property with respect to $(X^k,\mathbb{F})$ for all $k \in \{1,\dots,n\}$, then we say that X satisfies the weak Markov consistency property with respect to $\mathbb{F}$.

Definition 4.14 Let $\mathcal{Y} = \{Y^1,\dots,Y^n\}$ be a family of processes such that each Y^k is an $(\mathbb{F},\mathbb{F}^{Y^k})$-CMC with values in $\mathcal{X}_k$.

1. Let us fix $k \in \{1,\dots,n\}$. We say that X satisfies the weak Markov consistency property with respect to $(X^k,\mathbb{F},Y^k)$ if
   (i) The process X satisfies the weak Markov consistency property with respect to $(X^k,\mathbb{F})$.
   (ii) The conditional law of X^k given $\mathcal{F}_T$ coincides with the conditional law of Y^k given $\mathcal{F}_T$, i.e. (4.2) holds.
2. If X satisfies the weak Markov consistency property with respect to $(X^k,\mathbb{F},Y^k)$ for every $k \in \{1,\dots,n\}$ then we say that X satisfies the weak Markov consistency property with respect to $(\mathbb{F},\mathcal{Y})$.

It will be clear from Theorems 4.16 and 4.17, respectively, that the properties introduced in Definition 4.13 and Definition 4.14, respectively, can be characterized in terms of the structural properties, specifically, in terms of the $\mathbb{F}$-intensities, of all the processes referred to in these definitions.

4.2.1 Necessary and Sufficient Conditions for Weak Markov Consistency of a Multivariate $(\mathbb{F},\mathbb{F}^X)$-CDMC

We postulate in this subsection that the process X satisfies Assumption 4.4, and we aim here at providing a condition characterizing the weak Markov consistency of the process X.

Let us start by introducing

Condition (WM-k) There exist $\mathbb{F}$-adapted processes $\lambda^{k;x^k y^k}$, $x^k, y^k \in \mathcal{X}_k$, $x^k \neq y^k$, such that

$$\mathbb{1}_{\{X_t^k = x^k\}} \sum_{\substack{x^m, y^m \in \mathcal{X}_m \\ m=1,\dots,n, m \neq k}} \lambda_t^{(x^1,\dots,x^n)(y^1,\dots,y^n)} \times \mathbb{E}_{\mathbb{P}}\left(\mathbb{1}_{\{X_t^1 = x^1,\dots,X_t^{k-1}=x^{k-1}, X_t^{k+1}=x^{k+1},\dots,X_t^n = x^n\}} \Big| \mathcal{F}_t \vee \mathcal{F}_t^{X^k}\right) = \mathbb{1}_{\{X_t^k = x^k\}} \lambda_t^{k;x^k y^k} \quad dt \otimes d\mathbb{P}\text{-a.e.}, \tag{4.16}$$

for all $x^k, y^k \in \mathcal{X}_k$, $x^k \neq y^k$.

As in the case of Condition (SM-k) we have the following proposition, which is a simple consequence of Proposition D.8:

Proposition 4.15 *Let Λ, $\widehat{\Lambda}$ be $\mathbb{F}$-intensities of X. Then condition (WM-k) holds for Λ if and only if it holds for $\widehat{\Lambda}$.*

The next theorem characterizes weak Markov consistency in the present set-up.

Theorem 4.16 *A process X with $\mathbb{F}$-intensity Λ is weakly Markov consistent with respect to $(X^k, \mathbb{F})$ if and only if Condition (WM-k) is satisfied. Moreover, X^k admits an $\mathbb{F}$-intensity process*

$$\Lambda^k := [\lambda^{k;x^k y^k}]_{x^k, y^k \in \mathcal{X}_k}, \tag{4.17}$$

with $\lambda^{k;x^k x^k}$ given by

$$\lambda_t^{k;x^k x^k} = - \sum_{y^k \in \mathcal{X}_k, y^k \neq x^k} \lambda_t^{k;x^k y^k}, \qquad x^k \in \mathcal{X}_k, \qquad t \in [0,T].$$

Proof For simplicity of notation we give the proof for $k = 1$ and $n = 2$. In this case, (4.16) takes the following form $\left(\text{recall our notation: } H_t^{k;x^k} = \mathbb{1}_{\{X_t^k = x^k\}}\right)$

$$H_t^{1;x^1} \sum_{x^2, y^2 \in \mathcal{X}_2} \lambda_t^{(x^1 x^2)(y^1 y^2)} \mathbb{E}_{\mathbb{P}}\left(H_t^{2;x^2} \Big| \mathcal{F}_t \vee \mathcal{F}_t^{X^1}\right) = H_t^{1;x^1} \lambda_t^{1;x^1 y^1} \quad dt \otimes d\mathbb{P}\text{-a.e.} \tag{4.18}$$

for all $x^1, y^1 \in \mathcal{X}^1$, $x^1 \neq y^1$.

Step 1: Fix arbitrary $x^1, y^1 \in \mathcal{X}^1$, $x^1 \neq y^1$. In this step we prove that the $\mathbb{F} \vee \mathbb{F}^{X^1}$-optional projection of $K^{x^1 y^1}$, given in (4.5), is an $\mathbb{F} \vee \mathbb{F}^{X^1}$-local martingale, and we find its explicit form.

Towards this end we recall that in Step 1 of the proof of Theorem 4.6 we showed that the process $K^{x^1 y^1}$ is an $\mathbb{F} \vee \mathbb{F}^X$-local martingale. Now, let us denote by $\widetilde{K}^{x^1 y^1}$ the optional projection of $K^{x^1 y^1}$ onto the filtration $\mathbb{F} \vee \mathbb{F}^{X^1}$.[2] Observe that the sequence

[2] We note that for the existence of optional projections we do not need the right continuity of the filtration (see Ethier and Kurtz, 1986, Theorem 2.4.2).

$$\tau_\ell := \inf\left\{t \geq 0 : H_t^{1;x^1y^1} \geq \ell \text{ or } \int_0^t \left(\sum_{x^2,y^2 \in \mathcal{X}_2} \lambda_u^{(x^1,x^2)(y^1,y^2)} \right) du \geq \ell \right\},$$

for $\ell = 1, \ldots,$ is a sequence of $\mathbb{F} \vee \mathbb{F}^X$-stopping times, as well as $\mathbb{F} \vee \mathbb{F}^{X^1}$-stopping times, and that it is a localizing sequence for $K^{x^1y^1}$. So, by Theorem 3.7 in Föllmer and Protter (2011), the process $\widetilde{K}^{x^1y^1}$ is an $\mathbb{F} \vee \mathbb{F}^{X^1}$-local martingale.

Using Theorem 5.25 in He et al. (1992), with $A_t := \int_0^t H_u^{1;x^1} du$, we obtain that

$$\begin{aligned}
&\mathbb{E}\left(\int_0^t \sum_{x^2,y^2 \in \mathcal{X}_2} H_u^{2;x^2} \lambda_u^{(x^1,x^2)(y^1,y^2)} dA_u \,\middle|\, \mathcal{F}_t \vee \mathcal{F}_t^{X^1} \right) \\
&\quad = \int_0^t \mathbb{E}\left(\sum_{x^2,y^2 \in \mathcal{X}_2} H_u^{2;x^2} \lambda_u^{(x^1,x^2)(y^1,y^2)} \,\middle|\, \mathcal{F}_u \vee \mathcal{F}_u^{X^1} \right) dA_u.
\end{aligned}$$

From this and from (4.5) we conclude that the process $\widetilde{K}^{x^1y^1}$ satisfies

$$\widetilde{K}_t^{x^1y^1} = H_t^{1;x^1y^1} - \int_0^t H_u^{1;x^1} \mathbb{E}\left(\sum_{x^2,y^2 \in \mathcal{X}_2} H_u^{2;x^2} \lambda_u^{(x^1,x^2)(y^1,y^2)} \,\middle|\, \mathcal{F}_u \vee \mathcal{F}_u^{X^1} \right) du, \quad (4.19)$$

for $t \in [0,T]$.

Step 2: Now, suppose that (4.18) holds. We will prove that X^1 is an $(\mathbb{F}, \mathbb{F}^{X^1})$-CMC with $\mathbb{F}$-intensity $\Lambda_t^1 = [\lambda_t^{1;x^1y^1}]_{x^1,y^1 \in \mathcal{X}_1}$.

From (4.19) and (4.18) we have

$$\widetilde{K}_t^{x^1y^1} = H_t^{1;x^1y^1} - \int_0^t H_u^{1;x^1} \lambda_u^{1;x^1y^1} du, \quad t \in [0,T].$$

Thus, we can apply Theorem D.21 to process X^1 in order to conclude that Λ^1 is an $\mathbb{F}$-intensity of X^1, so that the $\mathbb{R}^d$-valued process $\widetilde{M}^1 = (\widetilde{M}^{1;x^1}; x^1 \in \mathcal{X}_1)^\top$, given as

$$\widetilde{M}_t^1 = H_t^1 - \int_0^t (\Lambda_u^1)^\top H_u^1 du, \quad t \in [0,T], \quad (4.20)$$

is an $\mathbb{F} \vee \mathbb{F}^{X^1}$-local martingale.

Next, using Theorem D.34 we will show that X^1 is an $(\mathbb{F}, \mathbb{F}^{X^1})$-CMC. Towards this end, we first observe that Assumption 4.4 combined with Corollary D.18 imply that $\mathbb{F}$ is immersed in $\mathbb{F} \vee \mathbb{F}^X$ and thus that $\mathbb{F}$ is immersed in $\mathbb{F} \vee \mathbb{F}^{X^1}$.

Moreover, as we will show now, all real-valued $\mathbb{F}$-local martingales, which, in view of the immersion of $\mathbb{F}$ in $\mathbb{F} \vee \mathbb{F}^{X^1}$, are also $\mathbb{F} \vee \mathbb{F}^{X^1}$-local martingales, are orthogonal to the $\mathbb{F} \vee \mathbb{F}^{X^1}$-local martingale M given as in (D.6) (with $\mathbb{F}^{X^1}$ in place of $\mathbb{G}$). Indeed, let us take an arbitrary real-valued $\mathbb{F}$-local martingale N. Then, by definition of M and the fact that M is a pure-jump local martingale, we have, for any $(x^1,x^2) \in \mathcal{X}$,

$$
\begin{aligned}
[N, M^{(x^1,x^2)}]_t &= \sum_{0<u\le t} \Delta N_u \Delta M_u^{(x^1,x^2)} \\
&= \sum_{0<u\le t} \Delta N_u \Delta H_u^{(x^1,x^2)}, \quad t \in [0,T].
\end{aligned} \tag{4.21}
$$

Now, since the jump times of N are $\mathbb{F}$-stopping times, by Proposition 6.1 in Jakubowski and Niewęgłowski (2010a) we conclude that N and X do not have common jump times or, equivalently, that N and M do not have common jump times. Therefore $[N, M^{(x^1,x^2)}] = 0$, so that N is orthogonal to the processes $M^{(x^1,x^2)}$, $(x^1,x^2) \in \mathcal{X}$. Consequently, the processes N and M are orthogonal.

Using the above, we will now deduce that all real-valued $\mathbb{F}$-local martingales are orthogonal to the process $\widetilde{M}^1$ defined in (4.20). In fact, taking N to be an arbitrary real-valued $\mathbb{F}$-local martingale, we see that the orthogonality of N and $\widetilde{M}^1$ is a consequence of the orthogonality of N and $\widetilde{M}^{1;x^1}$ for all all $x^1 \in \mathcal{X}_1$. The latter property follows from the following equalities:

$$
\begin{aligned}
[N, \widetilde{M}^{1;x^1}]_t &= \sum_{0<u\le t} \Delta N_u \Delta \widetilde{M}_u^{1;x^1} = \sum_{0<u\le t} \Delta N_u \Delta H_u^{1;x^1} \\
&= \sum_{0<u\le t} \sum_{x^2\in\mathcal{X}_2} \Delta N_u \Delta H_u^{(x^1,x^2)} = \sum_{x^2\in\mathcal{X}_2} \sum_{0<u\le t} \Delta N_u \Delta H_u^{(x^1,x^2)} \\
&= \sum_{x^2\in\mathcal{X}_2} [N, M^{(x^1,x^2)}]_t = 0, \quad t \in [0,T],
\end{aligned}
$$

where the penultimate equality follows from (4.21).

Consequently, we see that the assumptions of Theorem D.34 are fulfilled (taking $X = X^1$ and $\mathbb{G} = \mathbb{F}^{X^1}$), and thus we may conclude that X^1 is an $(\mathbb{F}, \mathbb{F}^{X^1})$-CMC with $\mathbb{F}$-intensity $\Lambda_t^1 = [\lambda_t^{1;x^1y^1}]_{x^1,y^1\in\mathcal{X}_1}$.

Step 3: Conversely, assume now that X^1 is an $(\mathbb{F}, \mathbb{F}^{X^1})$-CMC with $\mathbb{F}$-intensity $\Lambda_t^1 = [\lambda_t^{1;x^1y^1}]_{x^1,y^1\in\mathcal{X}_1}$. We will show that (4.18) holds.

Towards this end we fix $x^1, y^1 \in \mathcal{X}_1$, $x^1 \neq y^1$. Analogously to Step 3 in the proof of Theorem 4.6 we conclude that, taking $\widetilde{K}^{x^1y^1}$ as in (4.19) and taking $\widehat{K}^{x^1y^1}$ as given by

$$
\widehat{K}_t^{x^1y^1} = H_t^{1;x^1y^1} - \int_0^t H_u^{1;x^1} \lambda_u^{1;x^1y^1} du, \quad t \in [0,T],
$$

the difference $\widetilde{K}^{x^1y^1} - \widehat{K}^{x^1y^1}$ is a continuous $\mathbb{F} \vee \mathbb{F}^{X^1}$-local martingale of finite variation, starting at 0. Therefore it is equal to 0, which implies (4.18). The proof of the theorem is complete. □

The next theorem gives necessary and sufficient conditions for the weak Markov consistency of X with respect to $(\mathbb{F}, \mathcal{Y})$. The proof of the theorem is analogous to the proof of Theorem 4.7, and therefore it is omitted.

Theorem 4.17 *Let $\mathcal{Y} = \{Y^1, \ldots, Y^n\}$ be a family of processes such that each Y^k is an $(\mathbb{F}, \mathbb{F}^{Y^k})$-CDMC, with values in $\mathcal{X}_k$ and with $\mathbb{F}$-intensity $\Psi_t^k = [\psi_t^{k;x^k y^k}]_{x^k, y^k \in \mathcal{X}_k}$, $t \in [0,T]$. Let Λ be a version of the $\mathbb{F}$-intensity of process X. Then X satisfies the weak Markov consistency property with respect to $(\mathbb{F}, \mathcal{Y})$ if, and only if for all $k = 1, \ldots, n$, the following hold:*

(i) *Condition (WM-k) is satisfied with $\psi_t^{k;x^k y^k}$ in place of $\lambda_t^{k;x^k y^k}$.*
(ii) *The law of X_0^k given $\mathcal{F}_T$ coincides with the law of Y_0^k given $\mathcal{F}_T$.*

Discussion of Necessary Conditions for Weak Markov Consistency

Condition (WM-k) is difficult to verify since it entails computations of projections onto the filtration $\mathbb{F} \vee \mathbb{F}^{X^k}$. Here we will formulate an "algebraic-like" necessary condition for weak Markov consistency, which is easier to verify.

We start by imposing the following simplifying assumption on a process X:

Assumption 4.18 For each $k \in \{1, \ldots, n\}$ it holds that

$$\mathbb{P}\left(X_t^k = x^k \middle| \mathcal{F}_t\right) > 0 \quad dt \otimes d\mathbb{P}\text{-a.e.,} \quad \text{for all } x^k \in \mathcal{X}_k.$$

Clearly, this assumption imposes constraints on the initial distribution of the chain as well as on the structure of the intensity process of X. However, it allows us to simplify and streamline the discussion below. The general case can be dealt with in a similar way, with special attention paid to sets of ωs for which $\mathbb{P}(X_t^k = x^k | \mathcal{F}_t)(\omega) = 0$.

Before we proceed we observe that Assumption 4.18 implies that

$$\mathbb{P}\left(X_t^k = x^k\right) > 0 \quad dt\text{-a.e.,} \quad \text{for all } x^k \in \mathcal{X}_k.$$

We will also need a simple technical result regarding events $B(t,k,x^k)$ and $C(t,k,x^k)$ defined, for every $t \in [0,T]$, $x^k \in \mathcal{X}_k$, and $k \in \{1, \ldots, n\}$, as follows:

$$B(t,k,x^k) = \left\{\omega : X_t^k(\omega) = x^k\right\}, \quad C(t,k,x^k) = \left\{\omega : \mathbb{P}\left(X_t^k = x^k \middle| \mathcal{F}_t\right)(\omega) > 0\right\}.$$

Observe that (writing B and C in place of $B(t,k,x)$ and $C(t,k,x)$ to shorten the ensuing formulae)

$$\begin{aligned}\mathbb{P}(B \cap C) &= \mathbb{E}(\mathbb{1}_B \mathbb{1}_C) = \mathbb{E}(\mathbb{E}(\mathbb{1}_B | \mathcal{F}_t) \mathbb{1}_C) = \mathbb{E}\left(\mathbb{P}(B|\mathcal{F}_t) \mathbb{1}_{\{\mathbb{P}(B|\mathcal{F}_t)>0\}}\right) \\ &= \mathbb{P}(B) > 0.\end{aligned}$$

We are now ready to state and to prove

Proposition 4.19 *Assume that X satisfies Assumption 4.18. Fix $k \in \{1, \ldots, n\}$.*

Suppose that X is weakly Markov consistent with respect to $(X^k,\mathbb{F})$. Then, the $\mathbb{F}$-intensity Λ^k of X^k (cf. (4.17)) satisfies

$$\lambda_t^{k;x^k y^k}(\omega)=\sum_{\substack{x^m,y^m\in\mathcal{X}_m\\ m=1,\dots,n,m\neq k}}\lambda_t^{(x^1,\dots,x^n)(y^1,\dots,y^n)}(\omega)\times\frac{\mathbb{P}\left(X_t^1=x^1,\dots,X_t^n=x^n|\mathcal{F}_t\right)(\omega)}{\mathbb{P}\left(X_t^k=x^k|\mathcal{F}_t\right)(\omega)}\tag{4.22}$$

for all $x^k,y^k\in\mathcal{X}_k$, $y^k\neq x^k$, and $\omega\in B(t,k,x^k)\cap C(t,k,x^k)$, for almost every $t\in[0,T]$.

Proof Since weak Markov consistency with respect to $(X^k,\mathbb{F})$ holds, in view of Theorem 4.16 Λ^k satisfies (4.16). Taking conditional expectations in (4.16) with respect to $\mathcal{F}_t\vee\sigma(X_t^k)$ yields

$$\begin{aligned}
&\mathbb{1}_{\{X_t^k=x^k\}}\lambda_t^{k;x^k y^k}\\
&=\mathbb{E}\left(\mathbb{1}_{\{X_t^k=x^k\}}\lambda_t^{k;x^k y^k}\,\middle|\,\mathcal{F}_t\vee\sigma(X_t^k)\right)\\
&=\mathbb{E}\Bigg(\mathbb{1}_{\{X_t^k=x^k\}}\sum_{\substack{x^m,y^m\in\mathcal{X}_m\\ m=1,\dots,n,m\neq k}}\lambda_t^{(x^1,\dots,x^n)(y^1,\dots,y^n)}\\
&\qquad\times\mathbb{E}\left(\mathbb{1}_{\{X_t^1=x^1,\dots,X_t^{k-1}=x^{k-1},X_t^{k+1}=x^{k+1},\dots,X_t^n=x^n\}}|\mathcal{F}_t\vee\mathcal{F}_t^{X^k}\right)\Bigg|\mathcal{F}_t\vee\sigma(X_t^k)\Bigg)\\
&=\mathbb{1}_{\{X_t^k=x^k\}}\sum_{\substack{x^m,y^m\in\mathcal{X}_m\\ m=1,\dots,n,m\neq k}}\lambda_t^{(x^1,\dots,x^n)(y^1,\dots,y^n)}\\
&\qquad\times\mathbb{E}\left(\mathbb{1}_{\{X_t^1=x^1,\dots,X_t^{k-1}=x^{k-1},X_t^{k+1}=x^{k+1},\dots,X_t^n=x^n\}}\,\middle|\,\mathcal{F}_t\vee\sigma(X_t^k)\right).
\end{aligned}$$

Now, let us take an arbitrary $\omega\in B(t,k,x^k)\cap C(t,k,x^k)$. By Assumption 4.18, using Jakubowski and Niewęgłowski (2008b), Lemma 3, we have

$$\begin{aligned}
&\lambda_t^{k;x^k y^k}(\omega)\\
&\quad=\sum_{\substack{x^n,y^n\in\mathcal{X}_n\\ n=1,\dots,n,n\neq k}}\lambda_t^{(x^1,\dots,x^n)(y^1,\dots,y^n)}(\omega)\\
&\qquad\times\mathbb{E}\left(\mathbb{1}_{\{X_t^1=x^1,\dots,X_t^{k-1}=x^{k-1},X_t^{k+1}=x^{k+1},\dots,X_t^n=x^n\}}\,\middle|\,\mathcal{F}_t\vee\sigma(X_t^k)\right)(\omega)\\
&\quad=\sum_{\substack{x^n,y^n\in\mathcal{X}_n\\ n=1,\dots,n,n\neq k}}\lambda_t^{(x^1,\dots,x^n)(y^1,\dots,y^n)}(\omega)\frac{\mathbb{P}\left(X_t^1=x^1,\dots,X_t^n=x^n|\mathcal{F}_t\right)(\omega)}{\mathbb{P}\left(X_t^k=x^k|\mathcal{F}_t\right)(\omega)},
\end{aligned}$$

which shows that condition (4.22) is necessary for the weak Markov consistency of X with respect to $(X^k,\mathbb{F})$. □

As will be seen in Chapter 8, the next proposition is useful in the construction of weak CMC structures.

Proposition 4.20 *Let $\mathcal{Y} = \{Y^1, \ldots, Y^n\}$ be a family of processes such that each Y^k is an $(\mathbb{F}, \mathbb{F}^{Y^k})$-CDMC with values in $\mathcal{X}_k$ and with an $\mathbb{F}$-intensity $\Psi^k_t = [\psi^{k;x^k y^k}_t]_{x^k, y^k \in \mathcal{X}_k}$, $t \in [0,T]$. Assume that X satisfies Assumption 4.18, and let Λ be a version of its $\mathbb{F}$-intensity. In addition, suppose that X is weakly Markov consistent with respect to $(\mathbb{F}, \mathcal{Y})$. Then*

(i) *For almost every $t \in [0,T]$ we have*

$$\psi^{k;x^k y^k}_t(\omega) = \sum_{\substack{x^m, y^m \in \mathcal{X}_m \\ m=1,\ldots,n, m \neq k}} \lambda_t^{(x^1,\ldots,x^n)(y^1,\ldots,y^n)}(\omega) \times \frac{\mathbb{P}\left(X^1_t = x^1, \ldots, X^n_t = x^n | \mathcal{F}_t\right)(\omega)}{\mathbb{P}\left(X^k_t = x^k | \mathcal{F}_t\right)(\omega)} \tag{4.23}$$

for all $x^k, y^k \in \mathcal{X}_k$, $y^k \neq x^k$, and $\omega \in B(t,k,x^k) \cap C(t,k,x^k)$.

(ii) *The law of X^k_0 given $\mathcal{F}_T$ coincides with the law of Y^k_0 given $\mathcal{F}_T$.*

Proof Since X is weakly Markov consistent with respect to $(\mathbb{F}, \mathcal{Y})$, X is weakly Markov consistent with respect to $(X^k, \mathbb{F})$ for each k. Thus, in view of (4.22) and Lemma D.31 we conclude that (4.23) holds. This proves (i). Conclusion (ii) is clear from the weak Markov consistency of X with respect to $(\mathbb{F}, \mathcal{Y})$. □

Remark 4.21 Even though the above proposition gives a necessary, rather than a sufficient, condition for the weak Markov consistency of X with respect to $(\mathbb{F}, \mathcal{Y})$, it will be used in the construction of weak CMC structures in Section 8.2. To find a workable sufficient condition for the weak Markov consistency of X with respect to $(\mathbb{F}, \mathcal{Y})$ to hold is an open problem. Thus, for the time being, our strategy for constructing CMC structures will be to use the necessary condition (4.23) to construct process X, a candidate for a CMC structure, and then to verify that this process does indeed furnish a weak CMC structure. We refer to Section 8.2 for details.

4.2.2 When Does Weak Markov Consistency Imply Strong Markov Consistency?

It is clear that the strong Markov consistency of X with respect to $(X^k, \mathbb{F})$ implies the weak Markov consistency of X with respect to $(X^k, \mathbb{F})$. As it will be seen in Section 8.8, the process X may be weakly Markov consistent with respect to $(X^k, \mathbb{F})$ but may fail to satisfy the strong Markov consistency condition with respect to $(X^k, \mathbb{F})$. The following result provides sufficient conditions under which the weak Markov consistency of X with respect to $(X^k, \mathbb{F})$ implies the strong Markov consistency of X with respect to $(X^k, \mathbb{F})$.

Theorem 4.22 *Assume that X satisfies the weak Markov consistency condition with respect to $(X^k, \mathbb{F})$. If $\mathbb{F} \vee \mathbb{F}^{X^k}$ is $\mathbb{P}$-immersed in $\mathbb{F} \vee \mathbb{F}^X$ then X satisfies the strong Markov consistency condition with respect to $(X^k, \mathbb{F})$.*

Proof Suppose that $\mathbb{F} \vee \mathbb{F}^{X^k}$ is $\mathbb{P}$-immersed in $\mathbb{F} \vee \mathbb{F}^X$. Fix arbitrary $x_1^k, \ldots, x_m^k \in \mathcal{X}_k$ and $0 \leq t_1 \leq \cdots \leq t_m \leq T$. Let $A = \{X_{t_m}^k = x_m^k, \ldots, X_{t_1}^k = x_1^k\}$.

Since X^k is an $(\mathbb{F}, \mathbb{F}^{X^k})$-CMC, we have, for $s \leq t_1$,

$$\mathbb{P}(A|\mathcal{F}_s \vee \sigma(X_s^k)) = \mathbb{P}(A|\mathcal{F}_s \vee \mathcal{F}_s^{X^k}) = \mathbb{P}(A|\mathcal{F}_s \vee \mathcal{F}_s^X),$$

where in the second equality we have used the immersion of $\mathbb{F} \vee \mathbb{F}^{X^k}$ in $\mathbb{F} \vee \mathbb{F}^X$ (see Section 6.1.1 in Bielecki and Rutkowski (2004)). Thus X^k is an $(\mathbb{F}, \mathbb{F}^X)$-CMC. □

Remark 4.23 We note that the above theorem states only a sufficient condition for the weak Markov consistency of X to imply the strong Markov consistency of X (with respect to $(X^k, \mathbb{F})$). As it was shown in Theorem 3.18, in case of a trivial filtration $\mathbb{F}$, the condition that $\mathbb{F}^{X^k}$ is immersed in $\mathbb{F}^X$ is both necessary and sufficient for the weak Markov consistency of X to imply the strong Markov consistency of X (with respect to X^k).

5

Consistency of Multivariate Special Semimartingales

In this chapter we introduce and discuss various consistency concepts for multivariate special semimartingales. The results presented here are mainly based on Theorem 5.1, which generalizes to the case of semimartingales that are not special. Thus, these results themselves generalize in a straightforward manner to the case of semimartingales that are not special. We chose to work with special semimartingales in order to ease the presentation somewhat. For this reason we have also limited ourselves to the bivariate case. Towards this end we let $(\Omega, \mathcal{F}, \mathbb{F}, \mathbb{P})$ be a filtered canonical probability space over $\mathcal{X} = \mathbb{R}^{d_1} \times \mathbb{R}^{d_2}$ (see Definitions B.9 and B.15), which is assumed to satisfy the usual conditions.

A function $h : \mathbb{R}^d \to \mathbb{R}^d$ defined as $h(x) = x\mathbb{1}_{\{|x| \leq 1\}}$ will be called the standard truncation function of dimension d. Throughout this chapter the semimartingale truncation functions will be considered to be standard truncation functions of appropriate dimensions. In what follows, the semimartingale characteristics will always be computed with respect to the relevant standard truncation functions. Thus, the semimartingale characteristics for all the semimartingales appearing in the rest of this chapter are considered to be unique (as functions of the trajectories on the canonical space) once the filtration with respect to which the characteristics are computed is chosen.

We consider a canonical bivariate special semimartingale $X = (X^1, X^2)$ defined on the canonical space $(\Omega, \mathcal{F}, \mathbb{F}, \mathbb{P})$ and taking values in $\mathcal{X}$. This implies that the canonical decomposition of X is unique. Additionally, for simplicity we assume that $\mathcal{F}_0$ is trivial, so that X_0 is deterministic. We denote the (unique) $\mathbb{F}$-characteristics of X by (B, C, ν) recall that $\mathbb{F}^X = \mathbb{F}$.

5.1 Semimartingale Consistency and Semi-Strong Semimartingale Consistency

Note that, by definition, for $j = 1, 2$, the semimartingale X enjoys (trivially) the property of *semimartingale consistency with respect to* X^j: the coordinate X^j of X is an $\mathbb{R}^{d_j}$-valued semimartingale on $(\Omega, \mathcal{F}, \mathbb{F}, \mathbb{P})$, with its $\mathbb{F}$ characteristics given directly in terms of (B, C, ν). That is, for $j = 1, 2$, they are given by (B^j, C^j, ν^j), where B^j is the jth coordinate of B, $C^j = C_{j,j}$, and

$$\nu^1((0,t],A) = \nu((0,t],A \times \mathbb{R}^{d_2})), \quad A \in \mathcal{B}(\mathbb{R}^{d_1}),$$
$$\nu^2((0,t],B) = \nu((0,t],\mathbb{R}^{d_1} \times B)), \quad B \in \mathcal{B}(\mathbb{R}^{d_2}).$$

In view of Stricker's theorem (see Stricker (1977)) the semimartingale X also enjoys the property of *semi-strong semimartingale consistency* with respect to X^j: the coordinate X^j is a semimartingale on $(\Omega,\mathcal{F},\mathbb{F}^j,\mathbb{P})$ (see (B.5) and see (B.6) for the definition of $\mathbb{F}^j$). Moreover, it follows from Lemma 2.5 in Stricker (1977) that, since X^j is a special semimartingale on $(\Omega,\mathcal{F},\mathbb{F},\mathbb{P})$, if X_0^j is fixed then X^j is a special semimartingale on $(\Omega,\mathcal{F},\mathbb{F}^j,\mathbb{P})$, $j = 1,2$.

However, the $\mathbb{F}^j$-characteristics of X^j are not immediately available and, in general, they need to be computed. Computation of the $\mathbb{F}^j$-characteristics of X^j is frequently a daunting task, which involves the computation of appropriate projections of the $\mathbb{F}$-characteristics of X^j.

5.1.1 Computation of the $\mathbb{F}^j$-Characteristics of X^j

In this subsection we use the following notions and notation:

1. For a given process Z we denote by ${}^{o,\mathbb{F}^j}Z$ the optional projection of Z onto $\mathbb{F}^j$, defined in the sense of He et al. (1992), i.e. the unique $\mathbb{F}^j$-optional, finite-valued process ${}^{o,\mathbb{F}^j}Z$ such that, for every $\mathbb{F}^j$-stopping time τ, we have

$$\mathbb{E}(Z_\tau \mathbb{1}_{\tau<\infty}|\mathcal{F}_\tau^j) = {}^{o,\mathbb{F}^j}Z_\tau \mathbb{1}_{\tau<\infty}.$$

   Note that by Theorem V.5.1 in He et al. (1992) this optional projection exists if Z is a process such that $Z_\tau \mathbb{1}_{\tau<\infty}$ is σ-integrable with respect to $\mathcal{F}_\tau^j$ for every $\mathbb{F}^j$-stopping time τ.
2. We will also need a notion of predictable projection for functions $W\colon \widehat{\Omega} \to \mathbb{R}$ which are measurable with respect to $\widehat{\mathcal{F}}$, where

$$\widehat{\Omega} := \Omega \times \mathbb{R}_+ \times \mathbb{R}^{d_j}, \quad \widehat{\mathcal{F}} := \mathcal{F} \otimes \mathcal{B}(\mathbb{R}_+) \otimes \mathcal{B}(\mathbb{R}^{d_j}).$$

   The predictable projection of such a function W is defined as a measurable function ${}^{p,\mathbb{F}^j}W$ on $\widehat{\Omega}$ such that, for each $x \in \mathbb{R}^{d_j}$, the process ${}^{p,\mathbb{F}^j}W(\cdot,x)$ is a predictable projection of a process $W(\cdot,x)$[1].
3. We denote by $\mathcal{P}_{\mathbb{F}^j}$ the $\mathbb{F}^j$-predictable σ-field on $\Omega \times \mathbb{R}_+$, i.e. the sigma field generated by $\mathbb{F}^j$-adapted continuous processes.

   Analogously we introduce on $\widehat{\Omega}$ the σ-field $\widehat{\mathcal{P}}_{\mathbb{F}^j}$ defined by

$$\widehat{\mathcal{P}}_{\mathbb{F}^j} := \mathcal{P}_{\mathbb{F}^j} \otimes \mathcal{B}(\mathbb{R}^{d_j}).$$

4. A random measure π on $\mathcal{B}(\mathbb{R}_+) \otimes \mathcal{B}(\mathbb{R}^{d_j})$ is $\mathbb{F}^j$-predictable if for any $\widehat{\mathcal{P}}_{\mathbb{F}^j}$-measurable, positive, real-valued function W on $\widehat{\Omega}$, the real-valued process

$$V(\omega,t) = \int_{[0,t]\times\mathbb{R}^{d_j}} W(\omega,s,x^j)\pi(\omega;ds,dx^j)$$

[1] We refer to Theorem I.2.28 in Jacod and Shiryaev (2003) for the notion of the predictable projection of a process.

is $\mathbb{F}^j$-predictable or, equivalently, if for any positive, real, measurable function W on $\widehat{\Omega}$ we have

$$\mathbb{E}\left(\int_{\mathbb{R}_+\times\mathbb{R}^{d_j}} W(s,x^j)\pi(ds,dx^j)\right) = \mathbb{E}\left(\int_{\mathbb{R}_+\times\mathbb{R}^{d_j}} {}^{p,\mathbb{F}^j}W(s,x^j)\pi(ds,dx^j)\right).$$

5. For a random measure ν on $\mathcal{B}(\mathbb{R}_+)\times\mathcal{B}(\mathbb{R}^{d_j})$ we define its $\mathbb{F}^j$-dual predictable projection, which we denote by $\nu^{p,\mathbb{F}^j}$, as the unique $\mathbb{F}^j$-predictable measure such that, for every positive measurable function W on $\widehat{\Omega}$, we have

$$\mathbb{E}\left(\int_{\mathbb{R}_+\times\mathbb{R}^{d_j}} {}^{p,\mathbb{F}^j}W(t,x^j)\nu(dt,dx^j)\right) = \mathbb{E}\left(\int_{\mathbb{R}_+\times\mathbb{R}^{d_j}} W(t,x^j)\nu^{p,\mathbb{F}^j}(dt,dx^j)\right).$$

Since X^j is a special semimartingale in the filtration $\mathbb{F}$ and it is also a special semimartingale in the filtration $\mathbb{F}^j$, it admits two unique canonical decompositions:

$$\begin{aligned} X_t^j &= X_0^j + M_t^j + B_t^j + \int_0^t\int_{\mathbb{R}^{d_j}} x^j \mathbb{1}_{\{|x^j|>1\}}\mu^j(dt,dx^j) \\ &= X_0^j + X_t^{j,c} + \int_0^t\int_{\mathbb{R}^{d_j}} x^j(\mu^j(dt,dx^j)-\nu^j(dt,dx^j)) \\ &\quad + B_t^j + \int_0^t\int_{\mathbb{R}^{d_j}} x^j \mathbb{1}_{\{|x^j|>1\}}\nu^j(dt,dx^j) \end{aligned}$$

and

$$\begin{aligned} X_t^j &= X_0^j + \widetilde{M}_t^j + \widetilde{B}_t^j + \int_0^t\int_{\mathbb{R}^{d_j}} x^j \mathbb{1}_{\{|x^j|>1\}}\mu^j(dt,dx^j) \\ &= X_0^j + \widetilde{X}_t^{j,c} + \int_0^t\int_{\mathbb{R}^{d_j}} x^j(\mu^j(dt,dx^j)-\widetilde{\nu}^j(dt,dx^j)) \\ &\quad + \widetilde{B}_t^j + \int_0^t\int_{\mathbb{R}^{d_j}} x^j \mathbb{1}_{\{|x^j|>1\}}\widetilde{\nu}^j(dt,dx^j), \end{aligned}$$

where

- M^j is an $\mathbb{F}$-local martingale, B^j is an $\mathbb{F}$-predictable process of finite variation,
- $X^{j,c}$ is the continuous $\mathbb{F}$-local martingale part of X^j, and ν^j is the $\mathbb{F}$-compensator of the jump measure μ of X^j,

and where $\widetilde{M}^j$, $\widetilde{X}^{j,c}$, $\widetilde{B}^j$, and $\widetilde{\nu}^j$ are defined analogously but with respect to the filtration $\mathbb{F}^j$.

The $\mathbb{F}$-characteristic triple of X^j is (B^j,C^j,ν^j), where $C^j := \langle X^{j,c}\rangle^{\mathbb{F}}$, and the $\mathbb{F}^j$-characteristic triple of X^j is $(\widetilde{B}^j,\widetilde{C}^j,\widetilde{\nu}^j)$, where $\widetilde{C}^j := \langle \widetilde{X}^{j,c}\rangle^{\mathbb{F}^j}$.

In view of Proposition II.2.9 in Jacod and Shiryaev (2003), there exists an $\mathbb{F}$-predictable, locally integrable, increasing process, say A^j, such that (note that the multiplication point indicates a stochastic integral; see Appendix A)

$$B^j = b^j\cdot A^j, \quad C^j = c^j\cdot A^j, \quad \nu^j(dt,dx^j) = K_t^j(dx^j)dA_t^j,$$

where

- b^j is an $\mathbb{R}^{d^j}$-valued and $\mathbb{F}$-predictable process,

- c^j is an $\mathbb{F}$-predictable process taking values in the set of $d^j \times d^j$ symmetric non-negative definite matrices with real valued entries, and
- $K_t^j(\omega, dx^j)$ is a transition kernel from $(\Omega \times \mathbb{R}_+, \mathcal{P}_{\mathbb{F}})$ to $(\mathbb{R}^{d^j}, \mathcal{B}(\mathbb{R}^{d_j}))$, satisfying a condition analogous to condition II.2.11 in Jacod and Shiryaev (2003), and where $\mathcal{P}_{\mathbb{F}}$ is the $\mathbb{F}$-predictable σ-field on $\Omega \times \mathbb{R}_+$.

We assume that

$$A_t^j = \int_0^t a_u^j du, \quad t \geq 0. \tag{5.1}$$

In addition, we consider the following conditions:

K1 For every $T \geq 0$ we have

$$\mathbb{E} \int_0^T |b_u^j a_u^j| du < \infty.$$

K2 The process $b^j a^j$ admits an $\mathbb{F}^j$-optional projection.

K3 The process M^j is a true $\mathbb{F}$-martingale.

The following result is a consequence of Theorem 3.6 in Bielecki et al. (2018b).

Theorem 5.1 *Assume that conditions **K1–K3** are satisfied. Then an $\mathbb{F}^j$-characteristic triple of X^j is given by*

$$\widetilde{B}^j = \int_0^{\cdot} {}^{o,\mathbb{F}^j}(b^j a^j)_s ds, \quad \widetilde{C}^j = C^j, \quad \widetilde{\nu}^j(dt, dx^j) = \left(K_t^j(dx^j) a_t^j dt\right)^{p,\mathbb{F}^j}.$$

We close this section with an example which makes use of Theorem 5.1 and illustrates computation of the $\mathbb{F}^1$-characteristics of X^1.

Example 5.2 Let us consider $\mathcal{X} = \mathbb{R} \times \mathbb{R}$ with canonical process $X = (X^1, X^2)$. Assume that a probability measure $\mathbb{P}$ is such that X^1 and X^2 are one-point cádlág processes, that is, X^i, $i = 1, 2$, starts from 0 at time $t = 0$ and jumps to 1 at some random time. Thus, X can be identified with a pair of positive random variables T_1 and T_2 given by $T_i := \inf\{t > 0 : X_t^i = 1\}$, $i = 1, 2$. In other words, $X_t^i = \mathbb{1}_{\{T_i \leq t\}}$, $i = 1, 2$. Clearly, X is a special $\mathbb{F}$-semimartingale and thus X^1 is a special $\mathbb{F}$-semimartingale and a special $\mathbb{F}^1$-semimartingale.

We assume that, under $\mathbb{P}$, the probability distribution of (T_1, T_2) admits a continuous density function f. This implies that the $\mathbb{F}$-characteristics of X^1 are $(B^1, 0, \nu^1)$, where

$$B_t^1 = \int_0^t \kappa_s^1 ds \quad \text{and} \quad \nu^1(ds, dx^1) = \delta_1(dx^1) \kappa_s^1 ds;$$

here δ_1 is the Dirac measure at 1 and κ is given by

$$\kappa_s^1 = \frac{\int_s^\infty f(s,v)\, dv}{\int_s^\infty \int_s^\infty f(u,v)\, du\, dv} \mathbb{1}_{\{s \leq T_1 \wedge T_2\}} + \frac{f(s, T_2)}{\int_s^\infty f(u, T_2)\, du} \mathbb{1}_{\{T_2 < s \leq T_1\}}, \quad s \geq 0;$$

this result follows by, for example, an application of Last and Brandt (1995), Theorem 4.1.11.

Thus, according to Theorem 5.1, the $\mathbb{F}^1$-characteristics of X^1 are $(\widetilde{B}^1, 0, \widetilde{\nu}^1)$, where

$$\widetilde{B}^1_t = \int_0^t {}^{o,\mathbb{F}^1}(\kappa^1)_s ds, \quad \widetilde{\nu}^1 = (\nu^1)^{p,\mathbb{F}^1}.$$

We will provide a more explicit formula for $\widetilde{\nu}^1$. For that, we only need to compute $\widetilde{\nu}^1(dt,\{1\})$. This computation boils down to computing the $\mathbb{F}^1$-optional projection of the process κ^1. Indeed, for an arbitrary, $\mathbb{F}^1$-predictable, bounded function W on $\Omega \times \mathbb{R}_+ \times \mathbb{R}$ we have

$$\begin{aligned}\mathbb{E}\int_{\mathbb{R}_+\times\mathbb{R}} W(s,x^1)\nu^1(ds,dx^1) &= \mathbb{E}\int_{\mathbb{R}_+} W(s,1)\kappa^1_s ds = \mathbb{E}\int_{\mathbb{R}_+} {}^{p,\mathbb{F}^1}(W(\cdot,1)\kappa^1_\cdot)_s ds\\ &= \mathbb{E}\int_{\mathbb{R}_+} {}^{p,\mathbb{F}^1}(\kappa^1)_s W(s,1) ds = \mathbb{E}\int_{\mathbb{R}_+\times\mathbb{R}} {}^{p,\mathbb{F}^1}(\kappa^1)_s W(s,x^1)\delta_1(dx^1) ds,\end{aligned}$$

where ${}^{p,\mathbb{F}^1}(\kappa^1)$ denotes the $\mathbb{F}^1$-predictable projection of κ^1. Next, we note that the measure ρ defined by

$$\rho(dt,dx^1) := {}^{p,\mathbb{F}^1}(\kappa^1)_t \delta_1(dx^1)dt$$

is $\mathbb{F}^1$-predictable and thus, owing to the uniqueness of the dual predictable projections, we have $\rho = (\nu^1)^{p,\mathbb{F}^1}$ and so $\widetilde{\nu}^1 = {}^{p,\mathbb{F}^1}(\kappa^1)_t \delta_1(dx^1)dt$. Finally, we note that, in view of the continuity assumptions on f and the fact that κ^1 admits two jumps only, we have

$$\mathbb{E}\int_{\mathbb{R}_+\times\mathbb{R}} {}^{p,\mathbb{F}^1}(\kappa^1)_s W(s,x^1)\delta_1(dx^1) ds = \mathbb{E}\int_{\mathbb{R}_+\times\mathbb{R}} {}^{o,\mathbb{F}^1}(\kappa^1)_s W(s,x^1)\delta_1(dx^1) ds,$$

where ${}^{o,\mathbb{F}^1}(\kappa)$ denotes the $\mathbb{F}^1$-optional projection of κ^1. Consequently, $\widetilde{\nu}^1((0,t],\{1\})$ is given as follows:

$$\begin{aligned}&\widetilde{\nu}^1((0,t],\{1\})\\ &= \int_0^t \mathbb{E}\left(\frac{\int_s^\infty f(s,v)\,dv}{\int_s^\infty \int_s^\infty f(u,v)\,du\,dv} \mathbb{1}_{\{s \le T_1 \wedge T_2\}} + \frac{f(s,T_2)}{\int_s^\infty f(u,T_2)\,du} \mathbb{1}_{\{T_2 < s \le T_1\}} \middle| \mathcal{F}^1_s \right) ds\\ &= \int_0^t \frac{\int_0^\infty f(s,v)\,dv}{\int_s^\infty \int_0^\infty f(u,v)\,du\,dv} \mathbb{1}_{\{T_1 > s\}} ds,\end{aligned}$$

where the last equality follows from the celebrated Key Lemma (see e.g. Lemma 2.9 in Aksamit and Jeanblanc (2017)).

We note that the last result agrees with the classical computation of the intensity of T_1 in its own filtration, which is given as $\lambda^1_s = f^1(s)/(1-F^1(s))$ with $F^1(s) = \mathbb{P}(T_1 \le s)$ and $f^1(s) = \partial F^1(s)/\partial s$.

5.2 Strong Semimartingale Consistency

In this section we introduce and discuss the strong semimartingale consistency of the special semimartingale X with respect to X^j. As we will see, if the special semimartingale X is strong semimartingale consistent with respect to X^j then finding the $\mathbb{F}^j$-characteristics of X^j is immediate.

Definition 5.3 Let us fix $j \in \{1,2\}$. We say that the special semimartingale $X = (X^1,X^2)$ is *strongly semimartingale consistent with respect to* X^j if the $\mathbb{F}$-characteristic triple (B^j,C^j,ν^j) of X^j and the $\mathbb{F}^j$-characteristic triple $(\widetilde{B}^j,\widetilde{C}^j,\widetilde{\nu}^j)$ of X^j coincide (as functions of trajectories).

The concept of strong semimartingale consistency is highly nontrivial, as is demonstrated by the examples provided below.

5.2.1 Examples of Strongly Semimartingale-Consistent Multivariate Special Semimartingales

Example 5.4 Let $\mathcal{X} = \mathbb{R}\times\mathbb{R}$ and consider a Poisson process $X = (X^1,X^2)$. We will investigate the strong semimartingale consistency of X. There is a one-to-one correspondence between any time-homogeneous Poisson process with values in $\mathcal{X} = \mathbb{R}\times\mathbb{R}$ and a homogeneous Poisson measure, say μ, on $E := \{0;1\}^2 \setminus \{(0,0)\}$. See for instance the discussion in Bielecki et al. (2008b).

Let ν denote the $\mathbb{F}$-dual predictable projection of μ. The measure ν is a measure on a finite set, so it is uniquely determined by its values on the atoms in E. Therefore $X = (X^1,X^2)$ is uniquely determined by

$$\nu(dt,\{(1,0)\}) = \lambda_{1,0}dt, \quad \nu(dt,\{(0,1)\}) = \lambda_{0,1}dt, \quad \nu(dt,\{(1,1)\}) = \lambda_{1,1}dt \tag{5.2}$$

for some positive constants λ_{10}, λ_{01}, and λ_{11}. A time homogeneous Poisson process with values in $\mathcal{X} = \mathbb{R}\times\mathbb{R}$ is a special semimartingale, and the $\mathbb{F}$-characteristic triple of X is $(B,0,\nu)$, where

$$B_t = \begin{bmatrix} B_t^1 \\ B_t^2 \end{bmatrix} := \begin{bmatrix} (\lambda_{10}+\lambda_{11})t \\ (\lambda_{01}+\lambda_{11})t \end{bmatrix}.$$

The $\mathbb{F}$-characteristic triple of X^j is $(B^j,0,\nu^j)$, $j=1,2$, where

$$\nu^1(dt,\{1\}) = \nu(dt,\{(1,0)\}) + \nu(dt,\{(1,1)\}) = \lambda_{10}dt + \lambda_{11}dt$$

and $\nu^1(dt,\{0\}) = 0$, and similarly for $\nu^2(dt,\{x\}), x \in \{0,1\}$. Thus, by Theorem 5.1, the process X is strongly semimartingale consistent with respect to X^j, since X^j is a semimartingale on $(\Omega,\mathcal{F},\mathbb{F}^j,\mathbb{P})$ with deterministic $\mathbb{F}$-characteristic triple, $j=1,2$.

Example 5.5 Take $\mathcal{X} = \mathbb{R}\times\mathbb{R}$. Let $X = (X^1,X^2)$ be given as a strong solution to the stochastic differential equation

$$dX_t = m(X_t)dt + \Sigma(X_t)dW_t, \quad X(0) = \begin{bmatrix} 1 \\ 1 \end{bmatrix}, \tag{5.3}$$

where $W^\top = (W^1,W^2)$ is a two-dimensional standard Brownian motion process on the canonical space $(\Omega,\mathcal{F},\mathbb{F},\mathbb{P})$, and where

$$m(x^1,x^2) = (m_1(x^1,x^2), m_2(x^1,x^2))^\top,$$

$$\Sigma(x^1,x^2) = \begin{pmatrix} \sigma_{11}(x^1,x^2) & \sigma_{12}(x^1,x^2) \\ \sigma_{21}(x^1,x^2) & \sigma_{22}(x^1,x^2) \end{pmatrix}.$$

The process X is a special semimartingale, and the $\mathbb{F}$-characteristic triple of X is $(B,C,0)$ (see e.g. Theorem III.2.26 in Jacod and Shiryaev (2003)), where, for $t \geq 0$,

$$B_t = \int_0^t m(X_u)du \quad \text{and} \quad C_t = \int_0^t \Sigma(X_u)\Sigma^\top(X_u)du.$$

The process X^j is a special semimartingale on $(\Omega,\mathcal{F},\mathbb{F},\mathbb{P})$ and on $(\Omega,\mathcal{F},\mathbb{F}^j,\mathbb{P})$, $j = 1,2$. Suppose that the function Σ satisfies the two conditions

$$\sigma_{11}^2(x^1,x^2) + \sigma_{12}^2(x^1,x^2) = \sigma_1^2(x^1) \tag{5.4}$$

and

$$\sigma_{21}^2(x^1,x^2) + \sigma_{22}^2(x^1,x^2) = \sigma_2^2(x^2). \tag{5.5}$$

In addition, suppose that the function m satisfies

$$m_1(x^1,x^2) = \mu_1(x^1), \quad m_2(x^1,x^2) = \mu_2(x^2).$$

Then (5.3) takes the form

$$dX_t^j = \mu_j(X_t^j)dt + \sigma_1(X_t^j)dZ_t^j, \quad X^j(0) = 1, j = 1,2, \tag{5.6}$$

where, for $i,j = 1,2$, $i \neq j$, the process Z^j given by

$$Z_t^j = \int_0^t \frac{\sigma_{j,j}(X_u^1,X_u^2)}{\sigma_j(X_u^j)}dW_u^j + \int_0^t \frac{\sigma_{j,i}(X_u^1,X_u^2)}{\sigma_j(X_u^j)}dW_u^i, \quad t \geq 0, i \neq j$$

is a continuous $\mathbb{F}$-local martingale. Since in view of (5.4) and (5.5) the process $((Z_t^j)^2 - t)_{t\geq 0}$ is a continuous $\mathbb{F}$-local martingale, we obtain by Lévy's characterization theorem (cf. Jacod and Shiryaev (2003), Theorem II.4.4) that Z^j is a standard Brownian motion in the filtration $\mathbb{F}$, $j = 1,2$. Thus, using (5.6) we conclude that X^j has the $\mathbb{F}$-characteristic triple given as $(B^j,C^j,0)$, where

$$B_t^j = \int_0^t \mu_j(X_u^j)du, \quad C_t^j = \int_0^t \sigma_j^2(X_u^j)du, \quad t \geq 0, \ j = 1,2.$$

Clearly, in view of Theorem 5.1 we have $(B^j,C^j,0) = (\widetilde{B}^j,\widetilde{C}^j,0)$ and thus that $X = (X^1,X^2)$ is strongly semimartingale consistent with respect to $\mathbb{F}^j$, $j = 1,2$.

Example 5.6 Let us consider a finite bivariate Markov chain $X = (X^1,X^2)$, as in Section 3.1 (with $n = 2$). The process X is not a semimartingale on $(\Omega,\mathcal{F},\mathbb{F},\mathbb{P})$ in general, since X does not have to take values in a vector space. However, X can be identified with the Markov chain $Z = (Z^1,Z^2)$, where Z^j takes values in the so-called canonical state space for finite-state Markov chains, i.e. the set $U_j = \left\{e_k^j : k = 1,2,\ldots,d_j\right\}$, which is the canonical basis in $\mathbb{R}^{d_j}$, with d_j equal to the cardinality of $\mathcal{X}_j$. The identification of Z^j with X^j is made by an arbitrary bijective mapping $f_j : \mathcal{X}_j \to U_j$ and by setting

$$Z_t^j = f_j(X_t^j), \quad t \geq 0.$$

The infinitesimal generator of Z can be identified with the infinitesimal generator $\Lambda(t)$, $t \geq 0$, of X. The process Z is a special semimartingale, and we have

$$Z_t = Z_0 + M_t + K_t, \quad t \geq 0,$$

where

$$K_t = \int_0^t \Lambda^\top(s) Z_s ds, \quad t \geq 0,$$

and M is a $\mathbb{P}$-martingale in the filtration of the process Z.

So, without any loss of generality, we may take $X = (X^1, X^2)$ to be a Markov chain with values in $\mathcal{X} = \mathcal{X}_1 \times \mathcal{X}_2$, where $\mathcal{X}_j = \left\{ e_k^j : k = 1, 2, \ldots, d_j \right\}$ is the canonical basis in $\mathbb{R}^{d_j}$. In this case, X is both a Markov chain and a special semimartingale with values in $\mathcal{X} = \mathbb{R}^{d_1} \times \mathbb{R}^{d_2}$. Clearly, X^j is a special semimartingale on $(\Omega, \mathcal{F}, \mathbb{F}^j, \mathbb{P})$, $j = 1, 2$.

Let $\mathcal{J}_j := \{x_j - y_j,\ x_j \neq y_j : x_j, y_j \in \mathcal{X}_j\}$ be the set of jump values of X^j, $j = 1, 2$. Thus, for $j = 1, 2$, the jump measure of X^j is given as the random measure μ^j on $\mathbb{R}_+ \times \mathcal{J}_j$ given by

$$\mu^j(dt, A_j) = \sum_{y_j - x_j \in A_j} N^j_{x_j y_j}(dt), \quad A_j \subset \mathcal{J}_j,$$

where $N^j_{x_j y_j}(dt)$ is given in (C.9) for $j = 1$, and by an analogous formula for $j = 2$. Consequently, the $\mathbb{F}$-compensator of the random measure μ^j, say η^j, is given as

$$\eta^j(dt, A_j) = \sum_{y_j - x_j \in A_j} \nu^j_{x_j y_j}(dt), \quad A_j \subset \mathcal{J}_j,$$

where $\nu^j_{x_j y_j}$ is the $\mathbb{F}$ compensator of the random measure $N^j_{x_j y_j}$.

Let $\mathbb{F}$ denote the filtration of X. Assume that Condition (M) is satisfied (see Section 3.1). Then, as we know from the proof of Theorem 3.3 and from Remark 3.5, for any $x_1, y_1 \in \mathcal{X}^1$ the $\mathbb{F}$-compensator of the random measure $N^1_{x_1 y_1}$ is given by $\nu^1_{x_1 y_1}(dt) = \mathbb{1}_{\{X_t^1 = x^1\}} \lambda^1_{x^1 y^1}(t) dt$. Thus, we see that the special semimartingale X is strongly semimartingale consistent with respect to $\mathbb{F}^1$.

Analogously one can demonstrate that the special semimartingale X is strongly semimartingale consistent with respect to $\mathbb{F}^2$.

5.2.2 Examples of Multivariate Special Semimartingales That Are Not Strongly Semimartingale Consistent

Example 5.7 Consider the canonical stochastic basis $(\Omega, \mathcal{F}, \mathbb{F}, \mathbb{P})$, with $\mathcal{X} = \mathbb{R} \times \mathbb{R}^N$. Let X^2 be a Markov chain that takes values in $\{e_1, \ldots, e_N\}$, where $(e_i)_{i=1}^N$ is the canonical basis in $\mathbb{R}^N$. We assume that X^2 admits a constant generator matrix Λ. Consider the special semimartingale $X = (X^1, X^2)$, with X^1 given by

$$dX_t^1 = b^\top X_t^2 dt + \sigma dW_t, \quad X_0^1 = x^1 \in \mathbb{R},$$

where W is a real-valued standard Brownian motion, $\sigma \in \mathbb{R}_+$, and $b \in \mathbb{R}^N$, $b \neq 0$. The $\mathbb{F}$-characteristics of X^1 are

$$\left(\left(\int_0^t b^\top X_u^2 du\right)_{t\geq 0}, (\sigma^2 t)_{t\geq 0}, 0\right).$$

According to Stricker's theorem, X^1 is a special semimartingale on $(\Omega, \mathcal{F}, \mathbb{F}^1, \mathbb{P})$. Clearly, however, the $\mathbb{F}$-characteristics of X^1 do not coincide with its $\mathbb{F}^1$-characteristics. Indeed, Theorem 5.1 implies that the $\mathbb{F}^1$-characteristics of X^1 are

$$\left(\left(\int_0^t b^\top \mathbb{E}(X_u^2 | \mathcal{F}_u^1) du\right)_{t\geq 0}, (\sigma^2 t)_{t\geq 0}, 0\right).$$

Thus, $X = (X^1, X^2)$ is not strongly semimartingale consistent with respect to $\mathbb{F}^1$.

Example 5.8 Refer to Example 5.2. Clearly, in that example, in general $\nu^1 \neq \widetilde{\nu}^1$, so that $X = (X^1, X^2)$ is not strongly semimartingale consistent with respect to $\mathbb{F}^1$.

PART TWO

STRUCTURES

6

Strong Markov Family Structures

In this chapter we define and study strong Markov family structures for a collection of time-homogeneous nice $\mathbb{R}$-Feller–Markov families. For brevity, we will refer to strong Markov family structures as strong Markov structures. In this volume we do not study structures in the case of time-inhomogeneous Markov families.

Essentially, Markov structures are key objects of interest in modeling structured dependence of the Markovian type between stochastic dynamical systems of Markovian type such as Markov families or Markov processes. Accordingly, much of the discussion presented in this chapter is devoted to building Markov structures. Part of the input to any construction procedure is provided by the marginal data, which we refer to as the *marginal input*. Another part of the input is provided by data and/or postulates regarding the stochastic dependence between coordinates of the resulting Markov structure, which we refer to as the *dependence structure input*. These inputs have to be appropriately accounted for in constructions of Markov structures. In principle this can be done since, as discussed in this chapter, one has quite a substantial flexibility in building Markov structures; this allows for accommodating in a Markov structure model various dependence structures exhibited by the phenomena one wants to model.

The results and constructions presented here are rooted in the results of Chapter 2.

6.1 Strong Markov Structures for Nice Feller–Markov Families and Nice Feller–Markov Processes

We first introduce the strong Markov structures for nice Feller–Markov families, and then the strong Markov structures for nice Feller–Markov processes.

Strong Markov Structures for Nice Feller–Markov Families

Here, we adopt the set-up of Section 2.2.2. In particular, we refer here to the concept of the nice $\mathbb{R}^n$-Feller–Markov family $\mathcal{MMFH}$ (see Definition 2.23) and to the strong Markov consistency property of $\mathcal{MMFH}$ with respect to X^j relative to $\mathcal{MFH}^j$ (see Definition 2.25).

Now, let $\left\{\mathcal{MFH}^j := \left\{(\Omega^j, \mathcal{F}, \mathbb{F}^j, (Y_t^j)_{t\geq 0}, \mathbb{Q}_y^j, Q^j) : y \in \mathbb{R}\right\} :\ j = 1, \ldots, n\right\}$ be a collection of canonical (time-homogeneous) nice $\mathbb{R}$-Feller–Markov families. Let ρ_j be the symbol of the family $\mathcal{MFH}^j$ (refer to the convention adopted in Section 2.2.2), and let A^j be the corresponding PDO, $j = 1, \ldots, n$.

The next definition introduces the main object of interest in this section.

Definition 6.1 We call a nice $\mathbb{R}^n$-Feller–Markov family

$$\mathcal{MMFH} = \{(\Omega, \mathcal{F}, \mathbb{F}, X = (X_t^1, \ldots, X_t^n)_{t\geq 0}, \mathbb{P}_x, P) : x \in \mathbb{R}^n\}$$

a *strong Markov family structure* (or, a strong Markov structure, for short) for the collection $\mathcal{MFH}^j,\ j = 1, \ldots, n$, if

$$\mathcal{MMFH} \curvearrowright \mathcal{MFH}^j, \quad j = 1, \ldots, n,$$

that is, if $\mathcal{MMFH}$ satisfies the strong Markov consistency property with respect to X^j relative to $\mathcal{MFH}^j$ (see Definition 2.25). If this is the case then we call the families $\mathcal{MFH}^j,\ j = 1, \ldots, n$, the *predetermined margins* of $\mathcal{MMFH}$.

Remark 6.2 (i) We remark that the concept of a strong Markov structure should not be confused with the strong Markov property.
(ii) Let us note that the law of $(X_t)_{t\geq 0}$ under $\mathbb{P}_x$ is determined by the transition function P; thus the strong Markov structure is also a strong Markov copula structure.

In Section 6.2 we will study the following problem:

Given a collection of marginal data, represented by $\mathcal{MFH}^j,\ j = 1, \ldots, n$, corresponding to a collection of canonical time-homogeneous nice $\mathbb{R}$-Feller–Markov families, provide an algorithm for building various strong Markov structures for this collection. In other words, we ask how to construct nice $\mathbb{R}^n$-Feller–Markov $\mathcal{MMFH}$ families with predetermined margins.

Strong Markov Process Structures for Nice Feller–Markov Processes

Let $\mathcal{MMPH} = \{(\Omega, \mathcal{F}, \mathbb{F}, (X_t = (X_t^1 \ldots, X_t^n))_{t\geq 0}\mathbb{Q}, P)\}$ be a canonical time-homogeneous multivariate Markov process over $(\mathbb{R}^n, \mathcal{B}(\mathbb{R}^n))$.

For $t \geq 0$ define an operator T_t on $C_0(\mathbb{R}^n)$ by

$$T_t f(x) = \mathbb{E}_{\mathbb{Q}}(f(X_t)|X_0 = x) = \int_{\mathbb{R}^n} f(y)P(x,t,dy), \quad x \in \mathbb{R}^n. \tag{6.1}$$

If $T_t,\ 0 \leq t$, defined in (6.1) is a $C_0(\mathbb{R}^n)$-Feller semigroup then $\mathcal{MMPH}$ is called an $\mathbb{R}^n$-*Feller–Markov* process. If, in addition, $C_c^\infty(\mathbb{R}^n) \subseteq D(A)$, where A is the generator of $T_t,\ 0 \leq t$, then the process $\mathcal{MMPH}$ is called a *nice* $\mathbb{R}^n$-Feller–Markov process. An analogous definition of niceness applies to Markov processes that do not necessarily qualify as multivariate Markov processes. It is worth noting that, since the transition function of an $\mathbb{R}^n$-Feller–Markov process coincides with the transition function of an $\mathbb{R}^n$-Feller–Markov family associated with this Markov process, an $\mathbb{R}^n$-Feller–Markov family associated with a nice $\mathbb{R}^n$-Feller–Markov process is nice, and vice versa.

Definition 6.3 Let $\mathcal{MPH}^j$, $j = 1,\ldots,n$, be a given collection of nice $\mathbb{R}$-Feller–Markov processes and let $\mathcal{MFH}^j$, $j = 1,\ldots,n$, be the collection of the associated nice $\mathbb{R}$-Feller–Markov families. We call a nice $\mathbb{R}^n$-Feller–Markov process, say $\mathcal{MMPH}$, a strong Markov process structure for the collection $\mathcal{MPH}^j$, $j = 1,\ldots,n$, if the nice $\mathbb{R}^n$-Feller–Markov family $\mathcal{MMFH}$ associated with $\mathcal{MMPH}$ is a strong Markov family structure for the collection $\mathcal{MFH}^j$, $j = 1,\ldots,n$, of marginal Markov families associated with the collection $\mathcal{MPH}^j$, $j = 1,\ldots,n$.

Clearly the transition functions of the coordinates of a strong Markov process structure for the collection $\mathcal{MPH}^j$, $j = 1,\ldots,n$, coincide with the transition functions of $\mathcal{MPH}^j$, $j = 1,\ldots,n$. If one is interested in the equality of laws of coordinates then one should guarantee that the initial marginal laws of X_0 coincide with the initial laws of $Y_0^j, j = 1,\ldots n$, i.e.

$$\mathcal{L}(X_0^j) = \mathcal{L}(Y_0^j), \quad j = 1,\ldots,n.$$

This can be done via classical copula functions. In such a case we may call $\mathcal{MMPH}$ a strong Markov process copula structure for the collection $\mathcal{MPH}^j$, $j = 1,\ldots,n$.

Let us note that, in view of the discussion in Section 2.1.2, building a strong Markov process structure and a strong Markov process copula structure for an $\mathcal{MMPH}$ boils down to providing analogous constructions for the $\mathcal{MMFH}$ associated with this $\mathcal{MMPH}$.

6.2 Building Strong Markov Structures

We want to construct an $\mathcal{MMFH}$ that will constitute a strong Markov structure for $\mathcal{MFH}^j$, $j = 1,\ldots,n$. In view of Proposition 2.32 we will uniquely construct an $\mathcal{MMFH}$ if we construct a symbol, say q, that characterizes $\mathcal{MMFH}$. In this chapter we use the following convention: we say that condition **C1** holds if **C1**(j) holds for all $j = 1,\ldots,n$; an analogous convention applies to the conditions **C2**(j)–**C4**(j), $j = 1,\ldots,n$.[1]

Thus, building such a structure will, generically, proceed according to the following recipe:

(i) Begin by identifying the symbol ρ_j corresponding to each family $\mathcal{MFH}^j$, $j = 1,\ldots,n$. The collection $\{\rho_j, j = 1,\ldots,n\}$ is the collection of marginal inputs used in building a Markov structure.

(ii) Design a candidate symbol q that can serve as a solution of the following system of equations (corresponding to equation (2.53)):

$$q(x, \mathbf{e}_j\xi_j) = \rho_j(x_j, \xi_j), \quad x \in \mathbb{R}^n, \quad \xi_j \in \mathbb{R}, \quad j = 1,\ldots,n. \tag{6.2}$$

In particular, this implies that condition **C2** holds for q and that condition **C3** holds for $\tilde{q}_j$ given as $\tilde{q}_j(x_j, \xi_j) = q(x, \mathbf{e}_j\xi_j), j = 1,\ldots,n$. We need to choose a solution q to the system (6.2) such that it can serve as the symbol of an $\mathbb{R}^n$-Feller–Markov family, and such that condition **C1** is satisfied. For this purpose

[1] The conditions **C1**(j)–**C4**(j) are stated just before Remark 2.29.

Propositions 2.35 or 2.46 can be used. Of course, q needs to be such that the corresponding $\mathbb{R}^n$-Feller–Markov family is nice. This can be provided for using Proposition 2.40.

(iii) Once a symbol q solving the system (6.2) is constructed, one can construct the corresponding PDO, say A, the $C_0(\mathbb{R}^n)$-Feller semigroup $\mathcal{T} = (T_t,\ t \geq 0)$, and, consequently, the transition function P via $P(x,t,B) = T_t \mathbb{1}_B(x)$. Note that this requires the usual extension of the semigroup $\mathcal{T}$ from $C_0(\mathbb{R}^n)$ to the set of bounded Borel measurable functions. Having given the transition function P, we may construct a unique (canonical) multivariate Markov family

$$\mathcal{MMFH} = \{(\Omega, \mathcal{F}, \mathbb{F}, (X_t)_{t\geq 0}, \mathbb{P}_x, P) : x \in \mathbb{R}^n\};$$

see Theorem 1.3.5 in Gikhman and Skorokhod (2004).

(iv) Make sure that, for all $j = 1, \ldots, n$, it holds that the generator A^j corresponding to ρ_j satisfies condition **C4**, where the expectation $\mathbb{E}_x$ is computed under the measure $\mathbb{P}_x$ from (iii) above. For this purpose Corollary 2.43 or Proposition 2.47 can be used.

(v) Given (i)–(iv), we have at our disposal symbols $\rho_1, \ldots, \rho_n$ and q satisfying **C1**–**C4** and such that the corresponding Feller–Markov families are nice. So, we may use Theorem 2.33 to conclude that the $\mathcal{MMFH}$ constructed in (iii) constitutes a strong Markov structure for $\mathcal{MFH}^j$, $j = 1, \ldots, n$.

Remark 6.4 (i) We stress that, in general, there will be multiple solutions q to the system (6.2), satisfying all the requirements listed above; this leads to multiple strong Markov structures for $\mathcal{MFH}^j$, $j = 1, \ldots, n$. One can choose a particular structure depending on the additional operational criteria, which provide a dependence-structure input that is relevant to the particular modeler. This feature is key to applications of Markov structures: it allows one to build Markov structures with predetermined margins and a multitude of dependence structures between the coordinates X^j, $j = 1, \ldots, n$.
(ii) When constructing a Markov structure using the above recipe, the results of Section 2.2.2, such as Propositions 2.35 or 2.46 and Corollary 2.43 or Proposition 2.47, can be used to verify whether conditions **C1** and **C4** are satisfied for the symbol q that is chosen as a solution to the system (6.2). However, as noted above, conditions **C2** and **C3** are automatically satisfied for any symbol q satisfying the system (6.2). So, there is no need to verify whether condition **C3** is satisfied for such a q. In particular there is no need to use Propositions 2.38 or 2.41 for this purpose.

We will now discuss item (ii) of the above recipe in the context of the type I representation of symbols. Let

$$\begin{aligned}\rho_j(x_j, \xi_j) = & -i d_j(x_j)\xi_j + c_j(x_j)\xi_j^2 \\ & + \int_{\mathbb{R}\setminus\{0\}} \left(1 - e^{i z_j \xi_j} + \frac{i z_j \xi_j}{1 + |z_j|^2}\right) \mu_j(x_j, dz_j) \qquad (6.3)\end{aligned}$$

be the type I representation for ρ_j, $j = 1, \ldots, n$.

Next, recall from (2.36) that the type I representation of the symbol q, i.e

$$q(x,\xi) = -i\langle b(x),\xi\rangle + \langle \xi, a(x)\xi\rangle + \int_{\mathbb{R}^n\setminus\{0\}} \left(1 - e^{i\langle y,\xi\rangle} + \frac{i\langle y,\xi\rangle}{1+|y|^2}\right)\mu(x,dy), \quad x,\xi\in\mathbb{R}^n, \tag{6.4}$$

is given in terms of a triple of coefficients: an $\mathbb{R}^n$-valued Borel measurable function b, a Borel measurable function a with values in the set of symmetric positive-semidefinite matrices, and a Lévy kernel μ. Therefore, the symbol q in a representation of type I will satisfy the system of equations (6.2) if and only if its type I characteristic triple (b,a,μ) satisfies, for $j=1,\dots,n$, the following conditions:

1I b is a vector function such that

$$b_j(x) = d_j(x_j), \tag{6.5}$$

2I a is a symmetric non-negative-definite matrix function such that

$$a_{jj}(x) = c_j(x_j), \tag{6.6}$$

3I μ is a Lévy kernel such that

$$\int_{\mathbb{R}^n\setminus\{0\}} \left(1 - e^{iy_j\xi_j} + \frac{iy_j\xi_j}{1+|y|^2}\right)\mu(x,dy) = \int_{\mathbb{R}\setminus\{0\}} \left(1 - e^{iy_j\xi_j} + \frac{iy_j\xi_j}{1+|y_j|^2}\right)\mu_j(x_j,dy_j). \tag{6.7}$$

Remark 6.5 With regard to Remark 6.4 it is essential to observe that conditions (6.5)–(6.7) do not uniquely determine the triple (b,a,μ). In fact, only b is uniquely determined, but not a and μ.

Occasionally, we will be dealing with building of strong Markov structures in terms of symbols given via type II representations. To this end let us assume that ρ_j admits a type II representation, and let

$$\rho_j(x_j,\xi_j) = -id_j^0(x_j)\xi_j + c_j(x_j)\xi_j^2 + \int_{\mathbb{R}\setminus\{0\}} \left(1 - e^{iz_j\xi_j}\right)\mu_j(x_j,dz_j) \tag{6.8}$$

be this type II representation for $j=1,\dots,n$. The symbol q will satisfy the system of equations (6.2) if and only if its type II characteristic triple (b^0,a,μ) (see (2.39)) satisfies, for $j=1,\dots,n$:

1II b^0 is a vector function such that

$$b_j^0(x) = d_j^0(x_j), \tag{6.9}$$

2II a is a symmetric non-negative-definite matrix function such that

$$a_{jj}(x) = c_j(x_j), \tag{6.10}$$

3II μ is a Lévy kernel such that

$$\int_{\mathbb{R}^n\setminus\{0\}} \left(1 - e^{iy_j\xi_j}\right) \mu(x,dy) = \int_{\mathbb{R}\setminus\{0\}} \left(1 - e^{iy_j\xi_j}\right) \mu_j(x_j,dy_j). \tag{6.11}$$

In the next section we provide examples of the construction of strong Markov structures for various collections of $\mathbb{R}$-Feller–Markov families.

6.3 Examples

In all the examples below we set $x = (x_1,\ldots,x_n) \in \mathbb{R}^n$ and $\xi = (\xi_1,\ldots,\xi_n) \in \mathbb{R}^n$ for some $n > 1$.

Example 6.6 Independence Markov structure Let functions $\rho_1,\ldots,\rho_n$ be given as

$$\begin{aligned}\rho_j(z,\psi) &= -\,id_j(z)\psi + c_j(z)\psi^2 \\ &\quad + \int_{\mathbb{R}\setminus\{0\}} \left(1 - e^{iu\psi} + \frac{iu\psi}{1+u^2}\right) \mu_j(z,du), \quad z,\psi \in \mathbb{R},\end{aligned} \tag{6.12}$$

where the characteristic triples $(d_j,c_j,\mu_j)_{j=1}^n$ are assumed to satisfy conditions **S1**(1)–**S4**(1) and **H3**(1) (see Chapter 2). Then, in view of Proposition 2.39, there exist nice $\mathbb{R}$-Feller–Markov families, say $\{\mathcal{MFH}^j,\ j = 1,\ldots,n\}$, with corresponding symbols $\rho_1,\ldots,\rho_n$. Assume additionally that $\rho_1,\ldots,\rho_n$ satisfy (2.63).

Now let q be a function given by (6.4), where

$$b_j(x) := d_j(x_j), \quad a_{ij}(x) := c_j(x_j)\mathbb{1}_{\{i=j\}}, \quad i,j = 1,\ldots,n, \tag{6.13}$$

$$\mu(x,dy) := \sum_{j=1}^n \left(\bigotimes_{k=1,k\neq j}^n \delta_{\{0\}}(dy_k) \right) \otimes \mu_j(x_j,dy_j). \tag{6.14}$$

Here, by assumption, μ_j, $j = 1,\ldots,n$, are Lévy kernels from $(\mathbb{R},\mathcal{B}(\mathbb{R}))$ to $(\mathbb{R}\setminus\{0\},\mathcal{B}(\mathbb{R}\setminus\{0\}))$, so that

$$\int_{\mathbb{R}^n\setminus\{0\}} (|y|^2 \wedge 1)\mu(x,dy) = \sum_{j=1}^n \int_{\mathbb{R}\setminus\{0\}} (y_j^2 \wedge 1)\mu_j(x_j,dy_j) < \infty,$$

and thus μ is a Lévy kernel from $(\mathbb{R}^n,\mathcal{B}(\mathbb{R}^n))$ to $(\mathbb{R}^n\setminus\{0\},\mathcal{B}(\mathbb{R}^n\setminus\{0\}))$.

We note that the properties **S1**(1)–**S4**(1), **H3**(1) assumed for $\rho_1,\ \ldots,\rho_n$ imply that properties **S1**(n)–**S4**(n), **H3**(n) are satisfied for q. Thus, by application of Proposition 2.39, the above reasoning implies that q is a symbol of a nice $\mathbb{R}^n$-Feller–Markov family $\mathcal{MMFH}$ satisfying (2.63). Let us observe that

$$\begin{aligned}&\int_{\mathbb{R}^n\setminus\{0\}} \left(1 - e^{i\langle y,\xi\rangle} + \frac{i\langle y,\xi\rangle}{1+|y|^2}\right) \mu(x,dy) \\ &\quad = \sum_{j=1}^n \int_{\mathbb{R}\setminus\{0\}} \left(1 - e^{iy_j\xi_j} + \frac{iy_j\xi_j}{1+y_j^2}\right) \mu_j(x_j,dy_j)\end{aligned} \tag{6.15}$$

for $x, \xi \in \mathbb{R}^n$. This and (6.13) imply that

$$q(x,\xi) = \sum_{j=1}^{n} \rho_j(x_j, \xi_j). \tag{6.16}$$

Since the symbol q of $\mathcal{MMFH}$ satisfies (2.63), by Proposition 2.35 we see that condition **C1** is satisfied. From (6.16) and the fact that $\rho_j(x_j, 0) = 0$, for every $j = 1, \ldots, n$, we see that (6.2) holds. Thus conditions **C2** and **C3** hold. Putting the above observations together, and additionally assuming that condition **C4** holds,[2] we conclude from Theorem 2.33 that the family $\mathcal{MMFH}$ is a Markov structure for $\mathcal{MFH}^j$, $j = 1, \ldots, n$. We call this structure the *independence Markov structure*.

It is intuitively clear from (6.16), and can be established formally, that the independence Markov structure constructed here is a multivariate Markov family with independent coordinates.[3]

Example 6.7 Diffusion structure Consider a collection of functions $\rho_1, \ldots, \rho_n$ given by

$$\rho_j(z, \psi) = -i\psi d_j(z) + c_j(z)\psi^2, \quad z, \psi \in \mathbb{R}, \ j = 1, \ldots, n.$$

We will discuss two cases, representing two sets of conditions satisfied by c_j and d_j.

Case 1. We assume that d_j, c_j, $j = 1, \ldots, n$, are functions satisfying **S1**(1)–**S2**(1). In view of Remark 2.40 these conditions are sufficient for the existence of nice $\mathbb{R}$-Feller–Markov families $\{\mathcal{MFH}^j, \ j = 1, \ldots, n\}$ with corresponding symbols $\rho_1, \ldots, \rho_n$ satisfying (2.63). These $\mathbb{R}$-Feller–Markov families correspond to one-dimensional continuous diffusion processes starting from different initial positions, since the generator corresponding to ρ_j has, on $C_c^\infty(\mathbb{R})$, the form

$$A^j f(z) = d_j(z) \frac{\partial}{\partial z} f(z) + c_j(z) \frac{\partial^2}{\partial z^2} f(z).$$

We will now construct a Markov structure $\mathcal{MMFH}$ for $\mathcal{MFH}^j$, $j = 1, \ldots, n$. This is done by constructing a symbol q for $\mathcal{MMFH}$ in terms of $\rho_1, \ldots, \rho_n$. To this end we observe that it is natural to construct q as follows:

$$q(x,\xi) := -i\langle b(x), \xi \rangle + \langle \xi, a(x)\xi \rangle, \tag{6.17}$$

where the functions $b : \mathbb{R}^n \to \mathbb{R}^n$, $a : \mathbb{R}^n \to L(\mathbb{R}^n, \mathbb{R}^n)$ satisfy

$$b_j(x) = d_j(x_j), \quad a_{jj}(x) = c_j(x_j), \quad j = 1, \ldots, n, \tag{6.18}$$

and, moreover, a is taken to be a symmetric positive-semidefinite matrix-valued function chosen so that **S1**(n) holds. It is clear that **S2**(n) also holds for this construction.

[2] Let us recall that the fact that $C_c^\infty(\mathbb{R})$ is a core of each $A^j = -q^j(x_j, D)$, $j = 1, \ldots, n$, is sufficient for condition **C4** to hold; see Corollary 2.43.

[3] We say that the multivariate Markov family $\mathcal{MMFH} = \{(\Omega, \mathcal{F}, \mathbb{F}, (X_t)_{t \geq 0}, \mathbb{P}_x, P), x \in \mathbb{R}^n\}$ has independent coordinates if the coordinates of the process X are independent under every probability $\mathbb{P}_x$, $x \in \mathbb{R}^n$. In this regard see also Lemma 4.7 in Schnurr (2009).

Consequently, in view of Proposition 2.39 and Remark 2.40, q is a symbol of a PDO that has an extension that generates a nice $\mathbb{R}^n$-Feller–Markov family satisfying (2.63). Thus, in view of Proposition 2.35, condition **C1** holds here. We shall now verify that condition **C2** is also satisfied. Indeed, one can easily check that conditions (6.18) imply that

$$\begin{aligned} q(x, \mathbf{e}_j\xi_j) &= -i\langle b(x), \mathbf{e}_j\xi_j\rangle + \langle \mathbf{e}_j\xi_j, a(x)\mathbf{e}_j\xi_j\rangle \\ &= -ib_j(x)\xi_j + a_{jj}(x)\xi_j^2 \\ &= \rho_j(x_j, \xi_j), \quad x \in \mathbb{R}^n,\ \xi_j \in \mathbb{R}, \end{aligned} \tag{6.19}$$

which means that condition **C2**(j) is satisfied for $j = 1, \ldots, n$. Thus, since for each $j = 1, \ldots, n$ the function ρ_j is a symbol of a nice $C_0(\mathbb{R}^n)$-Feller family, we see that condition **C3** holds. Assuming that condition **C4** holds, which happens if one of the conditions of Proposition 2.45 holds, and using Theorem 2.33 we conclude that q defines a Markov structure $\mathcal{MMFH}$ for $\mathcal{MFH}^j$, $j = 1, \ldots, n$.

Case 2. We assume that d_j and c_j, $j = 1, \ldots, n$, are functions satisfying one of the assumptions of Proposition 2.45. Therefore there exist nice $\mathbb{R}$-Feller–Markov families $\{\mathcal{MFH}^j,\ j = 1, \ldots, n\}$ with corresponding symbols $\rho_1, \ldots, \rho_n$ satisfying (2.63). In this case, additionally we assume that a and b are chosen in such a way that (6.18) holds and that the conditions in Proposition 2.46 hold. In view of Proposition 2.46, q as given in (6.17) is a symbol for a pseudo-differential operator which generates a nice $\mathbb{R}^n$-Feller–Markov family satisfying condition **C1**. The properties postulated in (6.18) imply (6.19), so that **C2** is satisfied. Again, as in Case 1, this implies that **C3** holds. Proposition 2.45 and Corollary 2.43 imply condition **C4**, and we conclude, by Theorem 2.33, that q defines a Markov structure $\mathcal{MMFH}$ for $\mathcal{MFH}^j$, $j = 1, \ldots, n$.

Example 6.8 Bivariate Poisson structure In this example we take $n = 2$ and present a construction of a bivariate Poisson structure in the spirit of our recipe of Section 6.2. However, we slightly modify the procedure described there. Namely, we first construct a candidate for symbol q; next, we find $q(x, \mathbf{e}_j\xi_j)$ for $j = 1, 2$, and then we formulate conditions ensuring that (6.2) holds.

Let $\rho_j, j = 1, 2$, be given by

$$\rho_j(z, \psi) = (1 - e^{i\psi})\eta_j, \quad z, \psi \in \mathbb{R}, \tag{6.20}$$

with $\eta_j > 0$. It is clear that there exist $\mathbb{R}$-Feller–Markov families

$$\mathcal{MFH}^j = \left\{(\Omega^j, \mathcal{F}, \mathbb{F}^j, (Y_t^j)_{t \geq 0}, \mathbb{Q}_y^j, Q^j) : y \in \mathbb{R}\right\}, \quad j = 1, 2,$$

with corresponding symbols ρ_1 and ρ_2.

Let us fix $j \in \{1, 2\}$. Observe that ρ_j can be written in the form (2.36) with $a_j \equiv 0$, $b_j \equiv \frac{1}{2}\eta_j$, $\mu_j(x, dy) = \eta_j\delta_1(dy)$. Hence, using (2.37) we see that the compensator of the jump measure of Y^j is given by $\nu_j(du, dy) = \mu_j(Y_u^j, dy)du = \eta_j\delta_1(dy)du$.

Therefore the constants η_j are jump intensities of Y^j. Given the form of ν_j, it follows that, for any $A \in \mathcal{B}(\mathbb{R})$ such that, $1 \notin A$, we have

$$\mathbb{E}_{\mathbb{Q}^j_y}(\mu^{Y^j}((0,t] \times A)) = \mathbb{E}_{\mathbb{Q}^j_y}(\nu_j((0,t] \times A)) = 0, \quad t \geq 0,$$

where μ^{Y^j} is the jump measure of Y^j. Thus $\mu^{Y^j}((0,t] \times A) = 0$ for all $t \geq 0$ and hence we conclude that Y^j has jumps of size 1 only. Let us consider the following representation of the semimartingale Y^j:

$$\begin{aligned} Y^j_t = Y^j_0 + \frac{1}{2}\eta_j t + \int_0^t \int_{\mathbb{R}} h(y)\left(\mu^{Y^j}(ds,dy) - \nu_j(ds,dy)\right) \\ + \int_0^t \int_{\mathbb{R}} \left(y - h(y)\right)\mu^{Y^j}(ds,dy), \quad t \geq 0, \end{aligned}$$

where $h(y) = y/(1+y^2)$. Since

$$\int_0^t \int_{\mathbb{R}} (y - h(y))\nu_j(ds,dy) = \int_0^t \int_{\mathbb{R}} \frac{y^3}{1+y^2}\eta_j \delta_1(dy)ds = \frac{1}{2}\eta_j t, \quad t \geq 0,$$

we have, for every $t \geq 0$,

$$\begin{aligned} Y^j_t &= Y^j_0 + \eta_j t + \int_0^t \int_{\mathbb{R}} h(y)(\mu^{Y^j}(ds,dy) - \nu_j(ds,dy)) \\ &\quad + \int_0^t \int_{\mathbb{R}} (y - h(y))(\mu^{Y^j}(ds,dy) - \nu_j(ds,dy)) \\ &= Y^j_0 + \eta_j t + \int_0^t \int_{\mathbb{R}} y(\mu^{Y^j}(ds,dy) - \nu_j(ds,dy)). \end{aligned}$$

Now, from the fact that

$$\int_0^t \int_{\mathbb{R}} y\nu_j(ds,dy) = \eta_j t, \quad t \geq 0,$$

we conclude that

$$Y^j_t = Y^j_0 + \int_0^t \int_{\mathbb{R}} y\mu^{Y^j}(ds,dy), \quad t \geq 0.$$

From this and the fact that Y^j has jumps of size 1 only it follows that the process $N^j_t := Y^j_t - Y^j_0, t \geq 0$, has trajectories which are increasing step functions with jumps of size 1. From the above we also see that, for each $y \in \mathbb{R}$, $(N^j_t - \eta_j t)_{t \geq 0}$ is an $(\mathbb{F}^{Y^j}, \mathbb{Q}^j_y)$-local martingale. Thus, by Watanabe's characterization theorem (see the Remark following Theorem 2.3 in Watanabe (1964) and also see Brémaud (1975)), N^j is a one-dimensional Poisson process with intensity η_j under $\mathbb{Q}^j_y$. Consequently, Y^j is a one-dimensional Poisson process starting from y under $\mathbb{Q}^j_y$.[4] Accordingly, we call the family $\mathcal{MFH}^j$ *a one-dimensional Poisson family*.

[4] If N is a classical Poisson process, i.e. $N_0 = 0$, then the process $N_t + y$, $t \geq 0$, is a Poisson process starting from $y \in \mathbb{R}$.

A natural candidate for the symbol of the Markov structure with margins $\mathcal{MFH}^1$ and $\mathcal{MFH}^2$ is

$$q(x,\xi) = (1-e^{i\xi_1})\lambda_{(1,0)} + (1-e^{i\xi_2})\lambda_{(0,1)} + (1-e^{i(\xi_1+\xi_2)})\lambda_{(1,1)}, \tag{6.21}$$

where $\lambda_{(0,1)}, \lambda_{(1,0)}, \lambda_{(1,1)}$ are nonnegative constants. Since the conditions in Proposition 2.39 are satisfied, the function q is a symbol of a nice $\mathbb{R}^2$-Feller–Markov family $\mathcal{MMFH}$ satisfying (2.63). Let us consider the system of equations (6.2). Note that

$$q(x,\mathbf{e}_1\xi_1) = (1-e^{i\xi_1})(\lambda_{(1,0)} + \lambda_{(1,1)}),$$

$$q(x,\mathbf{e}_2\xi_2) = (1-e^{i\xi_2})(\lambda_{(0,1)} + \lambda_{(1,1)}).$$

Hence, and from the form of ρ_j (see (6.20)), we obtain that (6.2) is satisfied if and only if $\lambda_{(0,1)}, \lambda_{(1,0)}, \lambda_{(1,1)}$ satisfy the following system of linear equations:

$$\lambda_{(0,1)} + \lambda_{(1,1)} = \eta_2,$$

$$\lambda_{(1,0)} + \lambda_{(1,1)} = \eta_1.$$

The above system admits infinitely many solutions, which can be parameterized by $\lambda_{(1,1)}$. In this case $\lambda_{(0,1)}$, $\lambda_{(1,0)}$ are given by

$$\lambda_{(0,1)} = \eta_2 - \lambda_{(1,1)},$$

$$\lambda_{(1,0)} = \eta_1 - \lambda_{(1,1)}.$$

Since we are interested in nonnegative solutions, we restrict $\lambda_{(1,1)}$ to the interval $[0, \eta_1 \wedge \eta_2]$. Each choice of $\lambda_{(1,1)}$ will give us the Markov structure with margins $\mathcal{MFH}^1$ and $\mathcal{MFH}^2$.

Since q does not depend on x we see that condition **C2** holds, which together with the fact that ρ_j is a symbol of $\mathcal{MFH}^j$ for $j = 1,2$, implies **C3**. Since (2.63) holds, by Proposition 2.35 $\mathcal{MMFH}$ satisfies condition **C1**. Finally, for $j = 1,2$, Remark 2.44 implies that $C_c^\infty(\mathbb{R})$ is a core of A^j, the PDO associated with the symbol ρ_j, and thus by Corollary 2.43 we see that condition **C4** holds. Therefore, in view of Theorem 2.33 the function q in (6.21) is a symbol of a Markov structure

$$\mathcal{MMFH} = \left\{(\Omega, \mathcal{F}, \mathbb{F}, X = (X_t^1, X_t^2)_{t\geq 0}, \mathbb{P}_x, P), x \in \mathbb{R}^2\right\}$$

with margins $\mathcal{MFH}^1$ and $\mathcal{MFH}^2$. We call the structure $\mathcal{MMFH}$ a bivariate Poisson structure for $\mathcal{MFH}^1$ and $\mathcal{MFH}^2$.

The symbol q can be written in the form (2.36) with

$$b \equiv \left(\tfrac{1}{2}\eta_1 - \tfrac{1}{6}\lambda_{(1,1)}, \tfrac{1}{2}\eta_2 - \tfrac{1}{6}\lambda_{(1,1)}\right)', \quad a \equiv \begin{pmatrix} 0 & 0 \\ 0 & 0 \end{pmatrix},$$

$$\begin{aligned}\mu(x,dy_1,dy_2) = {} & \delta_{(1,0)}(dy_1,dy_2)(\eta_1 - \lambda_{(1,1)}) + \delta_{(0,1)}(dy_1,dy_2)(\eta_2 - \lambda_{(1,1)}) \\ & + \delta_{(1,1)}(dy_1,dy_2)\lambda_{(1,1)}.\end{aligned}$$

Using (2.37) we see that, for each $x \in \mathbb{R}^2$, the $(\mathbb{F}^X, \mathbb{P}_x)$-compensator of the jump measure of X is given by $\nu(du, dy_1, dy_2) = \mu(X_u, dy_1, dy_2)du$. So $\lambda_{(1,1)}$ is the intensity of common jumps, i.e. simultaneous jumps in the coordinates X^1 and X^2 of X, or, equivalently, the jumps of X equal to $(1,1)$. In accordance with the notion of the compensator of a jump measure of the process X, we see that $\lambda_{(1,1)}$, is the compensator of a process which counts the common jumps of X, is given as $\lambda_{(1,1)}t$, $t \geq 0$. In the same vein, the number $\eta_1 - \lambda_{(1,1)}$ is the intensity of the idiosyncratic jumps of X^1 only, i.e. the process $(\eta_1 - \lambda_{(1,1)})t$, $t \geq 0$, the compensator of the process that counts the jumps of the coordinate X^1 alone or, equivalently, the jumps of X equal to $(1,0)$. An analogous interpretation applies to the number $\eta_2 - \lambda_{(1,1)}$.

Generalization to the n-dimensional case is immediate. In Example 9.4 we discuss the bivariate Poisson structure in the context of the special semimartingale structures.

In Example 6.9 below we provide a generalization of Example 6.8 that allows the jump intensities to depend on x.

Example 6.9 Bivariate point process Markov structure We let $n = 2$ and we consider functions ρ_1 and ρ_2 given by

$$\rho_j(z, \psi) := \left(1 - e^{i\psi}\right) \eta_j(z), \quad z, \psi \in \mathbb{R}, j = 1, 2, \tag{6.22}$$

where η_1 and η_2 are assumed to be nonnegative continuous bounded functions. Since assumptions **S1**(1)–**S4**(1) and **H3**(1) are satisfied for ρ_1 and ρ_2, the existence of nice $\mathcal{MFH}^j$ with symbol ρ_j, $j = 1, 2$, follows from Proposition 2.39. Now, let us define the symbol q by

$$q(x, \xi) = \int_{\mathbb{R}^2 \setminus \{0\}} (1 - e^{i\langle y, \xi \rangle}) \mu(x, dy), \quad x, \xi \in \mathbb{R}^2,$$

with μ given by

$$\begin{aligned} \mu(x, dy_1, dy_2) = \delta_{(1,0)}(dy_1, dy_2)\lambda_{(1,0)}(x) + \delta_{(0,1)}(dy_1, dy_2)\lambda_{(0,1)}(x) \\ + \delta_{(1,1)}(dy_1, dy_2)\lambda_{(1,1)}(x), \end{aligned} \tag{6.23}$$

where $\lambda_{(0,1)}(x)$, $\lambda_{(1,0)}(x)$, $\lambda_{(1,1)}(x)$ are nonnegative continuous bounded functions of x. So, the corresponding symbol q is given as

$$q(x, \xi) = \left(1 - e^{i\xi_1}\right) \lambda_{(1,0)}(x) + \left(1 - e^{i\xi_2}\right) \lambda_{(0,1)}(x) + \left(1 - e^{i(\xi_1 + \xi_2)}\right) \lambda_{(1,1)}(x).$$

Let us note the analogy of its form with the form of q in Example 6.8. It is straightforward to verify that q satisfies **S1**(2)–**S4**(2) and **H3**(2). Therefore, by Proposition 2.39, there exists a nice $\mathbb{R}^2$-Feller–Markov family $\mathcal{MMFH}$ with symbol q satisfying (2.63). Now let us consider the system of equations (6.2). Note that

$$\begin{aligned} q(x, \mathbf{e}_1 \xi_1) &= \left(1 - e^{i\xi_1}\right) (\lambda_{(1,0)}(x) + \lambda_{(1,1)}(x)), \\ q(x, \mathbf{e}_2 \xi_2) &= \left(1 - e^{i\xi_2}\right) (\lambda_{(0,1)}(x) + \lambda_{(1,1)}(x)), \end{aligned} \tag{6.24}$$

for all $x \in \mathbb{R}^2$, $\xi \in \mathbb{R}^2$. Hence, and from the form of ρ_j (see (6.22)), we obtain that (6.2) is satisfied if and only if, for $x = (x_1, x_2) \in \mathbb{R}^2$, we have

$$\lambda_{(0,1)}(x) + \lambda_{(1,1)}(x) = \eta_2(x_2),$$

$$\lambda_{(1,0)}(x) + \lambda_{(1,1)}(x) = \eta_1(x_1).$$

In this case condition **C2** holds; together with the fact that, for $j = 1,\ldots,n$, ρ_j is a symbol of $\mathcal{MFH}^j$, it implies that **C3** holds. Given η_1 and η_2, there exist infinitely many continuous and bounded functions $\lambda_{(1,0)}, \lambda_{(0,1)}$ and $\lambda_{(1,1)}$ that satisfy the above equations. In fact, such triples can be parameterized by a $\lambda_{(1,1)}$ satisfying

$$0 \leq \lambda_{(1,1)}(x) \leq \inf_{x_1,x_2\in\mathbb{R}} (\eta_1(x_1) \wedge \eta_2(x_2)), \quad x = (x_1,x_2) \in \mathbb{R}^2.$$

Since (2.63) holds, by Proposition 2.35 $\mathcal{MMFH}$ satisfies condition **C1**. Finally, we check that condition **C4** is satisfied. We know that the operator A^j corresponding to ρ_j is well defined on $C_c^\infty(\mathbb{R})$ and has the form

$$A^j f(z) = \eta_j(z)(f(z+1) - f(z)), \quad z \in \mathbb{R}.$$

Note that $C_c^\infty(\mathbb{R})$ is dense in $C_0(\mathbb{R})$ and thus $C_c^\infty(\mathbb{R})$ is a core of A^j, $j = 1,2$. Thus, by Corollary 2.43, condition **C4** holds. Summing up, $\mathcal{MMFH}$ as constructed above with the symbol q constitutes a Markov structure for $\mathcal{MFH}^j$, $j = 1,2$.

In the previous example we constructed a Markov structure featuring the possibility of common jumps of its coordinates. Formula (6.23) gives the distribution of the common jump sizes. In the next example

Example 6.10 Markov jump structure Consider functions $\rho_1,\ldots,\rho_n$ on $\mathbb{R}^2$ of the form

$$\rho_j(z,\psi) = \eta_j(z) \int_{\mathbb{R}\setminus\{0\}} \left(1 - e^{iu\psi}\right) \nu_j(z,du), \quad z \in \mathbb{R},\ \psi \in \mathbb{R}, \tag{6.25}$$

where, for $j = 1,\ldots n$, η_j is a nonnegative, bounded, continuous function and ν_j is a probability kernel from $(\mathbb{R},\mathcal{B}(\mathbb{R}))$ to $(\mathbb{R},\mathcal{B}(\mathbb{R}))$, with the property that, for any bounded and measurable function f on $\mathbb{R}$ and for any sequence $z_k \in \mathbb{R}$ such that $\lim_k z_k = z$, we have

$$\lim_{k\to\infty} \int_{\mathbb{R}} f(u)\nu_j(z_k,du) = \int_{\mathbb{R}} f(u)\nu_j(z,du). \tag{6.26}$$

In particular, this implies that each function $\rho_j(z,\psi)$ is continuous in z.

The above assumptions imply that conditions **S1**(1)–**S4**(1) and **H3**(1) hold for $\rho_1,\ldots,\rho_n$ written in the type I representation, i.e. in the form (6.3). Thus, in view of Proposition 2.39 there exist nice Feller families, $\mathcal{MFH}^j$, $j = 1,\ldots,n$, with symbols $\rho_1,\ldots,\rho_n$ satisfying (2.63).

We will build a Markov structure

$$\mathcal{MMFH} = \left\{(\Omega,\mathcal{F},\mathbb{F},X = (X_t^1,\ldots,X_t^n)_{t\geq 0},\mathbb{P}_y,P) : y \in \mathbb{R}^n\right\}$$

for $\mathcal{MFH}^j$, $j = 1,\ldots,n$, by constructing a symbol q as follows

$$q(x,\xi) = \int_{\mathbb{R}^n\setminus\{0\}} \left(1 - e^{i\langle y,\xi\rangle}\right) \mu(x,dy), \quad x,\xi \in \mathbb{R}^n, \tag{6.27}$$

in terms of a measurable kernel μ yet to be specified.

In order to proceed we set $\mathcal{J} := \{S \subset \{1,\dots,n\} : S \neq \emptyset\}$. Next, to specify μ we denote by μ_S, for $S \in \mathcal{J}$ and $x \in \mathbb{R}^n$, a probability kernel with cumulative distribution function, say $F^S(x,\cdot)$, given by

$$F^S(x,y_1,\dots,y_n) = C^S(\nu_j(x_j,(-\infty,y_j]), j \in S) \prod_{i \in S^c} \delta_{\{0\}}((-\infty,y_i]),$$

where C^S is a copula function (see Appendix A) on $[0,1]^{|S|}$, and where $|S| = \mathrm{card}(S)$ is the cardinality of S.[5] We additionally assume that, for any $S \in \mathcal{J}$, for any bounded and measurable function h on $\mathbb{R}^n$, and for any sequence $x_k \in \mathbb{R}^n$ such that $\lim_k x_k = x$, we have

$$\lim_{k\to\infty} \int_{\mathbb{R}^n} h(u)\mu_S(x_k,du) = \int_{\mathbb{R}^n} h(u)\mu_S(x,du). \tag{6.28}$$

We now specify μ:

$$\mu(x,dy) = \sum_{S \in \mathcal{J}} \lambda_S(x)\mu_S(x,dy), \tag{6.29}$$

where λ_S, $S \in \mathcal{J}$, are nonnegative continuous bounded functions on $\mathbb{R}^n$. For a μ specified by (6.29), the symbol q introduced in (6.27) takes the form

$$q(x,\xi) = \sum_{S \in \mathcal{J}} \lambda_S(x) \int_{\mathbb{R}^n \setminus \{0\}} \left(1 - e^{i\langle y,\xi \rangle}\right) \mu_S(x,dy). \tag{6.30}$$

In view of the properties of the kernels μ_S, $S \in \mathcal{J}$, in particular, assumption (6.28), the symbol q specified in (6.30) satisfies **S1**(n)–**S4**(n). Moreover **H3**(n) holds, since the λ_S, $S \in \mathcal{J}$, are bounded. By Proposition 2.39 there exists a nice $\mathbb{R}^n$-Feller–Markov family with symbol q satisfying (2.63).

Now we will determine the conditions that we need to impose on such a constructed $\mathcal{MMFH}$ to guarantee that it is the desired structure. To this end we impose conditions ensuring that (6.2) holds for the q in (6.30). For this, given the form (6.30) of q, it is enough to demonstrate that equality (6.11) holds for all j. First, we note that, for each $S \in \mathcal{S}$ and fixed $k \in \{1,\dots,n\}$, we have

$$\begin{aligned}
&\int_{\mathbb{R}^n \setminus \{0\}} \left(1 - e^{i\langle y,\mathbf{e}_k \xi_k\rangle}\right) \mu_S(x,dy) \\
&= \int_{\mathbb{R}^n \setminus \{0\}} \left(1 - e^{iy_k\xi_k}\right) \mu_S(x,dy) \\
&= \mathbb{1}_{\{S\}}(k) \int_{\mathbb{R} \setminus \{0\}} \left(1 - e^{iy_k\xi_k}\right) \nu_k(x_k,dy_k),
\end{aligned}$$

where the last equality follows from the fact that, for each $x = (x_1,\dots,x_n) \in \mathbb{R}^n$ we have that μ_S is a probability measure with kth marginal equal to ν_k for $k \in S$ and equal to $\delta_{\{0\}}(dy_k)$ for $k \in S^c$.

[5] In particular, for $\mathrm{card}(S) = 1$ we have $C^S(x) = x$.

The above calculations and (6.25) imply that for all $k \in \{1,\dots,n\}$ we have

$$\begin{aligned} q(x, \mathbf{e}_k \xi_k) &= \sum_{S \in \mathcal{J}} \lambda_S(x) 1_{\{k \in S\}} \int_{\mathbb{R}\setminus\{0\}} \left(1 - e^{iy_k \xi_k}\right) \nu_k(x_k, dy_k) \\ &= \left(\sum_{S \in \mathcal{J}: k \in S} \lambda_S(x) \right) \int_{\mathbb{R}\setminus\{0\}} \left(1 - e^{iy_k \xi_k}\right) \nu_k(x_k, dy_k) \\ &= \rho_k(x_k, \xi_k), \end{aligned} \tag{6.31}$$

provided that

$$\sum_{S \in \mathcal{J}: k \in S} \lambda_S(x) = \eta_k(x_k), \quad x \in \mathbb{R}^n. \tag{6.32}$$

Condition (6.32) will be used in verifying that the $\mathcal{MMFH}$ corresponding to symbol q given in (6.30) is a Markov structure for $\mathcal{MFH}^j$, $j = 1,\dots,n$.

Functions λ_S satisfying (6.32) can be easily constructed. For example, one may take

$$\lambda_{\{k\}}(x) = \eta_k(x_k), \quad \lambda_S(x) \equiv 0 \quad \text{for } S \text{ such that } k \in S \text{ and } \mathrm{card}(S) > 1.$$

The construction of functions λ_S satisfying (6.32) and $\lambda_S \not\equiv 0$ for $\mathrm{card}(S) > 1$ can be done as well. Please see Example 6.9 for a particular case.

The above considerations show that, for an $\mathcal{MMFH}$ satisfying (6.32), condition **C2** is fulfilled; this, together with the fact that ρ_j is a symbol of $\mathcal{MFH}^j$, implies **C3**. Since (2.63) holds we see by Proposition 2.35 that $\mathcal{MMFH}$ satisfies **C1**. To construct an $\mathcal{MMFH}$ structure using Theorem 2.33, condition **C4**(j) has to hold for all ρ_j, $j = 1,\dots,n$. The sufficient condition is that $C_c^\infty(\mathbb{R})$ is a core of A^j corresponding to ρ_j. We know that the operator A^j is well defined on $C_0(\mathbb{R})$ and has the form

$$A^j f(z) = \eta_j(z) \int_{\mathbb{R}\setminus\{0\}} (f(z+u) - f(z)) \nu_j(z, du).$$

The space $C_c^\infty(\mathbb{R})$ is dense in $C_0(\mathbb{R})$ and thus $C_c^\infty(\mathbb{R})$ is a core of A^j.

Therefore, in view of Theorem 2.33 the symbol q given by (6.30) with functions λ_S satisfying (6.32) is a symbol of a Markov structure with margins $\mathcal{MFH}^j$, $j = 1,\dots,n$.

We complete the presentation of this example by providing some more insight into the properties of the Markov structure

$$\mathcal{MMFH} = \left\{ (\Omega, \mathcal{F}, \mathbb{F}, X = (X_t^1, \dots, X_t^n)_{t\geq 0}, \mathbb{P}_x, P) : x \in \mathbb{R}^n \right\}$$

constructed above. In view of Theorem 3.10 in Schnurr (2009), for any $y \in \mathbb{R}^n$ the $(\mathbb{F}, \mathbb{P}_y)$-compensator of μ^X, the jump measure of X, is the measure ν^X given by

$$\nu^X(dt, dy) = \mu(X_t, dy)dt = \sum_{S \in \mathcal{J}} \lambda_S(X_t) \mu_S(X_t, dy)dt. \tag{6.33}$$

Moreover, $\lambda_S(x)$ and the measure $\mu_S(x, \cdot)$ characterize common jumps of X, from the position x, such that all coordinate X^i with $i \in S$ jump together, but no coordinates X^i with $i \notin S$ jump. In fact, the random measure which counts such jumps, say μ_S^X, is given as

$$\mu_S^X(\omega, dt, dz) = \sum_{s>0} \mathbb{1}_{\{\Delta X_s^i(\omega) \neq 0, i \in S, \Delta X_s^i(\omega)=0, i \in S^c\}} \delta_{\{s, \Delta X_s(\omega)\}}(dt, dz).$$

Consequently,

$$\mu^X \equiv \sum_{S \in \mathcal{J}} \mu_S^X.$$

So, using (6.33) it can be shown that, for each $y \in \mathbb{R}^n$, the $(\mathbb{F}, \mathbb{P}_y)$-compensator of the measure μ_S^X is given by

$$\nu_S^X(dt, dz) = \lambda_S(X_t)\mu_S(X_t, dz)dt.$$

Example 6.11 Markov jump-diffusion structures with space-homogeneous jump size distribution Let functions $\rho_1, \ldots, \rho_n$ be of the form

$$\rho_j(z, \psi) := \rho_j^{(1)}(z, \psi) + \rho_j^{(2)}(z, \psi), \quad z, \psi \in \mathbb{R}, \tag{6.34}$$

with

$$\rho_j^{(1)}(z, \psi) = -id_j(z)\psi + c_j(z)\psi^2,$$

where $d_j, c_j \geq 0$, $j = 1, \ldots, n$, are functions satisfying one of the conditions of Proposition 2.45, and

$$\rho_j^{(2)}(z, \psi) = \eta_j(z) \int_{\mathbb{R}\setminus\{0\}} \left(1 - e^{iz\psi}\right) \nu_j(d\psi),$$

where η_j is a nonnegative bounded continuous function and ν_j is a probability measure, $j = 1, \ldots, n$. Since assumptions **S1**(1)–**S4**(1) and **H3**(1) are satisfied for $\rho_1, \ldots, \rho_n$, it follows from Proposition 2.39 that for each $j = 1, \ldots, n$ there exists a nice $\mathbb{R}$-Feller–Markov family $\mathcal{MFH}^j$ with symbol ρ_j. Each family $\mathcal{MFH}^j$ is a Markov jump-diffusion family with a jump size distribution that is independent of x. We construct a Markov structure $\mathcal{MMFH}$ for $\mathcal{MFH}^j$, $j = 1, \ldots, n$, by constructing a symbol q using formula (6.4). Accordingly, the symbol q is given by

$$q(x, \xi) = q_1(x, \xi) + q_2(x, \xi) \quad x, \xi \in \mathbb{R}^n,$$

where

$$q_1(x, \xi) = -i\langle b(x), \xi \rangle + \langle \xi, a(x)\xi \rangle,$$

and the functions $b : \mathbb{R}^n \to \mathbb{R}^n$, $a : \mathbb{R}^n \to L(\mathbb{R}^n, \mathbb{R}^n)$ satisfy

$$b_j(x) = d_j(x_j), \quad a_{jj}(x) = c_j(x_j), \quad j = 1, \ldots, n, \tag{6.35}$$

and moreover a is symmetric and chosen so that **S1**(n) holds. To specify q_2 we define a family of probability measures $\mu_S(dy)$, $S \in \mathcal{J}$, similar to that in Example 6.10 and having a cumulative distribution function $F^S(\cdot)$ on $\mathbb{R}^n$ given by

$$F^S(y_1,\dots,y_n) = C^S(\nu_j((-\infty,y_j]), j \in S) \prod_{i\in S^c} \delta_{\{0\}}((-\infty,y_i]),$$

where C^S is a copula on $[0,1]^{|S|}$. Now, let q_2 be given by

$$q_2(x,\xi) = \sum_{S\in\mathcal{J}} \lambda_S(x) \int_{\mathbb{R}^n\setminus\{0\}} \left(1 - e^{i\langle y,\xi\rangle}\right) \mu_S(dy), \quad x,\xi \in \mathbb{R}^n,$$

where $\mathcal{J} = \{S \subset \{1,\dots,n\} : S \neq \emptyset\}$ and the functions λ_S are nonnegative, continuous, bounded and satisfy

$$\sum_{S\in\mathcal{J}:k\in S} \lambda_S(x) = \eta_k(x_k), \quad x \in \mathbb{R}^n. \tag{6.36}$$

It is easy to check that the symbol q satisfies **S1**(n)–**S4**(n). Since

$$\sup_{x\in\mathbb{R}^n} |q(x,\xi)| \leq I_1(\xi) + I_2(\xi),$$

where

$$I_1(\xi) = \|b\|_\infty |\xi| + \|a\|_\infty |\xi|^2,$$
$$I_2(\xi) = \sum_{S\in\mathcal{J}} \|\lambda_S\|_\infty \left| \int_{\mathbb{R}^n\setminus\{0\}} \left(1 - e^{i\langle y,\xi\rangle}\right) \mu_S(x,dy)\right|$$

and $\mathcal{J}$ is finite, we see that **H3**(n) holds. Therefore, by Proposition 2.39 there exists a nice $\mathbb{R}^n$-Feller–Markov family $\mathcal{MMFH}$ with symbol q satisfying (2.63). Hence $\mathcal{MMFH}$ satisfies condition **C1** (by Proposition 2.35). Using (6.35) and (6.36) we conclude that $q(x,\mathbf{e}_j\xi_j) = \rho_j(x_j,\xi_j)$, in a similar way to that in the proofs of (6.19) and (6.31). So condition **C2** holds. This, together with the fact that ρ_j is a symbol of $\mathcal{MFH}^j$ for $j = 1,\dots,n$, implies **C3**. Lastly, we prove that condition **C4** holds. The space $C_c^\infty(\mathbb{R})$ is a core for the operator $-\rho_j^1(x_j,D)$ since c_j, d_j satisfy an assumption in Proposition 2.45. Moreover, since the operator $-\rho_j^2(x_j,D)$ is bounded, we conclude that $C_c^\infty(\mathbb{R})$ is also a core of $-(\rho_j^1+\rho_j^2)(x_j,D)$. Thus condition **C4** holds.

Summing up, and invoking Theorem 2.33, we conclude that the symbol q defines a Markov structure for $\mathcal{MFH}^j$, $j = 1,\dots,n$.

Example 6.12 Lévy structure This is a special case of Example 6.11. Specifically, let us consider a collection of functions $\rho_1,\dots,\rho_n$ of the form, for $z,\psi \in \mathbb{R}$,

$$\rho_j(z,\psi) := \rho_j^{(1)}(\psi) + \rho_j^{(2)}(\psi), \tag{6.37}$$

where

$$\rho_j^{(1)}(\psi) = -i\psi d_j + c_j\psi^2,$$
$$\rho_j^{(2)}(\psi) = \eta_j \int_{\mathbb{R}} \left(1 - e^{iu\psi}\right) \nu_j(du);$$

here $d_j \in \mathbb{R}$, $c_j \geq 0$, $\eta_j \geq 0$ are constants and ν_j is a probability measure, $j = 1,\dots,n$.

Thus, according to Theorem 2.2 in Böttcher et al. (2013), each ρ_j is a symbol of the generator of a conservative translation-invariant Feller semigroup. Moreover, in view of the discussion in Example 6.11, these semigroups correspond to time-homogeneous nice $\mathbb{R}$-Feller–Markov families, say

$$\mathcal{MFH}^j = \left\{(\Omega, \mathcal{F}, \mathbb{F}^j, (Y_t^j)_{t\geq 0}, \mathbb{Q}_y^j, Q^j) : y \in \mathbb{R}\right\}, \quad j = 1, \ldots, n.$$

Consequently, in view of Definition 2.5 and Theorem 2.6 in Böttcher et al. (2013), each family $\mathcal{MFH}^j$ is a collection of time-homogenous $\mathbb{R}$-valued Lévy processes (see Remark B.27).

As usual, we will construct a Markov structure for $\mathcal{MFH}^j$, $j = 1, \ldots, n$, by constructing a symbol q using an appropriate specification of the formula (6.4). To this end we postulate that the symbol q is given as

$$q(\xi) = q_1(\xi) + q_2(\xi), \quad \xi \in \mathbb{R}^n,$$

with functions q_1 and q_2 described in what follows. The function q_1 is given by

$$q_1(\xi) = -i\langle b, \xi\rangle + \langle \xi, a\xi\rangle, \quad \xi \in \mathbb{R}^n,$$

where $b = (d_1, \ldots, d_n)$, $a = [a_{ij}]$ is an $n \times n$ symmetric, non-negative-definite matrix such that $a_{jj} = c_j$. The function q_2 has the form

$$q_2(\xi) = \sum_{S \in \mathcal{J}} \lambda_S \int_{\mathbb{R}^n \setminus \{0\}} \left(1 - e^{i\langle y, \xi\rangle}\right) \mu_S(dy), \quad \xi \in \mathbb{R}^n,$$

where μ_S, $S \in \mathcal{J} = \{S \subset \{1, \ldots, n\} : S \neq \emptyset\}$, is a probability measure with cumulative distribution function F^S on $\mathbb{R}^n$ given by

$$F^S(y_1, \ldots, y_n) = C^S(\nu_j((-\infty, y_j]), j \in S) \prod_{i \in S^c} \delta_{\{0\}}((-\infty, y_i]),$$

C^S is a copula on $[0,1]^{|S|}$, and the constants λ_S satisfy

$$\sum_{S \in \mathcal{J} : j \in S} \lambda_S = \eta_j. \tag{6.38}$$

As in Example 6.11 we can show that conditions **C1**–**C4** are satisfied. Thus, the q constructed in this example produces a Markov structure, say $\mathcal{MMFH}$, for $\mathcal{MFH}^j$, $j = 1, \ldots, n$. Invoking again Theorem 2.2, Definition 2.5, and Theorem 2.6 from Böttcher et al. (2013), we conclude that the family $\mathcal{MMFH}$ is a collection of time-homogenous $\mathbb{R}^n$-valued Lévy processes. We call this Markov structure the Lévy structure.

We refer to Tankov (2003) and Kallsen and Tankov (2006) for a related study of dependence in the case of multivariate Lévy processes and in the case of general Lévy measures. The Lévy structures discussed in this example are in fact a special case of multivariate Lévy processes with given marginals, which are studied in Kallsen and Tankov (2006).

Example 6.13 Markov structure with Lévy marginals Let $\rho_1,\dots,\rho_n$ be given as in Example 6.12, so that they satisfy (6.37). As we know from Example 6.12, for each $j=1,\dots,n$, the function ρ_j is a symbol of a nice family $\mathcal{MFH}^j$, which is a collection of time-homogenous $\mathbb{R}$-valued Lévy processes.

As usual, we will build a Markov structure for $\mathcal{MFH}^j$, $j=1,\dots,n$, by constructing a symbol q using an appropriate specification of the formula (6.4). To this end we postulate that the symbol q is given as

$$q(x,\xi)=q_1(x,\xi)+q_2(x,\xi),\quad x,\xi\in\mathbb{R}^n,$$

but we choose functions q_1, q_2 different from those in Example 6.12: they depend on x as specified in what follows. We take the function q_1 to be of the form

$$q_1(x,\xi)=-i\langle b,\xi\rangle+\langle\xi,a(x)\xi\rangle,\quad x,\xi\in\mathbb{R}^n,$$

where $b=(d_1,\dots,d_n)$, $a=[a_{ij}]$ is a $n\times n$ symmetric, non-negative-definite matrix such that $a_{jj}(x)=c_j$. The function q_2 has the form

$$q_2(x,\xi)=\sum_{S\in\mathcal{J}}\lambda_S(x)\int_{\mathbb{R}^n\setminus\{0\}}\left(1-e^{i\langle y,\xi\rangle}\right)\mu_S(dy),\quad \xi\in\mathbb{R}^n,$$

where μ_S, $S\in\mathcal{J}=\{S\subset\{1,\dots,n\}:S\neq\emptyset\}$, is a probability measure with cumulative distribution function F^S on $\mathbb{R}^n$ given by

$$F^S(y_1,\dots,y_n)=C^S(\nu_j((-\infty,y_j]),j\in S)\prod_{i\in S^c}\delta_{\{0\}}((-\infty,y_i]),$$

C^S is a copula on $[0,1]^{|S|}$ for every $x\in\mathbb{R}^n$, and λ_S are nonnegative continuous bounded functions satisfying

$$\sum_{S\in\mathcal{J}:j\in S}\lambda_S(x)=\eta_j.$$

As in Example 6.11, we can show that conditions **C1–C4** are satisfied. Thus, the q constructed in this example produces a Markov structure, say $\mathcal{MMFH}$, for $\mathcal{MFH}^j$, $j=1,\dots,n$. The structure $\mathcal{MMFH}$ is a collection of $\mathbb{R}^n$-valued Feller–Markov processes that are not necessarily time-homogenous Lévy processes. However, the predetermined margins $\mathcal{MFH}^j$, $j=1,\dots,n$, of $\mathcal{MMFH}$ are time-homogenous $\mathbb{R}$-valued Lévy processes.

7

Markov Chain Structures

Here we adopt the framework of Chapter 3 and study the problem of building multivariate finite Markov chains whose coordinates are finite univariate Markov chains with given generator matrices. Specifically, we will be concerned here with building strong and weak Markov chain structures for a collection of finite Markov chains. We will use methods that are specific for Markov chains and that are based on the results derived in Chapter 3.

This problem was previously studied, in Bielecki et al. (2008b) and Bielecki et al. (2010) for example, in the context of strong Markov consistency (see Definition 3.1). Thus, essentially, these publications dealt with building *strong Markov chain structures*, which are nothing but strong Markov process structures for a family of finite Markov chains.

In this chapter we shall additionally be concerned with constructing *weak Markov chain structures*, which are related to the concept of weak Markov consistency (see Definition 3.8), in the context of finite Markov chains.

As in Chapter 3, in order to simplify the presentation we shall consider the bivariate case only.

Paraphrasing remarks from the beginning of Chapter 6, we claim that Markov chain structures are the key objects of interest in modeling structured dependence of the Markovian type between stochastic dynamical systems given in terms of Markov chains. Accordingly, much of the discussion presented in this chapter is devoted to the construction of Markov chain structures. Part of the input to any construction procedure is provided by the marginal data, which we refer to as the *Markov chain marginal inputs*, and which are given here in terms of the generator functions of the marginal processes. Another part of the input is provided by the data and/or postulates regarding the stochastic dependence between the coordinates of the resulting Markov chain structure, which we refer to as the *Markov chain dependence structure input*. These inputs have to be appropriately accounted for in constructions of Markov chain structures. This, in principle, can be done since, as is discussed in this chapter, there is quite substantial flexibility in constructing Markov chain structures; this allows for accommodating in a Markov structure model the various dependence structures exhibited by the phenomena one wants to model.

7.1 Strong Markov Chain Structures for Finite Markov Chains

The key observation leading to the concept and construction of a strong Markov chain structure for a collection of finite Markov chains is the following. We recall that we only consider conservative Markov chains.

Given two generator functions $\Lambda^1(t) = [\lambda^1_{x^1y^1}(t)]$ and $\Lambda^2(t) = [\lambda^2_{x^2y^2}(t)]$, for $t \geq 0$, suppose that there exists a valid generator function $\Lambda(t) = [\lambda^{x^1x^2}_{y^1y^2}(t)]_{x^1,y^1\in\mathcal{X}^1,x^2,y^2\in\mathcal{X}^2}$ for $t \geq 0$, satisfying

$$\lambda^1_{x^1y^1}(t) = \sum_{y^2\in\mathcal{X}^2} \lambda^{x^1x^2}_{y^1y^2}(t), \quad x^1,y^1 \in \mathcal{X}^1, x^1 \neq y^1 \tag{7.1}$$

for every $x^2 \in \mathcal{X}^2$, and satisfying

$$\lambda^2_{x^2y^2}(t) = \sum_{y^1\in\mathcal{X}^1} \lambda^{x^1x^2}_{y^1y^2}(t), \quad x^2,y^2 \in \mathcal{X}^2, x^2 \neq y^2 \tag{7.2}$$

for every $x^1 \in \mathcal{X}^1$. Then, Condition (M) (see Section 3.1.1) is clearly satisfied for $\Lambda(t)$, so that (see Remark 3.5) strong Markov consistency holds for the Markov chain, $X = (X^1,X^2)$, generated by $\Lambda(t)$. Moreover, the matrix function $\Lambda^i(t)$ is the generator function of X^i, $i = 1,2$.

This observation leads to the following definition.

Definition 7.1 Let Y^1 and Y^2 be two Markov chains with values in $\mathcal{X}^1$ and $\mathcal{X}^2$ and with generator functions $\Lambda^1(t) = [\lambda^1_{x^1y^1}(t)]$ and $\Lambda^2(t) = [\lambda^2_{x^2y^2}(t)]$, for $t \geq 0$, respectively. A *strong Markov chain structure* for the Markov chains Y^1 and Y^2 is any Markov chain $X = (X^1,X^2)$ which is generated by a matrix function $\Lambda(t) = [\lambda^{x^1x^2}_{y^1y^2}(t)]_{x^1,y^1\in\mathcal{X}^1,x^2,y^2\in\mathcal{X}^2}$, with $\lambda^{x^1x^2}_{x^1x^2}(t)$ given as

$$\lambda^{x^1x^2}_{x^1x^2}(t) = - \sum_{(z^1,z^2)\in\mathcal{X}^1\times\mathcal{X}^2,(z^1,z^2)\neq(x^1,x^2)} \lambda^{x^1x^2}_{z^1z^2}(t), \tag{7.3}$$

that satisfies (7.1) and (7.2) and such that it correctly defines the infinitesimal generator function of a conservative Markov chain with values in $\mathcal{X}^1 \times \mathcal{X}^2$. In this case we call Y^1 and Y^2 the predetermined margins of X.

If $X = (X^1,X^2)$ is a strong Markov chain structure for Y^1 and Y^2, then X^j is a Markov chain in the filtration of X, and the generator function of X^j is the same as the generator of the predetermined margin Y^j, $j = 1,2$. In addition, if the law of X^j_0 is the same as the law of Y^j_0 then the law of the process X^j coincides with the law of the process Y^j, $j = 1,2$, in which case the chain X becomes a strong Markov chain copula structure for Y^1 and Y^2. Various structures $X = (X^1,X^2)$ will of course feature various stochastic dependences between X^1 and X^2.

Remark 7.2 (i) It is clear that there exists at least one solution to (7.1) and (7.2) such that the matrix function $\Lambda(t) = [\lambda^{x^1x^2}_{y^1y^2}(t)]_{x^1,y^1\in\mathcal{X}^1,x^2,y^2\in\mathcal{X}^2}$ is a valid generator function. This solution corresponds to the case of independent Markov chains X^1 and X^2. In this case we have $\Lambda(t) = I^1 \otimes \Lambda^2(t) + \Lambda^1(t) \otimes I^2$, where $A \otimes B$ denotes the

Kronecker (tensor) product of matrices A and B,[1] and where I^i is the identity matrix with dimension equal to $|\mathcal{X}^i|$, the cardinality of $\mathcal{X}^i$. The function $\Lambda(t)$ can be also written more explicitly:

$$\lambda^{x^1x^2}_{y^1y^2}(t) = \begin{cases} \lambda^1_{x^1x^1}(t) + \lambda^2_{x^2x^2}(t), & y^1 = x^1, y^2 = x^2, \\ \lambda^1_{x^1y^1}(t), & y^1 \neq x^1, y^2 = x^2, \\ \lambda^2_{x^2y^2}(t), & y^2 \neq x^2, y^1 = x^1, \\ 0, & \text{otherwise.} \end{cases}$$

This is a special case of the structure considered below in Example 8.4.
(ii) Note that, typically, the system (7.1) and (7.2), considered as a system with given $\Lambda^1(t) = [\lambda^1_{x^1y^1}(t)]$ and $\Lambda^2(t) = [\lambda^2_{x^2y^2}(t)]$ and with unknown

$$\Lambda(t) = [\lambda^{x^1x^2}_{y^1y^2}(t)]_{x^1,y^1 \in \mathcal{X}^1, x^2,y^2 \in \mathcal{X}^2},$$

contains many more unknowns (i.e., $\lambda^{x^1x^2}_{y^1y^2}(t)$, x^1, $y^1 \in \mathcal{X}^1, x^2, y^2 \in \mathcal{X}^2$) than it contains equations. In fact, given that the cardinalities of $\mathcal{X}^1$ and $\mathcal{X}^2$ are K_1 and K_2, respectively, the system consists of $K_1(K_1 - 1) + K_2(K_2 - 1)$ equations in $K_1K_2(K_1K_2 - 1)$ unknowns. Thus, in principle, one can create several bivariate strong Markov chain structures for Y^1 and Y^2. This observation parallels Remark 6.5.

7.2 Weak Markov Chain Structures for Finite Markov Chains

The concept of weak Markov chain structures corresponds to the concept of weak Markov consistency. As we shall see, this concept is much more intricate than the concept of strong Markov chain structures.

Definition 7.3 1. Let Y^1 and Y^2 be two Markov chains with values in $\mathcal{X}^1$ and $\mathcal{X}^2$ and with generator functions $\Lambda^1(t) = [\lambda^1_{x^1y^1}(t)]$ and $\Lambda^2(t) = [\lambda^2_{x^2y^2}(t)]$, respectively. A *weak Markov chain structure* for Y^1 and Y^2 is any Markov chain $X = (X^1, X^2)$ which is generated by a matrix function $\Lambda(t) = [\lambda^{x^1x^2}_{y^1y^2}(t)]_{x^1,y^1 \in \mathcal{X}^1, x^2,y^2 \in \mathcal{X}^2}$, with $\lambda^{x^1x^2}_{x^1x^2}(t)$ given as

$$\lambda^{x^1x^2}_{x^1x^2}(t) = - \sum_{(z^1,z^2) \in \mathcal{X}^1 \times \mathcal{X}^2, (z^1,z^2) \neq (x^1,x^2)} \lambda^{x^1x^2}_{z^1z^2}(t), \tag{7.4}$$

that satisfies the following conditions:

WMC1 $\Lambda(\cdot)$ is a generator function of a bivariate Markov chain with values in $\mathcal{X}^1 \times \mathcal{X}^2$;

WMC2 Conditions (3.14) and (3.17) hold for X, so that X is weakly Markov consistent.

[1] For the definition of the tensor product of matrices see www.encyclopediaofmath.org/index.php/Tensor_product.

2. A *strictly weak Markov chain structure* for Y^1 and Y^2 is any Markov chain $X = (X^1, X^2)$ that is a weak Markov chain structure for Y^1 and Y^2 but is not a strong Markov chain structure for Y^1 and Y^2.

Thus, any weak Markov chain structure between the Markov chains Y^1 and Y^2 produces a bivariate Markov chain, say $X = (X^1, X^2)$, such that

- The coordinates X^1 and X^2 are Markovian in their own filtrations but not necessarily Markovian in the filtration of X.
- The transition law of X^i is the same as the transition law of Y^i, $i = 1,2$. In this case, we say that the process X satisfies the weak Markov consistency condition relative to Y^1 and Y^2.
- If, in addition, the initial law of X^i is the same as the initial law of Y^i then the law X^i is the same as the law of Y^i, $i = 1,2$, in which case the process X is a weak Markov chain copula structure for Y^1 and Y^2.

It is clear that any strong Markov chain structure for Y^1 and Y^2 is also a weak Markov chain structure for Y^1 and Y^2. We will see that there exist strictly weak Markov chain structures for Y^1 and Y^2.

In fact, part of our interest is in constructing strictly weak Markov chain structures. However, we do not have at our disposal any sufficient algebraic condition for strictly weak Markov consistency that would be analogous to Condition (M), which, as we know, is sufficient for the strong Markov consistency of a multivariate finite Markov chain. A possible way of constructing strictly weak Markov chain structures for Y^1 and Y^2 is to start with (3.19), which is a necessary condition for the weak Markov consistency of a multivariate finite Markov chain, and to find a generator function $\Lambda(t)$ that satisfies this condition with given $\Lambda^1(t)$ and $\Lambda^2(t)$ but does not satisfy Condition (M). Even though condition (3.19) is not sufficient for the weak Markov consistency of a multivariate finite Markov chain in general, it often is sufficient. Also, even though Condition (M) is not necessary for the strong Markov consistency of a multivariate finite Markov chain in general, it typically is necessary. So, typically, this method for constructing strictly weak Markov chain structures will work. This approach will be illustrated in Example 7.7 in Section 7.3.

Remark 7.4 (i) It needs to be said that the issue of constructing strictly weak Markov chain structures is very important from the practical point of view. For example, it is important in the context of credit risk management since weak Markov chain structures allow for the modeling of default contagion between individual obligors (debtors) and the rest of the credit pool (see Bielecki et al. (2014b) for a discussion); this kind of contagion is precluded, in general, in the context of strong Markov chain structures. Thus, strictly weak Markov chain structures make it possible to tackle a critical modeling requirement: They allow the modeling of contagion between credit events in credit portfolios; equally importantly, they enable the modeling of contagion between failure events in complex manufacturing systems.
(ii) In her doctoral dissertation Chang (2017) derived some sufficient conditions for strictly weak Markov consistency. However, these conditions are not easy to implement. More work is needed to make them useful in practice.

7.3 Examples

As before, we take $n=2$ in the examples below. We shall present examples illustrating:

- The building of a strong Markov chain structure (Example 7.5), i.e. the construction of a two-dimensional Markov chain $X=(X^1,X^2)$ with coordinates X^1 and X^2 that are Markovian in the filtration of X and such that the transition law of X^i agrees with the transition law of a given Markov chain Y^i, $i=1,2$.
- The building of a strictly weak Markov chain structure (Example 7.7), i.e. the construction of a two-dimensional Markov chain $X=(X^1,X^2)$ with coordinates X^1 and X^2 that are Markovian in their own filtrations but not Markovianin the filtration of X and are such that the transition law of X^i agrees with the transition law of a given Markov chain Y^i, $i=1,2$.
- The existence of a Markov chain for which weak Markov consistency does not hold, i.e. a Markov chain that cannot serve as a weak Markov chain structure (Example 7.8). In this example, the coordinate X^2 of the Markov chain $X=(X^1,X^2)$ is shown to be not Markovian in its own filtration.

Example 7.5 Let us consider two processes, Y^1 and Y^2, that are time-homogeneous Markov chains, each taking values in the state space $\{0,1\}$ and with respective generators

$$\Lambda^1 = \begin{array}{c} \\ 0 \\ 1 \end{array} \begin{array}{c} \begin{array}{cc} \quad 0 & \qquad 1 \end{array} \\ \begin{pmatrix} -(a+c) & a+c \\ 0 & 0 \end{pmatrix} \end{array} \tag{7.5}$$

and

$$\Lambda^2 = \begin{array}{c} \\ 0 \\ 1 \end{array} \begin{array}{c} \begin{array}{cc} \quad 0 & \qquad 1 \end{array} \\ \begin{pmatrix} -(b+c) & b+c \\ 0 & 0 \end{pmatrix} \end{array}, \tag{7.6}$$

for $a,b,c \geq 0$.

We shall first consider the version of equation (3.7), for $i=1,2$, that is relevant for this example. We identify $C^{i,*}$, $i=1,2$, with the matrices

$$C^{1,*} = \begin{pmatrix} 1 & 0 \\ 1 & 0 \\ 0 & 1 \\ 0 & 1 \end{pmatrix} \quad \text{and} \quad C^{2,*} = \begin{pmatrix} 1 & 0 \\ 0 & 1 \\ 1 & 0 \\ 0 & 1 \end{pmatrix}. \tag{7.7}$$

It can be easily checked that the matrix Λ given by

$$\Lambda = \begin{array}{c} \\ (0,0) \\ (0,1) \\ (1,0) \\ (1,1) \end{array} \begin{array}{c} \begin{array}{cccc} \quad (0,0) & \qquad (0,1) & \qquad (1,0) & \quad (1,1) \end{array} \\ \begin{pmatrix} -(a+b+c) & b & a & c \\ 0 & -(a+c) & 0 & a+c \\ 0 & 0 & -(b+c) & b+c \\ 0 & 0 & 0 & 0 \end{pmatrix} \end{array} \tag{7.8}$$

satisfies (3.7) for $i = 1,2$. Thus, according to the results presented in Section 3.1, the bivariate Markov chain $X = (X^1, X^2)$ with generator Λ is a strong Markov chain structure for Y^1 and Y^2.

Nevertheless, it will be instructive to verify this directly. To this end, let us consider the bivariate Markov chain $X = (X^1, X^2)$ on the state space

$$E = \{(0,0),(0,1),(1,0),(1,1)\}$$

generated by the matrix Λ given in (7.8). We first compute the transition probability matrix for X, for $t \geq 0$:

$$P(t) = \begin{pmatrix} e^{-(a+b+c)t} & e^{-(a+c)t}(1-e^{-bt}) & e^{-(b+c)t}(1-e^{-at}) & \kappa(t) \\ 0 & e^{-(a+c)t} & 0 & 1-e^{-(a+c)t} \\ 0 & 0 & e^{-(b+c)t} & 1-e^{-(b+c)t} \\ 0 & 0 & 0 & 1 \end{pmatrix},$$

where $\kappa(t) = e^{-(a+b+c)t} - e^{-(b+c)t} - e^{-(a+c)t} + 1$.

Thus, for any $t \geq 0$,

$$\lim_{h \to 0} \frac{P(X_{t+h}^2 = 0 | X_t^2 = 0) - 1}{h} = -(b+c).$$

Similarly, for any $t \geq 0$,

$$\lim_{h \to 0} \frac{P(X_{t+h}^1 = 0 | X_t^1 = 0) - 1}{h} = -(a+c).$$

Since both the processes X^1 and X^2 are absorbed in state 1, using arguments analogous to those used in Example 2.4 in Bielecki et al. (2015), we see that X^1 and X^2 are Markov chains in their own filtrations. From the calculations conducted above we conclude that the generator of X^i is Λ^i, $i = 1,2$.

To verify that Λ is a strong Markov chain structure for Y^1 and Y^2, it remains to show that coordinates X^1 and X^2 are Markovian in the filtration of X. This can be verified by direct computations: indeed,

$$\begin{aligned} &\lim_{h \to 0} \frac{P(X_{t+h}^1 = 0 | X_t^1 = 0, X_t^2 = 0) - 1}{h} \\ &\quad = \lim_{h \to 0} \frac{P(X_{t+h}^1 = 0 | X_t^1 = 0, X_t^2 = 1) - 1}{h} \\ &\quad = -(a+c) = \lim_{h \to 0} \frac{P(X_{t+h}^1 = 0 | X_t^1 = 0) - 1}{h}, \end{aligned}$$

or, equivalently,

$$\begin{aligned} P(X_{t+h}^1 = 0 | X_t^1 = 0, X_t^2 = 0) &= P(X_{t+h}^1 = 0 | X_t^1 = 0, X_t^2 = 1) \\ &= P(X_{t+h}^1 = 0 | X_t^1 = 0) = e^{-(a+c)h}, \end{aligned}$$

so that condition (3.2) is satisfied for X^1, and similarly for X^2.

Note that, in accordance with the concept of strong Markov consistency, the transition intensities and transition probabilities for X^1 do not depend on the state of X^2:

- No matter whether the state of X^2 is 0 or 1, the intensity of transition of X^1 from 0 to 1 is equal to $a+c$.
- No matter what the state of X^2 is, the transition probability of X^1 from 0 to 1 in t units of time is equal to

$$e^{-(b+c)t}(1-e^{-at})+e^{-(a+b+c)t}-e^{-(b+c)t}-e^{-(a+c)t}+1=1-e^{-(a+c)t}.$$

An analogous observation holds for X^2. Finally, note that X^1 and X^2 are independent if and only if $c=0$.

Remark 7.6 Note that the matrix Λ given in (7.8) admits the following decomposition:

$$\Lambda = B_0 + B_{12} - B_1 - B_2, \tag{7.9}$$

where

$$B_0 = I^1 \otimes \Lambda^2 + \Lambda^1 \otimes I^2,$$

$$B_{12} = \begin{array}{c} \\ (0,0) \\ (0,1) \\ (1,0) \\ (1,1) \end{array} \begin{array}{c} \begin{array}{cccc} (0,0) & (0,1) & (1,0) & (1,1) \end{array} \\ \begin{pmatrix} -c & 0 & 0 & c \\ 0 & 0 & 0 & 0 \\ 0 & 0 & 0 & 0 \\ 0 & 0 & 0 & 0 \end{pmatrix} \end{array},$$

$$B_1 = \begin{array}{c} \\ (0,0) \\ (0,1) \\ (1,0) \\ (1,1) \end{array} \begin{array}{c} \begin{array}{cccc} (0,0) & (0,1) & (1,0) & (1,1) \end{array} \\ \begin{pmatrix} -c & c & 0 & 0 \\ 0 & 0 & 0 & 0 \\ 0 & 0 & 0 & 0 \\ 0 & 0 & 0 & 0 \end{pmatrix} \end{array},$$

$$B_2 = \begin{array}{c} \\ (0,0) \\ (0,1) \\ (1,0) \\ (1,1) \end{array} \begin{array}{c} \begin{array}{cccc} (0,0) & (0,1) & (1,0) & (1,1) \end{array} \\ \begin{pmatrix} -c & 0 & c & 0 \\ 0 & 0 & 0 & 0 \\ 0 & 0 & 0 & 0 \\ 0 & 0 & 0 & 0 \end{pmatrix} \end{array}.$$

The pivot term B_0 corresponds to the independence structure. The other three terms in (7.9) model the dependence structure between X^1 and X^2.

Example 7.7 Let us consider two processes, Y^1 and Y^2, that are Markov chains, each taking values in the state space $\{0,1\}$, with respective generator functions

$$\Lambda^1(t) = \begin{pmatrix} -(a+c)+\alpha(t) & a+c-\alpha(t) \\ 0 & 0 \end{pmatrix}$$

and

$$\Lambda^2(t) = \begin{pmatrix} -(b+c)+\beta(t) & b+c-\beta(t) \\ 0 & 0 \end{pmatrix},$$

where

$$\alpha(t) = c\frac{e^{-at}(1-e^{-(b+c)t})b/(b+c)}{e^{-(a+b+c)t}+e^{-at}(1-e^{-(b+c)t})b/(b+c)},$$

$$\beta(t) = c\frac{e^{-bt}(1-e^{-(a+c)t})a/(a+c)}{e^{-(a+b+c)t}+e^{-bt}(1-e^{-(a+c)t})a/(a+c)}$$

for $a,b \geq 0$ and $c > 0$.

Here we seek a strictly weak Markov chain structure for Y^1 and Y^2. To this end we will investigate the necessary condition (3.19). We first note that in the present example the matrix representation of the operator Q_t^1 takes the form

$$Q_t^1 = \begin{pmatrix} P(X_t^1=0,X_t^2=0|X_t^1=0) & P(X_t^1=0,X_t^2=1|X_t^1=0) & P(X_t^1=1,X_t^2=0|X_t^1=0) & P(X_t^1=1,X_t^2=1|X_t^1=0) \\ P(X_t^1=0,X_t^2=0|X_t^1=1) & P(X_t^1=0,X_t^2=1|X_t^1=1) & P(X_t^1=1,X_t^2=0|X_t^1=1) & P(X_t^1=1,X_t^2=1|X_t^1=1) \end{pmatrix};$$

and similarly for Q_t^2. It turns out that the necessary condition (3.19) is satisfied by a valid generator matrix

$$\Lambda = \begin{pmatrix} -(a+b+c) & b & a & c \\ 0 & -a & 0 & a \\ 0 & 0 & -b & b \\ 0 & 0 & 0 & 0 \end{pmatrix}, \tag{7.10}$$

where $a,b \geq 0$ and $c > 0$. Verification of this is straightforward but computationally intensive.

Since the condition (3.19) is just a necessary condition for weak Markov consistency, the bivariate Markov process $X = (X^1, X^2)$ generated by the matrix Λ given in (7.10) may not be a weak Markov chain structure for Y^1 and Y^2. Whether it is needs to be verified by direct inspection.

Let us consider the bivariate Markov chain $X = (X^1, X^2)$ on the state space

$$E = \{(0,0),(0,1),(1,0),(1,1)\}$$

generated by the matrix Λ given by (7.10).

It is clear that the coordinates X^1 and X^2 are Markovian in their own filtrations. We shall show that:

- X^1 and X^2 are not Markovian in the filtration $\mathbb{F}^X$, and
- the generators of X^1 and X^2 are given by (7.11) and (7.12) below, respectively.

We first compute the transition probability matrix for X, for $t \geq 0$:

$$P(t) = \begin{pmatrix} e^{-(a+b+c)t} & e^{-at}(1-e^{-(b+c)t})\frac{b}{b+c} & e^{-bt}(1-e^{-(a+c)t})\frac{a}{a+c} & \gamma(t) \\ 0 & e^{-at} & 0 & 1-e^{-at} \\ 0 & 0 & e^{-bt} & 1-e^{-bt} \\ 0 & 0 & 0 & 1 \end{pmatrix},$$

where

$$\gamma(t) = 1 + e^{-(a+b+c)t}\left(\frac{a}{a+c} - \frac{c}{b+c}\right) - \frac{a}{a+c}e^{-bt} - \frac{b}{b+c}e^{-at}.$$

It follows that

$$\begin{aligned} P(X^1_{t+h} = 0 | X^1_t = 0, X^2_t = 0) &= e^{-(a+b+c)h} + e^{-ah}(1 - e^{-(b+c)h})\frac{b}{b+c} \\ &\neq P(X^1_{t+h} = 0 | X^1_t = 0, X^2_t = 1) = e^{-ah}, \end{aligned}$$

unless $c = 0$, which is the case of independent X^1 and X^2. Thus, in view of (3.2), in general, X^1 is not a Markov process in the filtration of X. Similarly for X^2.

We shall now compute the generator functions for X^2 and for X^1. For any $t \geq 0$ we have

$$\lim_{h\to 0} \frac{P(X^2_{t+h} = 0 | X^2_t = 0) - 1}{h} = -(b+c) + c\frac{P(X^1_t = 1, X^2_t = 0)}{P(X^2_t = 0)}.$$

Similarly, for any $t \geq 0$,

$$\lim_{h\to 0} \frac{P(X^1_{t+h} = 0 | X^1_t = 0) - 1}{h} = -(a+c) + c\frac{P(X^1_t = 0, X^2_t = 1)}{P(X^1_t = 0)}.$$

Thus, both X^1 and X^2 are time-inhomogeneous Markov chains with generator functions respectively given by

$$A^1(t) = \begin{pmatrix} -(a+c) + c\frac{P(X^1_t = 0, X^2_t = 1)}{P(X^1_t = 0)} & a + c - c\frac{P(X^1_t = 0, X^2_t = 1)}{P(X^1_t = 0)} \\ 0 & 0 \end{pmatrix} \quad (7.11)$$

and

$$A^2(t) = \begin{pmatrix} -(b+c) + c\frac{P(X^1_t = 1, X^2_t = 0)}{P(X^2_t = 0)} & b + c - c\frac{P(X^1_t = 1, X^2_t = 0)}{P(X^2_t = 0)} \\ 0 & 0 \end{pmatrix}. \quad (7.12)$$

If $X_0 = (0,0)$ then it is easily checked that $A^1(t) = \Lambda^1(t)$ and $A^2(t) = \Lambda^2(t)$ for any $t \geq 0$, as claimed. Consequently, the bivariate Markov process $X = (X^1, X^2)$ generated by the matrix Λ given in (7.10) is a weak Markov chain structure for Y^1 and Y^2 but is not a strong Markov chain structure for Y^1 and Y^2. So, it is a strictly weak Markov chain structure for Y^1 and Y^2.

Finally, note that the transition intensities and transition probabilities for X^1 depend on the state of X^2:

- When X is in the state $(0,0)$ at some point in time, the transition intensity of X^1 from 0 to 1 is equal to $a+c$; when X is in the state $(0,1)$ at some point in time, the transition intensity of X^1 from 0 to 1 is equal to a.
- When X is in the state $(0,0)$ at some point in time, the transition probability of X^1 from 0 to 1 in t units of time is

$$e^{-bt}(1-e^{-(a+c)t})\frac{a}{a+c}+1+e^{-(a+b+c)t}\left(\frac{a}{a+c}-\frac{c}{b+c}\right)$$
$$-\frac{a}{a+c}e^{-bt}-\frac{b}{b+c}e^{-at};$$

when X is in state $(0,1)$ at some point in time, the transition probability of X^1 from 0 to 1 in t units of time is

$$1-e^{-at}.$$

An analogous observation holds for X^2, i.e., the transition intensities and transition probabilities for X^2 depend on the state of X^1.

Example 7.8 Here we give an example of a bivariate Markov chain which is not weakly Markov consistent and thus cannot be strongly Markov consistent. Thus, it cannot serve as a Markov chain structure at all.

Let us consider the bivariate Markov chain $X=(X^1,X^2)$ on the state space

$$E=\{(0,0),(0,1),(1,0),(1,1)\}$$

generated by the matrix

$$A=\begin{pmatrix} -(a+b+c) & b & a & c \\ 0 & -(d+e) & d & e \\ 0 & 0 & -f & f \\ 0 & 0 & g & -g \end{pmatrix}. \tag{7.13}$$

We will show that X^2 is not Markovian in its own filtration. To this end we denote by N^2_{01} the process that counts the number of jumps of the coordinate X^2 from state 0 to state 1. Using (C.14) we see that the $\mathbb{F}^X$-intensity of such jumps is

$$\lambda_t=\mathbb{1}_{\{X^1_t=0,X^2_t=0\}}(b+c)+\mathbb{1}_{\{X^1_t=1,X^2_t=0\}}f.$$

So, the $\mathbb{F}^{X^2}$-intensity of N^2_{01}, which is the optional projection of the $\mathbb{F}^X$-intensity of N^2_{01} onto $\mathbb{F}^{X^2}$, has the form

$$(b+c)\mathbb{P}(X^1_t=0,X^2_t=0|\mathcal{F}^{X^2}_t)+f\,\mathbb{P}(X^1_t=1,X^2_t=0|\mathcal{F}^{X^2}_t). \tag{7.14}$$

Since $\{X^2_t=0,X^1_{t/2}=1\}\subseteq\{X^1_t=1\}$, we see that on the set

$$\{X^2_t=0,X^1_{t/2}=1\}$$

we have

$$\mathbb{P}(X_t^1 = 0, X_t^2 = 0 | X_t^2 = 0, X_{t/2}^1 = 1) = 0,$$
$$\mathbb{P}(X_t^1 = 1, X_t^2 = 0 | X_t^2 = 0, X_{t/2}^1 = 1) = 1.$$

Therefore, on the set $\{X_t^2 = 0, X_{t/2}^1 = 1\}$ the above optional projection is equal to

$$f\,\mathbb{P}(X_t^1 = 1, X_t^2 = 0 | X_t^2 = 0, X_{t/2}^1 = 1) = f.$$

However, on the set $\{X_t^2 = 0\}$ the optional projection in (7.14) is equal to

$$\begin{aligned}(b+c)\mathbb{P}(X_t^1 = 0, X_t^2 = 0 | X_t^2 = 0) + f\,\mathbb{P}(X_t^1 = 1, X_t^2 = 0 | X_t^2 = 0)\\ = (b+c-f)\mathbb{P}(X_t^1 = 0, X_t^2 = 0 | X_t^2 = 0) + f.\end{aligned}$$

Assuming that the process X starts from $(0,0)$ at time $t = 0$, it can be shown that $\mathbb{P}(X_t^1 = 0, X_t^2 = 0 | X_t^2 = 0) > 0$. The verification of this is straightforward but computationally intensive. Thus, if $b + c \neq f$ then the optional projection onto $\mathcal{F}_t^{X^2}$ of the $\mathbb{F}^X$-intensity of N_{01}^2 depends on the trajectory of X^2 up to time t, and not just on the state of X^2 at time t. This implies that the $\mathbb{F}^{X^2}$-intensity of N_{01}^2 depends on the trajectory of X^2 up to time t not only via state of X^2 at time t but also via other parts of the trajectory of X^2. Thus, in view of Corollary C.3, X^2 is not Markovian in its own filtration. It is obviously not Markovian in the filtration of the entire process X either.

8
Conditional Markov Chain Structures

Our goal in this chapter is to extend the theory of Markov structures from the universe of classical (finite) Markov chains to the universe of (finite) conditional Markov chains. Accordingly, we will use the term CMC structures. As it turns out, such an extension is not a trivial one. However, it is important both from the mathematical point of view and from the practical point of view. We will first discuss the so-called strong CMC structures and then weak CMC structures.

We let $T > 0$ be a fixed finite time horizon. Let $(\Omega, \mathcal{F})$ be an underlying measurable space endowed with the filtration $\mathbb{F} = (\mathcal{F}_t)_{t\in[0,T]}$. In what follows, $\mathcal{X}_k$ is a finite subset of some vector space, $k = 1,\dots,n$, and $\mathcal{X} := \times_{k=1}^{n} \mathcal{X}_k$. The processes considered in this chapter are defined on $(\Omega, \mathcal{F})$ and are restricted to the time interval $[0,T]$.

8.1 Strong CMC Structures

The next definition refers back to Definition 4.3.

Definition 8.1 Let $\mathcal{Y} = \{Y^1,\dots,Y^n\}$ be a family of processes, defined on the underlying probability space $(\Omega, \mathcal{F}, \mathbb{Q})$,[1] such that each Y^k is an $(\mathbb{F}, \mathbb{F}^{Y^k})$-CMC with values in $\mathcal{X}_k$. A *strong CMC structure* between the processes $Y^1,\dots,Y^n$ is any pair $(X, \mathbb{P})$, where $X = (X^1,\dots,X^n)$ is a multivariate process defined on $(\Omega, \mathcal{F})$ and with values in $\mathcal{X}$, and $\mathbb{P}$ is a probability measure on $(\Omega, \mathcal{F})$ such that

Condition 1 $\quad \mathbb{P}|_{\mathcal{F}_T \vee \mathcal{F}_T^{\mathcal{Y}}} = \mathbb{Q}|_{\mathcal{F}_T \vee \mathcal{F}_T^{\mathcal{Y}}}.$

Condition 2 $\quad X$ is an $(\mathbb{F}, \mathbb{F}^X)$-CMC on $(\Omega, \mathcal{F}, \mathbb{P})$ which satisfies the strong Markov consistency property with respect to $(\mathbb{F}, \mathcal{Y})$.[2]

The methodology developed in Section D.4 allows us to construct strong CMC structures between processes $Y^1,\dots,Y^n$ that are defined on the underlying

[1] It is always tacitly assumed that the probability space $(\Omega, \mathcal{F}, \mathbb{Q})$ is sufficiently rich to support all the stochastic processes and random variables that are considered.

[2] Note that Condition 1 above ensures that each Y^k considered on $(\Omega, \mathcal{F}, \mathbb{P})$ is also an $(\mathbb{F}, \mathbb{F}^{Y^k})$-CMC.

probability space $(\Omega, \mathcal{F}, \mathbb{Q})$ endowed with a reference filtration $\mathbb{F}$ and that are such that each Y^k is an $(\mathbb{F}, \mathbb{F}^{Y^k})$-CDMC with $(\mathbb{F}, \mathbb{Q})$-intensity, say, $\Psi^k = [\psi^{k;x^k y^k}]_{x^k, y^k \in \mathcal{X}_k}$. The additional feature of our construction is that, typically, the constructed CMC structures X are also $(\mathbb{F}, \mathbb{F}^X)$-DSMCs.

In view of Theorem D.40, Proposition 4.12, and Lemma D.31 a natural starting point for constructing a strong CMC structure between $Y^1, \ldots, Y^n$ is to determine a system of stochastic processes $[\lambda^{xy}]_{x,y \in \mathcal{X}}$ and an $\mathcal{X}$-valued random variable $\xi = (\xi^1, \ldots, \xi^n)$ on $(\Omega, \mathcal{F})$, which satisfy the following conditions:

CMC-1

$$\psi_t^{k;x^k y^k} = \sum_{\substack{y^m \in \mathcal{X}_m, \\ m=1,\ldots,n, m \neq k}} \lambda_t^{(x^1,\ldots,x^k,\ldots,x^m)(y^1,\ldots,y^k,\ldots,y^n)},$$

for $x^m \in \mathcal{X}_m$, $m = 1, \ldots, n$; $y^k \in \mathcal{X}_k$, $y^k \neq x^k$, $k = 1, \ldots, n$; $t \in [0, T]$.

CMC-2 The matrix process $\Lambda_t = [\lambda_t^{xy}]_{x,y \in \mathcal{X}}$ satisfies canonical conditions relative to the pair $(\mathcal{X}, \mathbb{F})$ (see Definition D.39).

CMC-3

$$\mathbb{Q}(\xi = y | \mathcal{F}_T) = \mathbb{Q}(\xi = y | \mathcal{F}_0), \quad y \in \mathcal{X}.$$

CMC-4

$$\mathbb{Q}(\xi^k = y^k | \mathcal{F}_T) = \mathbb{Q}(Y_0^k = y^k | \mathcal{F}_T), \quad y^k \in \mathcal{X}_k, \quad k = 1, \ldots, n.$$

We will call any pair (Λ, ξ) satisfying conditions **CMC-1–CMC-4**, a *strong* $(\mathbb{F}, \mathbb{Q})$-*CMC pre-structure* between the processes $Y^1, \ldots, Y^n$. The following theorem provides that, given a strong $(\mathbb{F}, \mathbb{Q})$-CMC pre-structure between the processes $Y^1, \ldots, Y^n$, we can construct on $(\Omega, \mathcal{F})$ a strong CMC structure $(X, \mathbb{P})$ between the processes $Y^1, \ldots, Y^n$.

Theorem 8.2 *Let $\mathcal{Y} = \{Y^1, \ldots, Y^n\}$ be a given family of processes on a probability space $(\Omega, \mathcal{F}, \mathbb{Q})$. Assume that for each $i = 1, \ldots, n$, the process Y^i is an $(\mathbb{F}, \mathbb{F}^{Y^i})$-CDMC with $(\mathbb{F}, \mathbb{Q})$-intensity $\Psi^i = [\psi^{i;x^i, y^i}]_{x^i, y^i \in \mathcal{X}_i}$. Let (Λ, ξ) be a strong $(\mathbb{F}, \mathbb{Q})$-CMC pre-structure between the processes $Y^1, \ldots, Y^n$. Then there exists a strong CMC structure $(X, \mathbb{P})$ between the processes $Y^1, \ldots, Y^n$.*

Proof We will construct a strong CMC structure $(X, \mathbb{P})$ using Theorem D.40. To this end let $\widetilde{\mathbb{F}} = \left\{\widetilde{\mathcal{F}}_t\right\}_{t \in [0,T]}$, where $\widetilde{\mathcal{F}}_t = \mathcal{F}_t \vee \mathcal{F}_t^Y$, $t \geq 0$, and $\widetilde{\mathcal{F}}_0 = \mathcal{F}_0$. Since Λ satisfies canonical conditions relative to the pair $(\mathcal{X}, \mathbb{F})$ it also satisfies canonical conditions relative to the pair $(\mathcal{X}, \widetilde{\mathbb{F}})$, since $\mathbb{F} \subseteq \widetilde{\mathbb{F}}$. Thus, invoking Theorem D.40 with $\mathbb{F}$ replaced by $\widetilde{\mathbb{F}}$ we obtain a probability measure $\mathbb{P}$ such that

$$\mathbb{P} = \mathbb{Q} \quad \text{on} \quad \widetilde{\mathcal{F}}_T,$$

so the measure $\mathbb{P}$ satisfies Condition 1 in the definition of a strong CMC structure.

Moreover, Theorem D.40 also provides an $(\widetilde{\mathbb{F}},\mathbb{G})$-CMC process X on $(\Omega,\mathcal{F},\mathbb{P})$ with $\widetilde{\mathbb{F}}$-intensity matrix process Λ, where $\mathbb{G}$ is given by (D.48), and with the initial distribution satisfying

$$\mathbb{P}(X_0 = x|\widetilde{\mathcal{F}}_T) = \mathbb{P}(X_0 = x|\widetilde{\mathcal{F}}_0) = \mathbb{Q}(X_0 = x|\widetilde{\mathcal{F}}_T), \quad x \in \mathcal{X}. \tag{8.1}$$

By Proposition D.42 the process X is also $(\widetilde{\mathbb{F}},\mathbb{G})$-DMC with $\widetilde{\mathbb{F}}$-intensity matrix process Λ.

Using Theorem D.30 with the pair $\mathbb{H},\mathbb{F}$ replaced with the pair $\mathbb{F},\widetilde{\mathbb{F}}$ we see that X is an $(\mathbb{F},\mathbb{F}^X)$-CDMC with $\mathbb{F}$-intensity Λ and initial condition satisfying

$$\mathbb{P}(X_0 = x|\mathcal{F}_T) = \mathbb{P}(X_0 = x|\mathcal{F}_0), \quad x \in \mathcal{X}. \tag{8.2}$$

Thus Assumption 4.4 is satisfied. From this and from the fact that (Λ,ξ) is a $(\mathbb{F},\mathbb{Q})$-CMC pre-structure between the processes $Y^1,\ldots,Y^n$ we conclude, using Proposition 4.12, that X satisfies Condition 2 in the definition of a strong CMC structure.

Thus, the pair $(X,\mathbb{P})$ is a CMC structure between processes $Y^1,\ldots,Y^n$. The proof is complete. □

Remark 8.3 (i) Note that in the definition of a strong CMC structure it is required that the $\mathcal{F}_T$-conditional distribution of X_0^k coincides with the $\mathcal{F}_T$-conditional distribution of Y_0^k, for $k \in 1,\ldots,N$, but otherwise the $\mathcal{F}_T$-conditional distribution of the multivariate random variable $X_0 = (X_0^1,\ldots,X_0^n)$ can be arbitrary.
(ii) In general, there exist numerous systems of stochastic processes that satisfy conditions **CMC-1** and **CMC-2**, so that there exist numerous strong pre-structures between conditional Markov chains $Y^1,\ldots,Y^n$ and, consequently, there exist numerous strong CMC structures between the conditional Markov chains $Y^1,\ldots,Y^n$.

Below we provide examples of strong CMC structures. The first example, dealing with conditionally independent univariate CMCs, does not really address the issue of modeling dependence between components of a multivariate CMC. Nevertheless, this example is a non-trivial, and therefore interesting, example from the mathematical point of view. Moreover, it provides a reality check for the theory of strong CMC structures: it would be not good for the theory if a multivariate conditional Markov chain $X = (X^1,\ldots,X^n)$ with conditionally independent components did not qualify as a strong CMC structure.

Example 8.4 Conditionally independent strong CMC structure This example generalizes the independence Markov structure presented in part (i) of Remark 7.2.

Let $Y^1,\ldots,Y^n$ be processes such that each Y^k is an $(\mathbb{F},\mathbb{F}^{Y^k})$-CDMC with values in $\mathcal{X}_k$ and with $\mathbb{F}$-intensity $\Psi_t^k = [\psi_t^{k;x^k y^k}]_{x^k,y^k\in\mathcal{X}_k}$. Assume that for each k the process Ψ^k satisfies canonical conditions relative to the pair $(\mathcal{X}_k,\mathbb{F})$. Additionally, assume that

$$\mathbb{Q}(Y_0^k = x^k|\mathcal{F}_T) = \mathbb{Q}(Y_0^k = x^k|\mathcal{F}_0), \quad x^k \in \mathcal{X}_k,\ k = 1,\ldots,n. \tag{8.3}$$

Consider a matrix-valued random process Λ given as the following Kronecker sum:

$$\Lambda_t = \sum_{k=1}^{n} I_1 \otimes \cdots \otimes I_{k-1} \otimes \Psi_t^k \otimes I_{k+1} \otimes \cdots \otimes I_n, \quad t \in [0,T], \tag{8.4}$$

where $\otimes$ is the Kronecker product and where I_k denotes the identity matrix of dimensions $|\mathcal{X}_k| \times |\mathcal{X}_k|$. Moreover, let us take an $\mathcal{X}$-valued random variable $\xi = (\xi^1, \ldots, \xi^n)$, which has $\mathcal{F}_T$-conditionally-independent coordinates, i.e.

$$\mathbb{Q}(\xi^1 = x^1, \ldots, \xi^n = x^n | \mathcal{F}_T) = \prod_{i=1}^{n} \mathbb{Q}(\xi^i = x^i | \mathcal{F}_T), \quad x = (x^1, \ldots, x^n) \in \mathcal{X}. \tag{8.5}$$

Additionally assume that the $\mathcal{F}_T$-conditional distributions of the coordinates of ξ and Y_0 coincide:

$$\mathbb{Q}(\xi^k = x^k | \mathcal{F}_T) = \mathbb{Q}(Y_0^k = x^k | \mathcal{F}_T), \quad x^k \in \mathcal{X}_k, \ k = 1, \ldots, n. \tag{8.6}$$

As shown in Theorem 3.1 in Bielecki et al. (2017b), Λ satisfies the conditions **CMC-1** and **CMC-2**. Furthermore, by (8.5) and (8.3), ξ satisfies **CMC-3** and, by (8.6) it also satisfies **CMC-4**. Thus, (Λ, ξ) is a strong $(\mathbb{F}, \mathbb{Q})$-CMC-pre-structure between the conditional Markov chains $Y^1, \ldots, Y^n$.

Consequently, in view of Theorem 8.2 there exists a strong CMC structure $(X, \mathbb{P})$ between the processes $Y^1, \ldots, Y^n$.

Finally, Theorem 3.3 in Bielecki et al. (2017b) demonstrates that the components of X are conditionally independent given $\mathcal{F}_T$. It is quite clear from (8.4) that the components X^i of X do not jump simultaneously; this, indeed, is the inherent feature of the conditionally independent strong CMC structure.

Next, we will present an example of a strong CMC structure whose components have common jumps.

Example 8.5 Common-jump strong CMC structure Let us consider two processes, Y^1 and Y^2, such that each Y^i is an $(\mathbb{F}, \mathbb{F}^{Y^i})$-CDMC taking values in the state space $\{0,1\}$. Suppose that their $\mathbb{F}$-intensities are

$$\Psi_t^1 = \begin{array}{c} \\ 0 \\ 1 \end{array} \begin{array}{c} \begin{array}{cc} 0 & 1 \end{array} \\ \begin{pmatrix} -a_t & a_t \\ 0 & 0 \end{pmatrix} \end{array}, \qquad \Psi_t^2 = \begin{array}{c} \\ 0 \\ 1 \end{array} \begin{array}{c} \begin{array}{cc} 0 & 1 \end{array} \\ \begin{pmatrix} -b_t & b_t \\ 0 & 0 \end{pmatrix} \end{array}$$

for $t \in [0,T]$, where a, b are nonnegative $\mathbb{F}$-progressively-measurable stochastic processes which have left limits and countably many jumps. Moreover, assume that $\mathbb{Q}(Y_0^1 = 0) = \mathbb{Q}(Y_0^2 = 0) = 1$. Next, let Λ be a matrix-valued process given by

$$\Lambda_t = \begin{array}{c} \\ (0,0) \\ (0,1) \\ (1,0) \\ (1,1) \end{array} \begin{array}{c} \begin{array}{cccc} (0,0) & (0,1) & (1,0) & (1,1) \end{array} \\ \begin{pmatrix} -(a_t + b_t - c_t) & b_t - c_t & a_t - c_t & c_t \\ 0 & -a_t & 0 & a_t \\ 0 & 0 & -b_t & b_t \\ 0 & 0 & 0 & 0 \end{pmatrix} \end{array}$$

for $t \in [0,T]$, where c is an $\mathbb{F}$-progressively-measurable stochastic processes which has left limits and countably many jumps and is such that

$$0 \leq c_t \leq a_t \wedge b_t, \quad t \in [0,T].$$

Moreover, let ξ be an $\mathcal{X}$-valued random variable satisfying $\mathbb{Q}(\xi = (0,0)) = 1$. It can be easily checked that (Λ, ξ) satisfies the conditions (CMC-1)–(CMC-4), so that it is a strong $(\mathbb{F},\mathbb{Q})$-CMC pre-structure between the conditional Markov chains Y^1, Y^2. Now, Theorem 8.2 provides that there exists a strong CMC structure $(X,\mathbb{P})$ between Y^1 and Y^2. We conclude the example by observing that the coordinates of the process X have common jumps only if $c > 0$.

Example 8.6 Perfect dependence strong CMC structure Let Y^1, …, Y^n be processes such that each Y^k is an $(\mathbb{F},\mathbb{F}^{Y^k})$ CMC and such that they have the same $\mathcal{F}_T$-conditional laws. Consider the process $X = (X^1,\ldots,X^n)$, where $X^k = Y^1, k = 1,\ldots,n$. It is clear that X furnishes a strong CMC structure $(X,\mathbb{Q})$ between the conditional Markov chains Y^1, …, Y^n.

8.2 Weak CMC Structures

The next definition refers back to Definition 4.14.

Definition 8.7 Let $\mathcal{Y} = \{Y^1,\ldots,Y^n\}$ be a family of processes, defined on the underlying probability space $(\Omega,\mathcal{F},\mathbb{Q})$, such that each Y^k is an $(\mathbb{F},\mathbb{F}^{Y^k})$-CMC with values in $\mathcal{X}_k$. A *weak CMC structure* between the processes $Y^1,\ldots,Y^n$ is any pair $(X,\mathbb{P})$, where $X = (X^1,\ldots,X^n)$ is a multivariate process given on $(\Omega,\mathcal{F})$ with values in $\mathcal{X}$ and $\mathbb{P}$ is a probability measure on $(\Omega,\mathcal{F})$, such that

Condition 1 $\mathbb{P}_{|_{\mathcal{F}_T \vee \mathcal{F}_T^{\mathcal{Y}}}} = \mathbb{Q}_{|_{\mathcal{F}_T \vee \mathcal{F}_T^{\mathcal{Y}}}}$.

Condition 2 X is an $(\mathbb{F},\mathbb{F}^X)$-CMC on $(\Omega,\mathcal{F},\mathbb{P})$ which satisfies the weak Markov consistency property with respect to $(\mathbb{F},\mathcal{Y})$.

In Section 8.1 we gave three examples of strong CMC structures. Consequently, they are also examples of weak CMC structures. Here, we will give an example of a weak-only CMC structure, that is, a weak CMC structure which is not a strong CMC structure. In particular, this feature implies that in the following example the immersion property formulated in Theorem 4.22 is not satisfied.

Example 8.8 Let us consider processes Y^1 and Y^2 defined on the probability space $(\Omega,\mathcal{F},\mathbb{Q})$ such that each Y^i is an $(\mathbb{F},\mathbb{F}^{Y^i})$-CDMC taking values in the state space $\mathcal{X}_i = \{0,1\}$. We assume that the $(\mathbb{F},\mathbb{Q})$-intensities of Y^1 and Y^2 are, respectively,

$$\Psi_t^1 = \begin{pmatrix} -a_t - c_t \dfrac{\delta_t}{\delta_t + \alpha_t} & a_t + c_t \dfrac{\delta_t}{\delta_t + \alpha_t} \\ 0 & 0 \end{pmatrix},$$

$$\Psi_t^2 = \begin{pmatrix} -b_t - c_t \dfrac{\delta_t}{\delta_t + \beta_t} & b_t + c_t \dfrac{\delta_t}{\delta_t + \beta_t} \\ 0 & 0 \end{pmatrix}$$

for $t \in [0,T]$, where

$$\alpha_t = \exp\left(-\int_0^t a_u du\right) \int_0^t b_u \exp\left(-\int_0^u (b_v + c_v) dv\right) du,$$

$$\beta_t = \exp\left(-\int_0^t b_u du\right) \int_0^t a_u \exp\left(-\int_0^u (a_v + c_v) dv\right) du,$$

$$\delta_t = \exp\left(-\int_0^t (a_u + b_u + c_u) du\right)$$

for $t \in [0,T]$, and a, b, c are positive $\mathbb{F}$-progressive stochastic processes which have left limits and countably many jumps. Moreover, suppose that $\mathbb{Q}(Y_0^i = 0) = 1$, for $i = 1, 2$, which implies that $\mathbb{Q}(Y_0^i = 0 | \mathcal{F}_T) = 1$.

Our goal is to find a weak CMC structure between Y^1 and Y^2. To this end we look for a probability measure $\mathbb{P}$ on $(\Omega, \mathcal{F})$ and an $(\mathbb{F}, \mathbb{F}^X)$-CMC process X defined on $(\Omega, \mathcal{F}, \mathbb{P})$ and taking values in $\mathcal{X} = \{(0,0),(0,1),(1,0),(1,1)\}$, such that the pair $(X, \mathbb{P})$ satisfies Conditions 1 and 2 in Definition 8.7.

To this end, let us consider a matrix-valued process Λ given by

$$\Lambda_t = \begin{matrix} \\ (0,0) \\ (0,1) \\ (1,0) \\ (1,1) \end{matrix} \begin{matrix} \begin{matrix} (0,0) & (0,1) & (1,0) & (1,1) \end{matrix} \\ \begin{pmatrix} -(a_t + b_t + c_t) & b_t & a_t & c_t \\ 0 & -a_t & 0 & a_t \\ 0 & 0 & -b_t & b_t \\ 0 & 0 & 0 & 0 \end{pmatrix} \end{matrix} \tag{8.7}$$

for $t \in [0,T]$, and a constant random variable ξ on $(\Omega, \mathcal{F})$ given by $\xi = (0,0)$.

Then, according to Theorem D.40 there exist a probability measure $\mathbb{P}$ on $(\Omega, \mathcal{F})$ satisfying Condition 1 in Definition 8.7 and an $(\mathbb{F}, \mathbb{F}^X)$-CMC process $X = (X^1, X^2)$ defined on $(\Omega, \mathcal{F}, \mathbb{P})$ taking values in

$$\mathcal{X} = \{(0,0),(0,1),(1,0),(1,1)\}$$

and such that $\mathbb{P}(X_0 = (0,0) | \mathcal{F}_0) = \mathbb{P}(X_0 = (0,0) | \mathcal{F}_T) = \mathbb{P}(X_0 = (0,0)) = 1$.

It remains to verify that the pair $(X, \mathbb{P})$ satisfies Condition 2 in Definition 8.7. For this we observe that the components X^1 and X^2 are processes with state space $\{0,1\}$ and such that state 1 is an absorbing state for both X^1 and X^2. Thus, by similar arguments to those in Example D.6, X^1 (resp. X^2) is an $(\mathbb{F}, \mathbb{F}^{X^1})$-CMC (resp. $(\mathbb{F}, \mathbb{F}^{X^2})$-CMC). Consequently, X is a weakly Markov consistent process relative to $(X^1, \mathbb{F})$ (resp. $(X^2, \mathbb{F})$).

We will now compute, using (4.22), an $\mathbb{F}$-intensity of X^1 and an $\mathbb{F}$-intensity of X^2. To this end we first solve the conditional Kolmogorov forward equation for $P(s,t) = [p_{xy}(s,t)]_{x,y \in S}$, $0 \le s \le t \le T$, i.e.

$$dP(s,t) = P(s,t)\Lambda(t)dt, \quad P(s,s) = \mathbb{I}. \tag{8.8}$$

This is done in order to compute the following conditional probabilities (see Theorem D.22):

$$\mathbb{P}(X_t = y|\mathcal{F}_T \vee \mathcal{F}_s^X) = \sum_{x\in S} \mathbb{1}_{\{X_s=x\}} p_{xy}(s,t), \quad 0 \le s \le t \le T,$$

which will be used in the computation of conditional probabilities of the form

$$\frac{\mathbb{P}(X_t^1 = x^1, X_t^2 = x^2|\mathcal{F}_t)}{\mathbb{P}(X_t^1 = x^1|\mathcal{F}_t)}$$

for $t \in [0,T]$. One can easily verify, by solving appropriate ODEs, that the unique solution of (8.8) is given by, for $0 \le s \le t \le T$,

$$P(s,t) =$$

$$\begin{pmatrix} \exp\left(-\int_s^t (a_u+b_u+c_u)du\right) & \alpha(s,t) & \beta(s,t) & \gamma(s,t) \\ 0 & \exp\left(-\int_s^t a_u du\right) & 0 & 1-\exp\left(-\int_s^t a_u du\right) \\ 0 & 0 & \exp\left(-\int_s^t b_u du\right) & 1-\exp\left(-\int_s^t b_u du\right) \\ 0 & 0 & 0 & 1 \end{pmatrix},$$

where

$$\alpha(s,t) = \exp\left(-\int_s^t a_u du\right)\int_s^t b_u \exp\left(-\int_s^u (b_v+c_v)dv\right) du,$$

$$\beta(s,t) = \exp\left(-\int_s^t b_u du\right)\int_s^t a_u \exp\left(-\int_s^u (a_v+c_v)dv\right) du,$$

$$\gamma(s,t) = 1-\exp\left(-\int_s^t (a_u+b_u+c_u)du\right) - \alpha(s,t) - \beta(s,t).$$

In the rest of the example we fix $t \in [0,T]$. Since X started at $(0,0)$, by Proposition D.17 we have

$$\begin{aligned}\mathbb{P}(X_t^1 = 0, X_t^2 = 0|\mathcal{F}_t) &= \mathbb{E}(\mathbb{P}(X_t^1 = 0, X_t^2 = 0|\mathcal{F}_T)|\mathcal{F}_t) \\ &= \mathbb{E}\left(\exp\left(-\int_0^t (a_u+b_u+c_u)du\right)|\mathcal{F}_t\right) \\ &= \exp\left(-\int_0^t (a_u+b_u+c_u)du\right) = \delta_t.\end{aligned}$$

In an analogous way we conclude that

$$\mathbb{P}(X_t^1 = 1, X_t^2 = 0|\mathcal{F}_t) = \beta(0,t),$$
$$\mathbb{P}(X_t^1 = 0, X_t^2 = 1|\mathcal{F}_t) = \alpha(0,t).$$

Thus

$$\frac{\mathbb{P}(X_t^1 = 1, X_t^2 = 0|\mathcal{F}_t)}{\mathbb{P}(X_t^2 = 0|\mathcal{F}_t)} = \frac{\beta_t}{\delta_t+\beta_t},$$

$$\frac{\mathbb{P}(X_t^1 = 0, X_t^2 = 1|\mathcal{F}_t)}{\mathbb{P}(X_t^1 = 0|\mathcal{F}_t)} = \frac{\alpha_t}{\delta_t+\alpha_t},$$

where for brevity of notation we set $\alpha_t = \alpha(0,t), \beta_t = \beta(0,t)$. Here (4.22) takes the form

$$
\begin{aligned}
&\mathbb{1}_{\{X_t^1=0\}}\lambda_t^{1;01} \\
&= \mathbb{1}_{\{X_t^1=0\}}\Bigg((\lambda_t^{(00)(10)} + \lambda_t^{(00)(11)})\frac{\mathbb{P}(X_t^1=0, X_t^2=0|\mathcal{F}_t)}{\mathbb{P}(X_t^1=0|\mathcal{F}_t)} \\
&\qquad + (\lambda_t^{(01)(10)} + \lambda_t^{(01)(11)})\frac{\mathbb{P}(X_t^1=0, X_t^2=1|\mathcal{F}_t)}{\mathbb{P}(X_t^1=0|\mathcal{F}_t)}\Bigg) \\
&= \mathbb{1}_{\{X_t^1=0\}}\left((a_t+c_t)\frac{\mathbb{P}(X_t^1=0, X_t^2=0|\mathcal{F}_t)}{\mathbb{P}(X_t^1=0|\mathcal{F}_t)} + a_t\frac{\mathbb{P}(X_t^1=0, X_t^2=1|\mathcal{F}_t)}{\mathbb{P}(X_t^1=0|\mathcal{F}_t)}\right) \\
&= \mathbb{1}_{\{X_t^1=0\}}\left((a_t+c_t) - c_t\frac{\alpha(0,t)}{\delta(0,t)+\alpha(0,t)}\right)
\end{aligned}
$$

and

$$
\begin{aligned}
&\mathbb{1}_{\{X_t^1=1\}}\lambda_t^{1;10} \\
&= \mathbb{1}_{\{X_t^1=1\}}\Bigg((\lambda_t^{(10)(00)} + \lambda_t^{(10)(01)})\frac{\mathbb{P}(X_t^1=1, X_t^2=0|\mathcal{F}_t)}{\mathbb{P}(X_t^1=1|\mathcal{F}_t)} \\
&\qquad + (\lambda_t^{(11)(00)} + \lambda_t^{(11)(01)})\frac{\mathbb{P}(X_t^1=1, X_t^2=1|\mathcal{F}_t)}{\mathbb{P}(X_t^1=1|\mathcal{F}_t)}\Bigg) \\
&= \mathbb{1}_{\{X_t^1=1\}}\left(0\times\frac{\mathbb{P}(X_t^1=0, X_t^2=0|\mathcal{F}_t)}{\mathbb{P}(X_t^1=0|\mathcal{F}_t)} + 0\times\frac{\mathbb{P}(X_t^1=0, X_t^2=1|\mathcal{F}_t)}{\mathbb{P}(X_t^1=0|\mathcal{F}_t)}\right) \\
&= \mathbb{1}_{\{X_t^1=1\}}0.
\end{aligned}
$$

Therefore, since $\mathbb{P}(X_t^1=0|\mathcal{F}_t) > 0$, the $\mathbb{F}$-intensity of X^1 satisfies

$$
\Lambda_t^1 = \begin{pmatrix} -a_t - c_t\frac{\delta_t}{\delta_t+\alpha_t} & a_t + c_t\frac{\delta_t}{\delta_t+\alpha_t} \\ 0 & 0 \end{pmatrix} = \Psi_t^1.
$$

Analogously, the $\mathbb{F}$-intensity of X^2 satisfies

$$
\Lambda_t^2 = \begin{pmatrix} -b_t - c_t\frac{\delta_t}{\delta_t+\beta_t} & b_t + c_t\frac{\delta_t}{\delta_t+\beta_t} \\ 0 & 0 \end{pmatrix} = \Psi_t^2.
$$

Consequently, using Lemma D.31 we deduce that X is a weak CMC structure for Y^1 and Y^2.

Finally, we will demonstrate that X is not a strong CMC structure for Y^1 and Y^2. We have, for $0 \le s \le t$,

$$
\begin{aligned}
&\mathbb{P}(X_t^1=0|\mathcal{F}_T \vee \sigma(X_s))\mathbb{1}_{\{X_s^1=0, X_s^2=0\}} \\
&= \mathbb{1}_{\{X_s^1=0, X_s^2=0\}}(p_{(0,0)(0,0)}(s,t) + p_{(0,0)(0,1)}(s,t)) \\
&= \mathbb{1}_{\{X_s^1=0, X_s^2=0\}}\Bigg(\exp\left(-\int_s^t (a_u+b_u+c_u)du\right) \\
&\qquad + \exp\left(-\int_s^t a_u du\right)\int_s^t b_u \exp\left(-\int_s^u (b_v+c_v)dv\right)du\Bigg)
\end{aligned}
$$

and

$$
\begin{aligned}
&\mathbb{P}(X_t^1 = 0 | \mathcal{F}_T \vee \sigma(X_s)) \mathbb{1}_{\{X_s^1=0, X_s^2=1\}} \\
&= \mathbb{1}_{\{X_s^1=0, X_s^2=1\}} (p_{(0,1)(0,0)}(s,t) + p_{(0,1)(0,1)}(s,t)) \\
&= \mathbb{1}_{\{X_s^1=0, X_s^2=1\}} \exp\left(-\int_s^t a_u du\right).
\end{aligned}
$$

Clearly

$$
\begin{aligned}
&\left(\exp\left(-\int_s^t (a_u + b_u + c_u) du\right)\right. \\
&\left. + \exp\left(-\int_s^t a_u du\right) \int_s^t b_u \exp\left(-\int_s^u (b_v + c_v) dv\right) du\right) \neq \exp\left(-\int_s^t a_u du\right), \quad (8.9)
\end{aligned}
$$

unless $c \equiv 0$ on $[s,t]$. In this case (8.9) implies that

$$
\mathbb{P}(\mathbb{P}(X_t^1 = 0 | \mathcal{F}_s \vee \mathcal{F}_s^X) \neq \mathbb{P}(X_t^1 = 0 | \mathcal{F}_t \vee \sigma(X_s^1))) > 0.
$$

Thus process X is not strongly Markov consistent, so X is a weak-only CMC structure between Y^1 and Y^2 unless $c \equiv 0$. For $c \equiv 0$, it follows from Example 8.5 that process X is a strong CMC structure between Y^1 and Y^2.

Remark 8.9 Note that Λ_t admits the following representation:

$$
\Lambda_t = \Psi_t^1 \otimes I_2 + I_1 \otimes \Psi_t^2 + B_t^{12} - B_t^1 - B_t^2, \tag{8.10}
$$

where the terms $\Psi_t^1 \otimes I_2 + I_1 \otimes \Psi_t^2$ give the conditionally independent structure between Y^1 and Y^2 (see Example 8.4), and the remaining terms

$$
B_t^{12} = \begin{pmatrix} -c_t & 0 & 0 & c_t \\ 0 & 0 & 0 & 0 \\ 0 & 0 & 0 & 0 \\ 0 & 0 & 0 & 0 \end{pmatrix},
$$

$$
B_t^1 = \begin{pmatrix} -c_t \frac{\delta_t}{\delta_t + \beta_t} & c_t \frac{\delta_t}{\delta_t + \beta_t} & 0 & 0 \\ 0 & 0 & 0 & 0 \\ 0 & 0 & -c_t \frac{\delta_t}{\delta_t + \beta_t} & c_t \frac{\delta_t}{\delta_t + \beta_t} \\ 0 & 0 & 0 & 0 \end{pmatrix},
$$

$$
B_t^2 = \begin{pmatrix} -c_t \frac{\delta_t}{\delta_t + \alpha_t} & 0 & c_t \frac{\delta_t}{\delta_t + \alpha_t} & 0 \\ 0 & -c_t \frac{\delta_t}{\delta_t + \alpha_t} & 0 & c_t \frac{\delta_t}{\delta_t + \alpha_t} \\ 0 & 0 & 0 & 0 \\ 0 & 0 & 0 & 0 \end{pmatrix}
$$

introduce the dependence structure between Y^1 and Y^2.

9

Special Semimartingale Structures

In this brief chapter we discuss semimartingale structures for a collection of special semimartingales. As in Chapter 5, we confine ourselves to the bivariate case and we consider semimartingale characteristics with respect to the standard truncation function. We start with a definition of the semimartingale structure, and then we follow up with examples.

9.1 Semimartingale Structures: Definition

Let Y^i be an $\mathbb{R}^{d_i}$-valued special semimartingale defined on a canonical probability space $(\Omega^i, \mathcal{G}^i, \mathbb{G}^i, \mathbb{P}^i)$, $i = 1,2$. Let $(\widehat{B^i}, \widehat{C^i}, \widehat{\nu}^i)$ denote the $\mathbb{G}^i$-canonical characteristics[1] of Y^i, $i = 1,2$.

It is useful in applications to assume that the law of Y^i, $i = 1,2$, is uniquely determined by its characteristic triple. This uniqueness property can be verified in terms of the uniqueness of the solution to the related martingale problem (see Jacod and Shiryaev (2003), Section III.2) for details). It is satisfied in the examples presented in this section.

Definition 9.1 We say that an $\mathbb{R}^{d_1} \times \mathbb{R}^{d_2}$-valued process $X = (X^1, X^2)$ on the canonical probability space $(\Omega, \mathcal{F}, \mathbb{F}, \mathbb{P})$ is a *semimartingale structure* for $Y^i, i = 1,2$, if the following conditions hold:

- The process $X = (X^1, X^2)$ is a special semimartingale.
- The $\mathbb{F}^i$-characteristics of X^i, say $(\widetilde{B^i}, \widetilde{C^i}, \widetilde{\nu}^i)$, are equal (as functions of trajectories) to $(\widehat{B^i}, \widehat{C^i}, \widehat{\nu}^i)$ for $i = 1,2$.

If, in addition, the process X is strongly semimartingale-consistent with respect to X^i for each $i = 1,2$ (see Definition 5.3) then we call X a strong semimartingale structure for $Y^i, i = 1,2$. Otherwise, we call X a semi-strong semimartingale structure for $Y^i, i = 1,2$.

[1] That is, considered as functions of trajectories.

Working with just standard truncation functions is important for us, as it provides for the uniqueness of the canonical characteristics (as discussed at the beginning of Chapter 5), which makes comparison of the triples $(\widetilde{B}^i,\widetilde{C}^i,\widetilde{\nu}^i)$ and $(\widehat{B}^i,\widehat{C}^i,\widehat{\nu}^i)$ straightforward: there is no ambiguity as to which characteristics of Y^i and X^i to compare.

9.2 Semimartingale Structures: Examples

We first present an example of a semi-strong semimartingale structure and then we proceed with examples of strong semimartingale structures.

Example 9.2 Let us consider the set-up of Example 5.2. The coordinate X^1 of X is a special semimartingale on $(\Omega,\mathcal{F},\mathbb{F}^1,\mathbb{P})$, and the $\mathbb{F}^1$-jump-characteristic of X^1 is given as

$$\begin{aligned}\widetilde{\nu}^1(ds,dx_1) = \delta_1(dx_1)\mathbb{E}\left(\frac{\int_s^\infty f(s,v)\,dv}{\int_s^\infty\int_s^\infty f(u,v)\,du\,dv}\mathbb{1}_{\{s\le T_1\wedge T_2\}}\middle|\mathcal{F}_s^1\right)ds\\ +\mathbb{E}\left(\frac{f(s,T_2)}{\int_s^\infty f(u,T_2)\,du}\mathbb{1}_{\{T_2<s\le T_1\}}\middle|\mathcal{F}_s^1\right)ds.\end{aligned}$$

Thus, in order to compute $\widetilde{\nu}^1(ds,dx_1)$ it is enough to compute $\widetilde{\nu}^1(ds,\{1\})$. Now, we have the following:

$$\begin{aligned}\widetilde{\nu}^1(ds,\{1\}) &= \frac{\int_s^\infty f(s,v)\,dv}{\int_s^\infty\int_s^\infty f(u,v)\,du\,dv}\mathbb{P}(T_2\ge s|T_1\ge s)\mathbb{1}_{\{T_1\ge s\}}\,ds\\ &\quad+\mathbb{E}\left(\frac{f(s,T_2)}{\int_s^\infty f(u,T_2)\,du}\mathbb{1}_{\{T_2<s\le T_1\}}\middle|\{T_1\ge s\}\right)\mathbb{1}_{\{T_1\ge s\}}\,ds\\ &= \frac{\int_s^\infty f(s,v)\,dv}{\int_s^\infty\int_s^\infty f(u,v)\,du\,dv}\frac{\int_s^\infty\int_s^\infty f(u,v)\,du\,dv}{\int_s^\infty\int_0^\infty f(u,v)\,du\,dv}\mathbb{1}_{\{T_1\ge s\}}\,ds\\ &\quad+\int_0^s\frac{f(s,v)}{\int_s^\infty f(u,v)\,du}\frac{\int_s^\infty f(u,v)\,du}{\int_s^\infty\int_0^\infty f(u,v)\,du\,dv}\,dv\mathbb{1}_{T_1\ge s}\,ds\\ &= \frac{\int_0^\infty f(s,v)\,dv}{\int_s^\infty\int_0^\infty f(u,v)\,du\,dv}\mathbb{1}_{\{T_1\ge s\}}\,ds\\ &= \frac{\int_0^\infty f(s,v)\,dv}{\int_s^\infty\int_0^\infty f(u,v)\,du\,dv}(1-\mathbb{1}_{\{X_s^1=1\}})\,ds.\end{aligned}$$

Analogous results hold for the coordinate X^2 of X.

Now let Y^i be a real-valued one-jump semimartingale defined on a canonical probability space $(\Omega^i,\mathcal{G}^i,\mathbb{G}^i,\mathbb{P}^i)$, $i=1,2$. Let $(\widehat{B}^i,0,\widehat{\nu}^i)$ denote the $\mathbb{G}^i$-canonical characteristics of Y^i, $i=1,2$.

The process $X=(X^1,X^2)$ is a semi-strong semimartingale structure for Y^1 and Y^2 if $\widetilde{\nu}^i=\widehat{\nu}^i$ for $i=1,2$.

The next three examples are devoted to strong semimartingale structures.

Example 9.3 Let Y^i be an $\mathbb{R}^{d_i}$-valued special semimartingale defined on a canonical probability space $(\Omega^i, \mathcal{G}^i, \mathbb{G}^i, \mathbb{P}^i)$, $i = 1,2$. Let $(\widehat{B}^i, \widehat{C}^i, \widehat{\nu}^i)$ denote the $\mathbb{G}^i$-canonical characteristics of Y^i, $i = 1,2$.

Let an $\mathbb{R}^{d_1} \times \mathbb{R}^{d_2}$-valued canonical process $X = (X^1, X^2)$, defined on the canonical probability space $(\Omega, \mathcal{F}, \mathbb{F}, \mathbb{P})$, be a special semimartingale with the canonical characteristic triple (B, C, ν), where

- $B(x^1_\cdot, x^2_\cdot) = (\widehat{B}^1(x^1_\cdot), \widehat{B}^2(x^2_\cdot))$,
- $C(x^1_\cdot, x^2_\cdot) = \mathrm{diag}(\widehat{C}^1(x^1_\cdot), \widehat{C}^2(x^2_\cdot))$,
- $\nu(x^1_\cdot, x^2_\cdot, dt, dy) = \widehat{\nu}^1(x^1_\cdot, dt, dy_1) \otimes \delta_0(dy_2) + \delta_0(dy_1) \otimes \widehat{\nu}^2(x^2_\cdot, dt, dy_2)$;

the law of X_0 is the product of the laws of Y_0^1 and Y_0^2. Then (B, C, ν) is a strong semimartingale structure for Y^1 and Y^2. In fact, it is an independence structure in the sense that X^1 and X^2 are independent.

Example 9.4 Let us consider the set up of Example 5.4 and the process $X = (X^1, X^2)$ therein.

Take two Poisson processes Y^1 and Y^2, each with values in $\mathbb{R}^1$, with intensities λ_1 and λ_2, respectively. They are special semimartingales.

Suppose that λ_{10}, λ_{01}, and λ_{11} are such that

$$\lambda_{10} + \lambda_{11} = \lambda_1, \quad \lambda_{01} + \lambda_{11} = \lambda_2;$$

then, in view of the discussion given in Example 5.4, we see that process X is a strong semimartingale structure for Y^1 and Y^2. We call X a bivariate Poisson structure for Y^1 and Y^2. It is worth noting that the present example and Example 6.8 provide two alternative methods for construction of a Poisson structure.

Example 9.5 Let us return to the set-up of Example 5.5.

Consider two real-valued special semimartingales Y^1 and Y^2 with corresponding canonical characteristic triples $(\widehat{B}^i, \widehat{C}^i, 0)$, $i = 1,2$, where

$$\widehat{B}^i_t = \int_0^t \mu_i(Y^i_u)du, \quad \widehat{C}^i_t = \int_0^t \sigma_i^2(Y^i_u)du, \quad t \geq 0.$$

Then, in view of the discussion in Example 5.5, we see that the process X therein is a strong semimartingale structure for Y^1 and Y^2. It is indeed a bivariate diffusion structure.

PART THREE

FURTHER DEVELOPMENTS

10

Archimedean Survival Processes, Markov Consistency, ASP Structures

Archimedean Survival Processes (ASPs), which are interesting from the theoretical point of view, originate in some financial applications. It turns out that applications of ASPs and ASP structures go beyond finance. We refer to Section 12.3 for a discussion of practical applications of ASP structures.

This chapter builds upon results presented in Hoyle et al. (2011), Hoyle and Mengütürk (2013), and Jakubowski and Pytel (2016). We will borrow some results from these papers but will skip the proofs, which, for the most part, are quite technical and would obscure the main message that we intend to convey in this chapter.

10.1 Archimedean Survival Processes and ASP Structures

Archimedean Survival Processes (ASPs) were introduced by Hoyle and Mengütürk (2013) to model the realized variance of two assets. It turns out that ASPs are very important objects to study in the context of stochastic structures, from both the theoretical and applied perspectives.

10.1.1 Preliminaries

We begin by recalling the definitions of the gamma process and the gamma random bridge process. The underlying probability space is $(\Omega, \mathcal{F}, \mathbb{P})$.

Let us fix two positive constants α and λ and a time interval $[0,T]$, $T \in [1/\alpha, \infty)$.[1]

Definition 10.1 A (real-valued) Lévy process $\gamma = (\gamma_t)_{t\in[0,T]}$ whose increments are gamma random variables is called a *gamma process*. If γ_1 is a gamma random variable with shape parameter α and rate parameter λ then we say that γ has parameters α and λ.

[1] The restriction of T to the interval $[1/\alpha, \infty)$ is a technical requirement that is needed for the validity of some results proved in papers to which we refer in this chapter.

We will consider the gamma process $\gamma = (\gamma_t)_{t\in[0,T]}$ starting from 0, i.e. $\gamma_0 = 0$. The gamma process is increasing. For $t \in (0,T]$, the density of γ_t is given by

$$f_t(x) := \mathbb{1}_{(0,+\infty)}(x)\frac{\lambda^{\alpha t}x^{\alpha t-1}}{\Gamma(\alpha t)}e^{-\lambda x}, \tag{10.1}$$

where Γ is the gamma function.

For simplicity of presentation we assume in the rest of the section that $\alpha = 1$ and $\lambda = 1$.

The gamma bridge $\rho = (\rho_t)_{t\in[0,T]}$ is defined, for $t \in [0,T]$, by the formula

$$\rho_t := \frac{\gamma_t}{\gamma_T}.$$

So, it is equivalent in law to a conditioned gamma process that starts from 0 and is equal to 1 at time T.

Definition 10.2 A *gamma random bridge* is a process $M = (M_t)_{t\in[0,T]}$ defined as

$$M_t = R\rho_t \quad t \in [0,T],$$

where ρ is a gamma bridge on $[0,T]$ and R is a strictly positive random variable that is independent of ρ. The law of R is called a *generating law of the gamma random bridge M* and we denote it by ψ.

It was proved in Hoyle et al. (2011) that a gamma random bridge is a time-homogeneous Markov process and that

$$\mathbb{P}(M_t \in dy \mid M_s = x) = \frac{\Psi_t(y)}{\Psi_s(x)}f_{t-s}(y-x)dy, \tag{10.2}$$

$$\mathbb{P}(M_T \in dy \mid M_s = x) = \frac{\psi_s(dy,x)}{\Psi_s(x)} \tag{10.3}$$

for $0 \le s < t < T$, where $\{\psi_t(\cdot,x), t \in [0,T), x \in (0,\infty)\}$, with $\psi_t(\cdot,x)$ defined as

$$\psi_0(B,x) = \psi(B),$$
$$\psi_t(B,x) = \int_B \frac{f_{T-t}(z-x)}{f_T(z)}\psi(dz)$$

for any $B \in \mathcal{B}(0,\infty)$, is a family of Borel measures and

$$\Psi_t(x) = \psi_t((0,+\infty),x). \tag{10.4}$$

10.1.2 Archimedean Survival Processes

We are ready to define the Archimedean survival process.

Definition 10.3 An $\mathbb{R}^n$-valued process $X = (X_t)_{t\in[0,1]}$ is called an n-dimensional *Archimedean survival process* (ASP) if

$$X_t = \begin{bmatrix} X_t^1 \\ X_t^2 \\ \vdots \\ X_t^i \\ \vdots \\ X_t^n \end{bmatrix} := \begin{bmatrix} M_t - M_0 \\ M_{1+t} - M_1 \\ \vdots \\ M_{(i-1)+t} - M_{i-1} \\ \vdots \\ M_{(n-1)+t} - M_{n-1} \end{bmatrix}, \quad t \in [0,1],$$

where $(M_t)_{t\in[0,n]}$ is a gamma random bridge with generating law ψ.[2] We call ψ *the generating law of* X.

The law of X_1 is given by the Archimedean survival copula; i.e. the law of the random vector X_1 is obtained from the marginal laws of its components via a survival copula C of the form

$$C(u) = \varphi(\varphi^{-1}(u_1), \varphi^{-1}(u_2) \ldots, \varphi^{-1}(u_n))$$

for $u \in [0,1]^n$, where φ: $[0,\infty)$ is a decreasing (and strictly decreasing on $[0, \inf_{\varphi(u)=0} u))$ and continuous function that satisfies the conditions $\varphi(0) = 1$, $\lim_{u\to\infty} \varphi(u) = 0$ (see Nelsen (2006)). We refer to Hoyle and Mengütürk (2013) for the exact form of the generator φ. In general, φ depends on the generating law ψ and on the shape parameter α. So, these two inputs model the dependence between the X_1^i, $i = 1,\ldots,n$. It was shown in Proposition 3.7 in Hoyle and Mengütürk (2013) that the dependences between the X_t^i, $i = 1,\ldots,n$, for $t \in [0,1)$ are given in terms of a Liouville copula, which, in general, is derived from ψ and α as well.

The coordinates of an ASP are nonnegative $\mathbb{P}$-a.s. for every $t \in [0,1]$, since a gamma bridge process has increasing paths $\mathbb{P}$-a.s.

We now have the following important result,

Proposition 10.4 *(Hoyle et al. (2011), Corollary 3.12) Let M be a gamma random bridge with generating law ψ. Fix s_1, T_1 satisfying $0 < T_1 \le T - s_1$. Then the process*

$$\Xi_t = M_{s_1+t} - M_{s_1}, \quad t \le T_1,$$

is a gamma random bridge on $[0, T_1]$ with generating law

$$\psi^*(dx) = \frac{x^{T_1-1}}{\beta(T_1, T-T_1)} \int_{(0,\infty)} \frac{(z-x)^{T-T_1-1}}{z^{T-1}} \psi(dz)dx, \tag{10.5}$$

where β is the beta function.

[2] Recall that we have set $\alpha = \lambda = 1$.

From Proposition 10.4 we see that every coordinate X^i of process X is a gamma random bridge (with $\alpha = \lambda = 1$ and with generating law ψ^*). Moreover, the laws of the components X^i, $i = 1,\ldots,n$, are all the same.

Remark 10.5 It is possible to generalize the construction of an ASP in such a way that the components X^i, $i = 1,\ldots,n$, are gamma random bridges with different laws. We refer to Hoyle and Mengütürk (2013) for a discussion.

10.1.3 ASP Structures

Now let $(Y_t^i)_{t\in[0,1]}$, $i = 1,\ldots,n$, be gamma random bridges with $\alpha = \lambda = 1$ and with generating law ϕ^*. From the above discussion we see that the ASP X constitutes a stochastic structure for the family $(Y_t^i)_{t\in[0,1]}$, $i = 1,\ldots,n$. This justifies the following definition.

Definition 10.6 Let $(Y_t^i)_{t\in[0,1]}$, $i = 1,\ldots,n$, be gamma random bridges with $\alpha = \lambda = 1$ and with generating law ϕ^*. We call the ASP X given in Definition 10.3 the *ASP structure* for $(Y_t^i)_{t\in[0,1]}$, $i = 1,\ldots,n$, if for the law ψ^* given in (10.5) it holds that $\phi^* = \psi^*$.

The above definition naturally generalizes to the case of arbitrary positive α and λ.

In view of Remark 10.5 we see that one can create structures for a family Y^i, $i = 1,\ldots,n$, of gamma random bridges that evolve according to different laws.

10.2 Weak and Strong Markov Consistency for ASPs

In this section we discuss the strong and weak Markov consistency properties of ASPs. We start with the weak Markov consistency property.

Definition 3.8 of weak Markov consistency was stated in the case of a finite Markov chain. It carries verbatim to the case of an ASP X.

We already know that that every coordinate X^i of X is a gamma random bridge. As stated just below Definition 10.2, Hoyle et al. (2011) showed that a gamma random bridge is a time-homogeneous Markov process. Thus, we obtain the following important result.

Theorem 10.7 *Let X be an ASP. Then X is weakly Markov consistent.*

We now proceed with a discussion of the strong Markov consistency of an ASP. To this end we recall Definition 2.15, giving the strong Markov consistency of a multivariate Markov process. The following two theorems come from Jakubowski and Pytel (2016).

Theorem 10.8 *Let X be an ASP. If its generating law has a distribution with support different from* $(0,\infty)$ *or if it has a discrete distribution then X is not strongly Markov consistent.*

We see from Theorems 10.7 and 10.8 that if the generating law of an ASP X has a distribution with the support different from $(0,\infty)$ or if it has a discrete distribution then X is strictly weakly Markov consistent (see Definition 3.10).

An ASP can be strongly Markov consistent in some cases. The next theorem gives a necessary and sufficient condition for an ASP to be strongly Markov consistent, under the assumption that R has a continuous density with support equal to $(0,\infty)$. In the statement of the theorem we let f_n^λ denote the density of the gamma distribution with shape parameter n and rate parameter λ, i.e.

$$f_n^\lambda(x) = \frac{\lambda^n x^{n-1}}{\Gamma(n)} e^{-\lambda x} \mathbb{I}_{(0,\infty)}(x). \tag{10.6}$$

Theorem 10.9 *Suppose that X is an n-dimensional ASP. Let* $(0,\infty)$ *be the support of R with a continuous density g. Then X is strongly Markov consistent if and only if*

$$g = f_n^\lambda \quad \text{a.e.}$$

for some $\lambda > 0$.

In view of the above theorems we see that the ASP structure defined in Definition 10.6 may be of strictly weak form or of strong form, depending on whether the ASP X exhibits strictly weak Markovian consistency or whether it exhibits strong Markovian consistency.

11
Generalized Multivariate Hawkes Processes

A very interesting class of stochastic processes was introduced in Hawkes (1971a, b). These processes, called now Hawkes processes, are intended to model self-exciting and mutually exciting random phenomena that evolve in time. The self-exciting phenomena are modeled as univariate Hawkes processes, and the mutually exciting phenomena are modeled as multivariate Hawkes processes. Hawkes processes belong to the family of marked point processes and, of course, a univariate Hawkes process is just a special case of a multivariate process.

In this chapter we define and study generalized multivariate Hawkes processes, as well as the related consistencies and structures. Generalized multivariate Hawkes processes are multivariate marked point processes that add an important feature to the family of the (classical) multivariate Hawkes processes: they allow for the explicit modeling of the simultaneous occurrence of excitation events coming from different sources, i.e. caused by different coordinates of the multivariate process. The importance of this feature is rather intuitive, and it will be illustrated by some examples in Section 12.4.

We need to stress that we limit ourselves here to the case of linear generalized multivariate Hawkes processes, the counterpart of linear classical Hawkes processes. That is to say, we do not study here what would be the counterpart of nonlinear classical Hawkes processes. We refer, for example, to Chapter 1 in Zhu (2013) for a comparison of linear and nonlinear Hawkes processes. We also note that the generalized Hawkes processes introduced here should not be confused with those studied in Vacarescu (2011). In particular, we do not introduce any additional random factors, such as Brownian motion, into the compensators of the multivariate marked point process N appearing in Definition 11.5 below.

11.1 Generalized $\mathbb{G}$-Hawkes Process

Let $(\Omega, \mathcal{F}, \mathbb{P})$ be a probability space on which we consider a filtration $\mathbb{G}$ and let $(\mathcal{X}, \boldsymbol{\mathcal{X}})$ be a Borel measurable space. We postulate that the filtration $\mathbb{G}$ satisfies the usual conditions and that $\mathcal{G}_0$ is the trivial σ-field.

We take ∂ to be a point external to $\mathcal{X}$, and we let $\mathcal{X}^\partial := \mathcal{X} \cup \partial$. On $(\Omega, \mathcal{F}, \mathbb{P})$ we consider a *marked point process* N with mark space $\mathcal{X}$, i.e. a sequence of random elements

$$N = ((T_n, X_n))_{n \geq 1}, \tag{11.1}$$

where, for each n,

(i) $T_n > 0$ is a $\mathbb{G}$-stopping time,
(ii) X_n is a $\mathcal{G}_{T_n}$-measurable random variable with values in $\mathcal{X}^\partial$,
(iii) $T_n \leq T_{n+1}$ and if $T_n < \infty$ then $T_n < T_{n+1}$,
(iv) $X_n = \partial$ if and only if $T_n = \infty$.

The explosion time of N, say T_∞, is defined as

$$T_\infty := \lim_{n \to \infty} T_n.$$

Following typical techniques used in the theory of marked point processes (MPPs), in particular, following Section 1.3 in Last and Brandt (1995), we associate with the process N an integer-valued random measure on $(\mathbb{R}_+ \times \mathcal{X}, \mathcal{B}(\mathbb{R}_+) \otimes \boldsymbol{\mathcal{X}})$, also denoted by N and defined as

$$N(dt, dx) := \sum_{n \geq 1} \delta_{(T_n, X_n)}(dt, dx) \mathbb{1}_{\{T_n < \infty\}}, \tag{11.2}$$

so that

$$N((0,t], A) = \sum_{n \geq 1} \mathbb{1}_{\{T_n \leq t\}} \mathbb{1}_{\{X_n \in A\}}$$

for $A \in \boldsymbol{\mathcal{X}}$.

Typically, we take $\mathbb{G}$ to be the natural filtration of N, denoted by $\mathbb{F}^N$ and defined as $\mathbb{F}^N := (\mathcal{F}_t^N,\ t \geq 0)$, where $\mathcal{F}_t^N$ is the $\mathbb{P}$-completed σ-field $\sigma(N((s,r] \times A) : 0 \leq s < r \leq t,\ A \in \mathcal{X}),\ t \geq 0$. In view of Theorem 2.2.4 in Last and Brandt (1995) the filtration $\mathbb{F}^N$ satisfies the usual conditions.

In particular, N is $\mathbb{F}^N$-optional, so, using Proposition 4.1.1 in Last and Brandt (1995) we conclude that the T_n are $\mathbb{F}^N$-stopping-times and X_n are $\mathcal{F}_{T_n}$-measurable. Thus, given N we see that requirements (i) and (ii) are satisfied for $\mathbb{G} = \mathbb{F}^N$.

We now consider a random measure ν on $(\mathbb{R}_+ \times \mathcal{X}, \mathcal{B}(\mathbb{R}_+) \otimes \boldsymbol{\mathcal{X}})$ defined as

$$\nu(\omega, dt, dy) := \mathbb{1}_{]\!]0, T_\infty(\omega)[\![}(t) \kappa(\omega, t, dy) dt, \tag{11.3}$$

where, for $A \in \boldsymbol{\mathcal{X}}$,

$$\kappa(t, A) = \eta(t, A) + \int_{(0,t) \times \mathcal{X}} f(t, s, x, A) N(ds, dx), \tag{11.4}$$

η is a finite kernel from $(\Omega \times [0, \infty), \mathcal{P}(\mathbb{G}))$ to $(\mathcal{X}, \boldsymbol{\mathcal{X}})$, and f is a kernel from $(\Omega \times \mathbb{R}_+ \times \mathbb{R}_+ \times \mathcal{X}, \mathcal{F} \otimes \mathcal{B}(\mathbb{R}_+) \otimes \mathcal{B}(\mathbb{R}_+) \otimes \boldsymbol{\mathcal{X}})$ to $(\mathcal{X}, \boldsymbol{\mathcal{X}})$. For fixed $t \geq 0$ and $A \in \boldsymbol{\mathcal{X}}$, the integral $\int_{(0,t) \times \mathcal{X}} f(t, s, x, A) N(ds, dx)$ in (11.4) is understood as

$$(W * N)_t,$$

where

$$W(s,x) = \mathbb{1}_{(0,t)}(s) f(t,s,x,A);$$

see Appendix A. We assume also that f is a kernel satisfying the following:

(i) $f(t,s,x,A) = 0$ for $s > t$,
(ii) θ defined as

$$\theta(t,A) := \int_{(0,t)\times\mathcal{X}} f(t,s,x,A) N(ds,dx), \quad t \geq 0,\ A \in \boldsymbol{\mathcal{X}},$$

is a kernel from $(\Omega \times [0,\infty), \mathcal{P}(\mathbb{G}))$ to $(\mathcal{X}, \boldsymbol{\mathcal{X}})$ which is finite for $t < T_\infty$.

Remark 11.1 Clearly, we have

$$\theta(t,A) = \sum_{n:\ T_n < t} f(t,T_n,X_n,A). \tag{11.5}$$

Note that $\kappa(t,\mathcal{X})$ is finite for any $t < T_\infty$. We assume additionally that $\kappa(t,\mathcal{X}) > 0$ for all $t \geq 0$ and that the integral $\int_{[0,t]} \kappa(s,A)ds$ is finite for any $A \in \boldsymbol{\mathcal{X}}$ and any $t < T_\infty$. This last assumption is satisfied under mild boundedness conditions imposed on η and f.

Note that the process $\nu((0,\cdot],A) = \int_{(0,\cdot]} \mathbb{1}_{]\!]0,T_\infty(\omega)[\![}(s)\kappa(s,A)ds$ is continuous for any set $A \in \boldsymbol{\mathcal{X}}$ and thus it is $\mathbb{G}$-predictable. Consequently, ν is a $\mathbb{G}$-predictable random measure.

We are ready to state the underlying definition in this chapter.

Definition 11.2 Let N be the marked point process introduced in (11.1) with corresponding random measure N defined in (11.2). We call N a *generalized* $\mathbb{G}$*-Hawkes process* on $(\Omega, \mathcal{F}, \mathbb{P})$, if the $(\mathbb{G},\mathbb{P})$-compensator of N, say ν, is of the form (11.3). The kernel κ is called the $\mathbb{G}$*-Hawkes kernel* for N. If N is a *generalized* $\mathbb{F}^N$*-Hawkes process*, we simply call it a *generalized Hawkes process* with Hawkes kernel κ.

Remark 11.3 (i) A necessary condition for generalized Hawkes processes to feature self-excitation and mutual excitation is that $f \neq 0$.
(ii) Recall that the compensator of a random measure is unique (up to equivalence). Thus, the compensator ν of N is unique. However, the representation (11.3), (11.4) is not unique in general. What we mean by this is that, for any given η and f in the representation (11.3), (11.4), one can always find $\widetilde{\eta} \neq \eta$ and $\widetilde{f} \neq f$ such that

$$\kappa(t,dy) = \widetilde{\eta}(t,dy) + \int_{(0,t)\times\mathcal{X}} \widetilde{f}(t,s,x,dy) N(ds,dx). \tag{11.6}$$

The uniqueness of the representation (11.3), (11.4) will be ensured if, for example, η is additionally postulated to be deterministic (which is the case in most of the examples that follow).
(iii) With a slight abuse of terminology we will refer to κ as to the Hawkes intensity kernel of N. Accordingly, we will refer to the quantity $\kappa(t,A)$ as the intensity at time t of the event regarding the process N that amounts to the marks of N taking values in

the set A or, for short, as the intensity at time t of the marks of N taking values in A.
(iv) We refer to Example 11.22 for an interpretation of the components η and f of the kernel κ in the case of a generalized multivariate Hawkes process, which is introduced in Definition 11.5.

Definition 11.2 is quite encompassing and has been stated as such with a view towards future applications. The only application of Definition 11.2 in this book will be in the case $\mathbb{G} = \mathbb{F}^N$. It will be demonstrated in Section 11.2.2 that a generalized $\mathbb{F}^N$-Hawkes process, with given compensator ν of the form (11.3), exists on a relevant canonical space. Specifically, it will be shown that there exists a probability measure $\mathbb{P}_\nu$ on this canonical space such that N, given as the relevant canonical mapping, admits an $(\mathbb{F}^N, \mathbb{P}_\nu)$-compensator given by ν. So, N is a generalized Hawkes process. In Section 11.2.4 we present a different construction of a generalized Hawkes process with a given compensator, which does not refer to any canonical space; the additional feature of this construction is that it facilitates simulations of the trajectories of Hawkes processes.

Remark 11.4 We note that in our definition of the generalized $\mathbb{G}$-Hawkes process the integral in (11.4) was taken over the interval $(0,t)$. In the definition of the classic Hawkes process, the corresponding integral is taken over the interval $(-\infty,t)$; see e.g. Embrechts et al. (2011). The "$(0,t)$" convention is used, however, in many applications of classic Hawkes processes (see Example 11.17). We will use this convention in order to simplify the study presented in this chapter.

11.2 Generalized Multivariate Hawkes Processes

This section is split into several subsections to enhance the logic of the presentation.

11.2.1 Definition of a Generalized Multivariate Hawkes Process

We now introduce the concept of a generalized multivariate Hawkes process, which is a particular case of a generalized $\mathbb{G}$-Hawkes process. We will take $\mathbb{G} = \mathbb{F}^N$; this, in part, motivates the use of the term "generalized multivariate Hawkes process" rather than "generalized multivariate $\mathbb{G}$-Hawkes process".

We first construct an appropriate mark space. Specifically, we fix an integer $d \geq 1$ and let $(E_i, \mathcal{E}_i)$, $i = 1, \ldots, d$, be some non-empty Borel measurable spaces and Δ be a dummy mark, the meaning of which will be explained below. Very often, in practical modeling, the spaces E_i are discrete. The instrumental rationale for considering a discrete mark space is that in most applications of the Hawkes processes that we are familiar with and/or can imagine, a discrete mark space is sufficient to account for the intended features of the modeled phenomenon.

We set $E_i^\Delta := E_i \cup \Delta$, and denote by $\mathcal{E}_i^\Delta$ the sigma algebra on E_i^Δ generated by $\mathcal{E}_i$. Then we define a mark space, say E^Δ, as follows:

$$E^\Delta := E_1^\Delta \times E_2^\Delta \times \cdots \times E_d^\Delta \setminus (\Delta, \Delta, \ldots, \Delta). \tag{11.7}$$

By $\mathcal{E}^\Delta$ we denote a trace sigma algebra of $\otimes_{i=1}^d \mathcal{E}_i^\Delta$ on E^Δ, i.e.

$$\mathcal{E}^\Delta := \left\{ A \cap E^\Delta : A \in \otimes_{i=1}^d \mathcal{E}_i^\Delta \right\}.$$

Moreover, denoting by ∂_i a point which is external to E_i^Δ, we define $E_i^\partial := E_i^\Delta \cup \{\partial_i\}$ and denote by $\mathcal{E}_i^\partial$ the sigma algebra generated by $\mathcal{E}_i$ and $\{\partial_i\}$. Analogously, we define

$$E^\partial = E^\Delta \cup \partial,$$

where $\partial = (\partial_1, \ldots, \partial_d)$ is a point external to $E_1^\Delta \times E_2^\Delta \times \cdots \times E_d^\Delta$, and by $\mathcal{E}^\partial$ we denote the sigma field generated by $\mathcal{E}^\Delta$ and $\{\partial\}$.

Definition 11.5 A generalized Hawkes process $N = ((T_n, X_n))_{n \geq 1}$ with mark space $\mathcal{X} = E^\Delta$ given by (11.7), and with $\mathcal{X}^\partial = E^\partial$, is called a *generalized multivariate Hawkes process (of dimension d).*

We interpret $T_n \in (0, \infty)$ and $X_n \in E^\Delta$ as the event times of N and as the corresponding mark values, respectively. Thus, if $T_n < \infty$ we have[1]

$$X_n = (X_n^i, i = 1, \ldots, d), \quad \text{where} \quad X_n^i \in E_i^\Delta.$$

Also, we interpret the X^i as the marks associated with the ith coordinate of N (see Definition 11.7). With this interpretation, the equality $X_n^i(\omega) = \Delta$ means that there is no event taking place with regard to the ith coordinate of N at the (general) event time $T_n(\omega)$.

Definition 11.6 We say that $T_n(\omega)$ is a common event time for a generalized multivariate Hawkes process N if there exist i and j, $i \neq j$, such that $X_n^i(\omega) \in E_i$ and $X_n^j(\omega) \in E_j$. We say that the process N admits common event times if

$$\mathbb{P}(\omega \in \Omega : \exists n \text{ such that } T_n(\omega) \text{ is a common event time }) > 0.$$

Otherwise, we say that the process N admits no common event times.

The ith Coordinate of a Generalized Multivariate Hawkes Process N

Definition 11.7 We define the ith coordinate N^i of N as

$$N^i((0,t], A) := \sum_{n \geq 1} \mathbb{1}_{\{T_n \leq t\}} \mathbb{1}_{\{X_n \in A^i\}}, \tag{11.8}$$

for $A \in \mathcal{E}_i$, and

$$A^i = \Big(\bigtimes_{j=1}^{i-1} E_j^\Delta \Big) \times A \times \Big(\bigtimes_{j=i+1}^{d} E_j^\Delta \Big). \tag{11.9}$$

[1] Note that here d is the number of components in X_n and n is the index of the nth element in the sequence $(X_n)_{n \geq 1}$.

Clearly, N^i is an MPP and

$$N^i((0,t],A) = N((0,t],A^i).$$

Indeed, the ith coordinate process N^i can be represented in terms of a sequence $(T^i_k, Y^i_k)_{k\geq 1}$, which is related to the sequence $(T_n, X^i_n)_{n\geq 1}$ as follows:

$$(T^i_k, Y^i_k) = \begin{cases} (T_{m^i_k}, X^i_{m^i_k}) & \text{if } m^i_k < \infty, \\ (T_{m^i_{K^i}+k-\hat{k}^i}, \Delta) & \text{if } m^i_k = \infty \text{ and } T_\infty < \infty, \\ (\infty, \partial^i) & \text{if } m^i_k = \infty \text{ and } T_\infty = \infty, \end{cases} \tag{11.10}$$

where $K^i = \max\{n : m^i_n < \infty\}$, with the m^i defined as

$$\begin{aligned} m^i_1 &= \inf\{n \geq 1 : X^i_n \in E_i\}, \\ m^i_k &= \inf\{n > m^i_{k-1} : X^i_n \in E_i\} \quad \text{for } k > 1. \end{aligned}$$

We clearly have

$$N^i((0,t],A) = \sum_{k\geq 1} \mathbb{1}_{\{T^i_k \leq t\}} \mathbb{1}_{\{Y^i_k \in A\}}, \quad A \in \mathcal{E}_i. \tag{11.11}$$

In particular this means that, for the ith coordinate N^i, the times $T_n(\omega)$ such that $X^i_n(\omega) = \Delta$ are disregarded as event times for this coordinate since the events occurring with regard to the entire N at these times do not affect the ith coordinate.

We define the completed filtration $\mathbb{F}^{N^i} = (\mathcal{F}^{N^i}_t, t \geq 0)$ generated by N^i in analogy to $\mathbb{F}^N$: specifically, $\mathcal{F}^{N^i}_t$ is the $\mathbb{P}$-completion of the σ-field $\sigma(N^i((s,r] \times A) : 0 \leq s < r \leq t,\ A \in \mathcal{E}_i)$, $t \geq 0$. In view of Theorem 2.2.4 in Last and Brandt (1995), the filtration $\mathbb{F}^{N^i}$ satisfies the usual conditions.

We define the explosion time T^i_∞ of N^i as

$$T^i_\infty := \lim_{n\to\infty} T^i_n.$$

Clearly, $T^i_\infty \leq T_\infty$.

The random measure N^i is both $\mathbb{F}^N$-optional and $\mathbb{F}^{N^i}$-optional. In what follows we will need to know the compensator of N^i with respect to $\mathbb{F}^N$ and the compensator of N^i with respect to $\mathbb{F}^{N^i}$. The following two propositions come in handy in this regard.

Proposition 11.8 *Let N be a generalized multivariate Hawkes process with Hawkes kernel κ. Then the $(\mathbb{F}^N, \mathbb{P})$-compensator, say ν^i, of the measure N^i defined in* (11.8) *is given as*

$$\nu^i(\omega, dt, dy_i) = \mathbb{1}_{]\!]0;T^i_\infty(\omega)[\![}(t)\kappa^i(\omega, t, dy_i)dt, \tag{11.12}$$

where

$$\kappa^i(t,A) := \kappa(t,A^i), \quad t \geq 0,\ A \in \mathcal{E}_i, \tag{11.13}$$

with A^i as defined in (11.9).

Proof According to Theorems 4.1.11 and 4.1.7 in Last and Brandt (1995), the ith coordinate N^i admits a unique $\mathbb{F}^N$-compensator, say v^i, with the property that $v^i([\![T^i_\infty;\infty[\![\times E_i) = 0$. For every n and $A \in \mathcal{E}_i$, the processes $M^{i,n,A}$ and $\widehat{M}^{i,n,A}$ given as

$$M^{i,n,A}_t = N^i((0,t\wedge T_n]\times A) - \int_0^{t\wedge T_n} \mathbb{1}_{]\!]0;T_\infty[\![}(u)\kappa^i(u,A)du, \quad t \geq 0,$$

and

$$\widehat{M}^{i,n,A}_t = N^i((0,t\wedge T_n]\times A) - v^i((0,t\wedge T_n]\times A), \quad t \geq 0,$$

are $(\mathbb{F}^N,\mathbb{P})$-martingales. Hence the process

$$\int_0^{t\wedge T_n} \left(\mathbb{1}_{]\!]0;T_\infty[\![}(u)\kappa^i(u,A)du - v^i(du,A)\right), \quad t \geq 0,$$

is an $\mathbb{F}^N$-predictable martingale. Since it is of integrable variation and null at $t=0$, it is null for all $t \geq 0$ (see e.g. Theorem VI.6.3 in He et al. (1992)). From the above discussion and the fact that $T^i_\infty \leq T_\infty$, we deduce that

$$\int_0^{t\wedge T_n} \mathbb{1}_{]\!]0;T^i_\infty[\![}(u)\kappa^i(u,A)du = v^i((0,t\wedge T_n]\times A), \quad t \geq 0.$$

This proves the proposition. □

Remark 11.9 Note that for each i, the function κ^i defined in (11.13) is a measurable kernel from $(\Omega\times\mathbb{R}_+,\mathcal{P}\otimes\mathcal{B}(\mathbb{R}_+))$ to $(E_i,\mathcal{E}_i)$. It is important to observe that, in general, there is no one-to-one correspondence between the Hawkes kernel κ and all the marginal kernels κ^i, $i=1,\ldots,d$. We mean by this that there may exist another Hawkes kernel, say $\widehat{\kappa}$, such that $\widehat{\kappa} \neq \kappa$ and

$$\kappa^i(t,A) = \widehat{\kappa}(t,A^i), \quad t \geq 0,\ A \in \mathcal{E}_i,\ i=1,\ldots,d. \tag{11.14}$$

The following important result gives the $(\mathbb{F}^{N^i},\mathbb{P})$-compensator of the measure N^i.

Proposition 11.10 *Let N be a generalized multivariate Hawkes process with Hawkes kernel κ. Then the $(\mathbb{F}^{N^i},\mathbb{P})$-compensator of the measure N^i, say $\widetilde{v}^i$, is given as*

$$\widetilde{v}^i(\omega,dt,dy_i) = (v^i)^{p,\mathbb{F}^{N^i}}(\omega,dt,dy_i), \tag{11.15}$$

where $(v^i)^{p,\mathbb{F}^{N^i}}$ is the dual predictable projection of v^i onto $\mathbb{F}^{N^i}$ under $\mathbb{P}$.

Proof Using Theorems 4.1.9 and 3.4.6 in Last and Brandt (1995), as well as the uniqueness of the compensator, it is enough to show that, for any $A \in \mathcal{E}^i$ and any $n \geq 1$, the process $(v^i)^{p,\mathbb{F}^{N^i}}((0,t\wedge T^i_n],A)$, where $(v^i)^{p,\mathbb{F}^{N^i}}$ is the dual predictable projection of v^i onto $\mathbb{F}^{N^i}$ under $\mathbb{P}$, is the $(\mathbb{F}^{N^i},\mathbb{P})$-compensator of the increasing process $N^i((0,t\wedge T^i_n],A)$, $t \geq 0$. This, however, follows from Theorem 3.6 in Bielecki et al. (2018b). □

Remark 11.11 (i) It has to be stressed that N^i may be neither a generalized $\mathbb{F}^N$-Hawkes process nor a generalized $\mathbb{F}^{N^i}$-Hawkes process in the sense of Definition 11.2. The reason is that the compensator ν^i (resp. $\widetilde{\nu}^i$) may not admit representations of the form (11.3), (11.4).
(ii) If κ^i is $\mathbb{F}^{N^i}$-predictable then $\widetilde{\nu}^i = \nu^i$.[2] We will discuss this case in Proposition 11.29. In particular, Proposition 11.29 will provide structural conditions on η and f under which κ^i is $\mathbb{F}^{N^i}$-predictable and, moreover, the process N^i is a generalized $\mathbb{F}^{N^i}$-Hawkes process.

An Idiosyncratic Group of I Coordinates of a Generalized Multivariate Hawkes Process N

In what follows (see Examples 11.18, 11.19, 11.21 and 11.22) we will need the concept of an idiosyncratic group of I coordinates of a generalized multivariate Hawkes process (MHP) N.

Let us fix $\mathcal{I} = \{d_1,\ldots,d_I\}$, where $I \in \{1,\ldots,d\}$ and $1 \le d_1 < \cdots < d_I \le d$. We define

$$A^{idio,\mathcal{I}} = \Big(\bigtimes_{j=1}^{d_1-1}\{\Delta\}\Big) \times A_{d_1} \times \Big(\bigtimes_{j=d_1+1}^{d_2-1}\{\Delta\}\Big) \times A_{d_2} \times \Big(\bigtimes_{j=d_2+1}^{d_3-1}\{\Delta\}\Big) \times \cdots \times \Big(\bigtimes_{j=d_{I-1}+1}^{d_I-1}\{\Delta\}\Big) \times A_{d_I} \times \Big(\bigtimes_{j=d_I+1}^{d}\{\Delta\}\Big), \qquad (11.16)$$

where $A_{d_i} \in \mathcal{E}_{d_i}$ for $d_i \in \mathcal{I}$. In the above we use the convention that for $d_{i+1} = d_i + 1$ we have $A_{d_i} \times \Big(\bigtimes_{j=d_i+1}^{d_{i+1}-1}\{\Delta\}\Big) \times A_{d_{i+1}} = A_{d_i} \times A_{d_{i+1}}$.

Definition 11.12 We define the $\mathcal{I}$th idiosyncratic group of coordinates $N^{idio,\mathcal{I}}$ of N as

$$N^{idio,\mathcal{I}}((0,t],A_{\mathcal{I}}) := N((0,t],A^{idio,\mathcal{I}}),$$

where

$$A_{\mathcal{I}} = \bigtimes_{d_j \in \mathcal{I}} A_{d_j}, \quad A_{d_i} \in \mathcal{E}_{d_i},\ d_i \in \mathcal{I},\ t \ge 0.$$

We will sometimes write $N^{idio,d_1,\ldots,d_I}$ in place of $N^{idio,\{d_1,\ldots,d_I\}}$.

Clearly, $N^{idio,\mathcal{I}}$ is an MPP. For example, $N^{idio,i}$ is an MPP which records idiosyncratic events occurring with regard to X^i, i.e. events that only involve X^i, so that $X_n^j = \Delta$ for $j \neq i$ at the times T_n at which these events take place. Let us note that

$$N^i = \sum_{A \subseteq \{1,\ldots,d\}, i \in A} N^{idio,A}.$$

[2] Roughly speaking, this situation parallels the case of multivariate Markov chains satisfying Condition (M) (see Section 3.1).

Likewise, for $i \neq j$, $N^{idio.i.j}$ is an MPP which records idiosyncratic events occurring with regard to X^i and X^j, i.e. events that only involve X^i and X^j simultaneously, so that $X^k_n = \Delta$ for $k \notin \{i, j\}$ at the times T_n at which these events take place. And so on.

In this volume we will not engage in a detailed study of $N^{idio.\mathcal{I}}$. Suffice it to say that such a study would parallel the study of N^i.

Relation with the Existing Literature

We end this section by briefly commenting on the relation of generalized multivariate Hawkes processes to the multivariate Hawkes processes studied previously in the literature.

In Brémaud and Massoulié (1996) a multivariate Hawkes process H was defined as a family of point processes $H_1, \dots, H_d$ that do not admit common jumps and such that the $\mathbb{F}^H$-intensity of H_k, $k = 1, \dots, d$, is given by

$$\lambda_k(t) = \nu_k + \sum_{j=1}^{d} \int_{(0,t)} h_{j,k}(t-s) H_j(ds), \quad t \geq 0.$$

Let us note that in our set-up such a process H can be viewed as a generalized multivariate Hawkes process N with mark space

$$E^\Delta = \left(\bigtimes_{k=1}^{d} \{\{1\} \cup \Delta\} \right) \setminus (\Delta, \dots, \Delta)$$

and with Hawkes intensity kernel

$$\kappa(t, dy) = \sum_{k=1}^{d} \widehat{\lambda}_k(t) \delta_{e_k}(dy), \quad t \geq 0,$$

where

$$\widehat{\lambda}_k(t) = \nu_k + \int_{(0,t) \times E^\Delta} \sum_{j=1}^{d} h_{j,k}(t-s) \mathbb{1}_{\{1\}}(x_j) N(ds, dx)$$

and $e_i = (e_{i,1}, \dots, e_{i,d})$ for $i = 1, \dots, d$, with

$$e_{i,j} := \begin{cases} 1, & j = i, \\ \Delta, & j \neq i. \end{cases}$$

It is straightforward to verify that for each k we have

$$\widehat{\lambda}_k(t) = \nu_k + \sum_{j=1}^{d} \int_{(0,t)} h_{j,k}(t-s) N^j(ds), \quad t \geq 0,$$

where N^j is the jth coordinate of N (see Definition 11.7). Thus H_j and N^j have intensities of the same form.

Liniger (2009) defined two types of multivariate Hawkes processes. The first type is the *pseudo-multivariate Hawkes* process, which is defined as an MPP N with mark space $\mathcal{X} = \mathbb{R}^d$ and with independent marks which are identically distributed according to a probability density on $\mathbb{R}^d$ given by

$$f(x) = f_1(x_1) \cdot \ldots \cdot f_d(x_d),$$

where $f_1, \ldots, f_d$ are probability densities on $\mathbb{R}$; the $\mathbb{F}^N$-intensity of this MPP of the form

$$\lambda(t) = \eta + \vartheta \int_{(-\infty,t)\times\mathbb{R}^d} w(t-s)\Big(\prod_{k=1}^{d} f_k(x_k)\Big) N(ds,dx), \quad t \geq 0,$$

where $w : \mathbb{R}_+ \to \mathbb{R}_+$ is the so-called *decay function*. Such a process is a generalized multivariate Hawkes process N with mark space

$$E^\Delta = \left(\bigtimes_{k=1}^{d} \{\mathbb{R} \cup \Delta\} \right) \setminus (\Delta, \ldots, \Delta) \tag{11.17}$$

and with Hawkes kernel of the form

$$\kappa(t,dy) = \lambda(t)\Big(\prod_{i=1}^{d} f_i(y_i)\mathbb{1}_{\mathbb{R}}(y_i)\Big) dy_1 \cdots dy_d,$$

where

$$\lambda(t) = \left(\eta + \vartheta \int_{(-\infty,t)\times E^\Delta} w(t-s)\Big(\prod_{k=1}^{d} f_k(x_k)\mathbb{1}_{\mathbb{R}}(x_k)\Big) N(ds,dx) \right).$$

The second type, defined by Liniger (2009), is the *genuine-multivariate Hawkes process* H, which is given as a family of MPPs $H_1, \ldots, H_d$ admitting no common jumps. Additionally, it is postulated that, for each $k = 1, \ldots, d$, the process H_k has real-valued marks distributed according to the probability density f_k and that the $\mathbb{F}^H$-intensity of H_k is

$$\gamma_k(t) = \eta_k + \sum_{j=1}^{d} \vartheta_{k,j} \int_{(0,t)\times\mathbb{R}} w_k(t-s) f_j(x) H_j(ds,dx).$$

Such a process is a generalized multivariate Hawkes process N with mark space given by (11.17) and with Hawkes kernel

$$\kappa(t,dy) = \sum_{k=1}^{d} \widehat{\gamma}_k(t) \left(\Big(\bigotimes_{i=1}^{k-1} \delta_\Delta(dy_i) \Big) \otimes (f_k(y_k)dy_k) \otimes \Big(\bigotimes_{j=k+1}^{d} \delta_\Delta(dy_j) \Big) \right), \quad t \geq 0,$$

where

$$\widehat{\gamma}_k(t) = \eta_k + \int_{(0,t)\times E^\Delta} \sum_{j=1}^{d} \vartheta_{k,j} w_k(t-s) f_j(x_j) \mathbb{1}_{\mathbb{R}}(x_j) N(ds,dx), \quad t \geq 0.$$

It is straightforward to verify that, for each k, we have

$$\widehat{\gamma}_k(t) = \eta_k + \sum_{j=1}^{d} \vartheta_{k,j} \int_{(0,t)\times\mathbb{R}} w_k(t-s) f_j(x) N^j(ds,dx).$$

Let us note that the generalized multivariate Hawkes process N corresponding to a genuine-multivariate Hawkes process admits no common event times. This follows from the fact that κ does not charge subsets of E^{Δ} which are associated with common event times. However, for the generalized multivariate Hawkes process N corresponding to a pseudo-multivariate Hawkes process, all event times are common. We see, then, that our concept of the generalized multivariate Hawkes process allows for the modeling of these two extreme situations regarding the common event times, and everything in between; in this regard see Example 11.21.

11.2.2 Existence of Generalized Multivariate Hawkes Process: The Canonical Probability Space Approach

In this book we present two methods for justifying the correctness of Definition 11.5. The first, presented here, refers to the concept of canonical space. The second is given in Section 11.2.4.

In order to proceed, let $(\Omega, \mathcal{F})$ be an underlying canonical space. Specifically, we take $(\Omega, \mathcal{F})$ to be the canonical space of multivariate marked point processes with marks taking values in E^{∂}. That is, Ω consists of elements $\omega = ((t_n, x_n))_{n \geq 1}$, satisfying $(t_n, x_n) \in (0, \infty] \times E^{\partial}$ and:

$$t_n \leq t_{n+1};$$
$$\text{if } t_n < \infty \text{ then } t_n < t_{n+1};$$
$$t_n = \infty \text{ iff } x_n = \partial.$$

The σ-field $\mathcal{F}$ is defined to be the smallest σ-field on Ω such that the mappings $\pi_n : \Omega \to ((0, \infty], \mathcal{B}((0, \infty])$, $\chi_n : \Omega \to (E^{\partial}, \mathcal{E}^{\partial})$ defined by

$$\pi_n(\omega) := t_n, \quad \chi_n(\omega) := x_n$$

are measurable for every n.

Note that the canonical space considered in this section agrees with the definition of a canonical space considered in Last and Brandt (1995) (see Remark 2.2.5 therein).

On this space we consider a sequence of measurable mappings

$$N = ((T_n, X_n))_{n \geq 1}, \tag{11.18}$$

given as

$$T_n(\omega) = t_n, \quad X_n(\omega) = \chi_n(\omega), \quad \omega \in \Omega.$$

Clearly, these mappings satisfy

(i) $T_n \leq T_{n+1}$ and if $T_n < +\infty$ then $T_n < T_{n+1}$,
(ii) $X_n = \partial$ iff $T_n = \infty$.

We call such an N a canonical mapping.

The following result provides the existence of a probability measure $\mathbb{P}_{\nu}$ on $(\Omega, \mathcal{F})$ such that the canonical mapping N becomes a generalized Hawkes process with a given Hawkes kernel κ, which in a unique way determines the compensator ν.

Theorem 11.13 *Consider the canonical space $(\Omega, \mathcal{F})$ and the canonical mapping N given by* (11.18)*. Let measures N and ν be associated with this canonical mapping through* (11.2) *and* (11.3), (11.4)*, respectively. Then there exists a unique probability measure $\mathbb{P}_\nu$ on $(\Omega, \mathcal{F})$ such that the measure ν is an $(\mathbb{F}^N, \mathbb{P}_\nu)$-compensator of N. Thus N is a generalized multivariate Hawkes process on $(\Omega, \mathcal{F}, \mathbb{P}_\nu)$.*

Proof We will make use of Theorem 8.2.1 in Last and Brandt (1995) with $\mathbf{X} = E^\Delta$, $\varphi = \omega$, and

$$\bar{\alpha}(\omega, dt) := \nu(\omega, dt, E^\Delta) = \mathbb{1}_{]\!]0, T_\infty(\omega)[\![}(t)\kappa(\omega, t, E^\Delta)dt, \tag{11.19}$$

from which we can conclude the assertion of the theorem. To this end, we will verify that all the assumptions of the said theorem are satisfied in the present case. As already observed, the random measure ν is $\mathbb{F}^N$-predictable. Next, let us fix $\omega \in \Omega$. Given (11.19) we see that $\bar{\alpha}$ satisfies the following equalities:

$$\bar{\alpha}(\omega, \{0\}) = 0, \quad \bar{\alpha}(\omega, \{t\}) = 0 \leq 1, t \geq 0,$$

which correspond to conditions (4.2.6) and (4.2.7) in Last and Brandt (1995), respectively. It remains to show that their condition (4.2.8) holds as well; i.e.

$$\bar{\alpha}(\omega, [\![\pi_\infty(\omega), \infty[\![) = 0, \tag{11.20}$$

where

$$\pi_\infty(\omega) := \inf\{t \geq 0 : \bar{\alpha}(\omega, (0, t]) = \infty\}.$$

To see this, we first note that (11.19) implies

$$\bar{\alpha}(\omega, [\![T_\infty(\omega), \infty[\![) = 0.$$

Thus it suffices to show that $\pi_\infty(\omega) \geq T_\infty(\omega)$. By the definition of $\bar{\alpha}$ we can write

$$\bar{\alpha}(\omega, (0, t]) = \begin{cases} \int_0^t \kappa(\omega, s, E^\Delta)ds, & t < T_\infty(\omega), \\ \int_0^{T_\infty(\omega)} \kappa(\omega, s, E^\Delta)ds, & t \geq T_\infty(\omega). \end{cases}$$

If $T_\infty(\omega) = \infty$ then clearly we have $\pi_\infty(\omega) = \infty = T_\infty(\omega)$.

Next, if $T_\infty(\omega) < \infty$ then $\lim_{t \uparrow T_\infty(\omega)} \bar{\alpha}(\omega, (0, t]) = a$. We need to consider two cases now: $a = \infty$ and $a < \infty$.

If $a = \infty$ then $\bar{\alpha}(\omega, (0, t]) = \infty$ for $t \geq T_\infty(\omega)$ and $\bar{\alpha}(\omega, (0, t]) < \infty$ for $t < T_\infty(\omega)$, in view of our assumptions imposed on κ at the beginning of this section. This implies that $\pi_\infty(\omega) = T_\infty(\omega)$.

If $a < \infty$ then $\bar{\alpha}(\omega, (0, t]) = a < \infty$ for $t \geq T_\infty(\omega)$; hence $\pi_\infty(\omega) = \infty \geq T_\infty(\omega)$. Thus $\pi_\infty(\omega) \geq T_\infty(\omega)$, which implies that (11.20) holds.

Since ω is arbitrary, we conclude that, for all $\omega \in \Omega$, conditions (4.2.6)–(4.2.8) in Last and Brandt (1995) are satisfied. So, applying Theorem 8.2.1 in Last and Brandt (1995) with $\beta = \nu$, we obtain that there exists a unique probability measure $\mathbb{P}_\nu$ such that ν is a $\mathbb{F}^N$-compensator of N under $\mathbb{P}_\nu$. □

Thanks to Theorem 11.13 we may now offer the following definition:

Definition 11.14 Let $\mathbb{P}_\nu$ be the probability measure on $(\Omega, \mathcal{F})$ whose existence was asserted in Theorem 11.13. Then the canonical mapping N on $(\Omega, \mathcal{F}, \mathbb{P}_\nu)$ is called the *canonical generalized multivariate Hawkes process*. The kernel κ is called the *(canonical) Hawkes kernel* for N.

Remark 11.15 Since $\mathbb{F}_0^N$ is a completed trivial σ-field, it is a consequence of Theorem 3.6 in Jacod (1974/75) that the compensator ν determines the law of N under $\mathbb{P}_\nu$ and, consequently, the Hawkes kernel determines the law of N under $\mathbb{P}_\nu$. Since $\mathbb{P}_\nu$ is unique, the Hawkes kernel κ determines the law of N. However, in view of Remark 11.9, the kernel κ^i may not determine the law of N^i; consequently, the kernel $\widetilde{\kappa}^i$ may not determine the law of N^i. It remains an open problem, for now, to determine sufficient conditions under which the law of N^i is determined by κ^i or by $\widetilde{\kappa}^i$. This problem is a special case of a more general problem: what are general sufficient conditions under which the characteristics of a semimartingale determine the law of the semimartingale?

Remark 11.16 Let us consider the ith coordinate of our canonical generalized multivariate Hawkes process N given by (11.18). It is interesting to see the values that the sequence $N^i = (T_k^i, Y_k^i)_{k \geq 1}$ attains in this canonical set-up. We first observe that here, for any $\omega \in \Omega$,

$$m_1^i(\omega) = \inf\{n \geq 1 : \pi_n(\omega) < \infty, \chi_n^i(\omega) \in E_i\}.$$

Thus, if ω is such that $m_1^i(\omega) < \infty$ then

$$T_1^i(\omega) = \pi_{m_1^i(\omega)}(\omega) = t_{m_1^i(\omega)}, \quad X_1^i(\omega) = \chi^i_{m_1^i(\omega)}(\omega).$$

Analogous derivations can be made for T_n^i and X_n^i, $n > 1$.

11.2.3 Examples

We will give examples of generalized multivariate Hawkes processes, which refer to our canonical measurable space $(\Omega, \mathcal{F})$.

For $t \geq 0$, $\omega = ((t_n, x_n))_{n \geq 1}$ and $A \in \mathcal{E}^\Delta$ we set

$$N(\omega, (0,t], A) := \sum_{n \geq 1} \mathbb{1}_{\{t_n \leq t, x_n \in A\}}. \tag{11.21}$$

In all the examples below we define a kernel κ of the form (11.4), with η and f properly chosen so that we may apply Theorem 11.13 to the effect that there exists a probability measure $\mathbb{P}_\nu$ on $(\Omega, \mathcal{F})$ such that the process N given by (11.21) is a Hawkes process with Hawkes kernel equal to κ. In other words, there exists a probability measure $\mathbb{P}_\nu$ on $(\Omega, \mathcal{F})$ such that the ν given in (11.3), (11.4) is the $\mathbb{F}^N$-compensator of N under $\mathbb{P}_\nu$.

For a Hawkes process N with mark space E^Δ we introduce the following notation:

$$N_t = N((0,t], E^\Delta), \quad t \geq 0.$$

Likewise we denote, for $i = 1, \ldots, d$,

$$N_t^i = N^i((0,t], E_i), \quad t \geq 0.$$

Example 11.17 Classical univariate Hawkes process We take $d = 1$ and $E_1 = \{1\}$, so that $E^\Delta = E_1 = \{1\}$. As usual, and in accordance with (11.2), we identify N with a point process $(N_t)_{t\geq 0}$. Now we take

$$\eta(t, \{1\}) = \lambda(t),$$

where λ is a positive locally integrable function, and, for $0 \leq s \leq t$, we take

$$f(t, s, 1, \{1\}) = w(t - s)$$

for some nonnegative function w defined on $\mathbb{R}_+$ (recall that $f(t, s, 1, \{1\}) = 0$ for $s > t \geq 0$). Using these objects we define κ by

$$\kappa(t, dy) = \bar{\kappa}(t)\delta_{\{1\}}(dy),$$

where

$$\bar{\kappa}(t) = \lambda(t) + \int_{(0,t)} w(t - s)dN_s.$$

In the case of a classical univariate Hawkes process, sufficient conditions under which the explosion time is a.s. infinite, i.e.

$$T_\infty = \infty, \quad \mathbb{P}_\nu\text{-a.s.},$$

are available in terms of the Hawkes kernel. Specifically, sufficient conditions for the no-explosion case are given in Bacry et al. (2013b):

$$\lambda \text{ is locally bounded} \quad \text{and} \quad \int_0^\infty w(u)du < \infty.$$

Example 11.18 Generalized multivariate Hawkes process with no common event times For simplicity we present only a generalized bivariate Hawkes process N. So $d = 2$ and the mark space is given as

$$E^\Delta = E_1^\Delta \times E_2^\Delta \setminus \{(\Delta, \Delta)\} = \{(\Delta, y_2), (y_1, \Delta), (y_1, y_2) : y_1 \in E_1, y_2 \in E_2\}.$$

Here, to define $\kappa(t, dy)$ we take

$$\eta(t, dy) = \eta_1(t, dy_1) \otimes \delta_\Delta(dy_2) + \delta_\Delta(dy_1) \otimes \eta_2(t, dy_2),$$

where η_i is a kernel from $(\mathbb{R}_+, \mathcal{B}(\mathbb{R}_+))$ to $(E_i, \mathcal{E}_i)$ and δ_Δ is a Dirac measure. Moreover, for $0 \leq s \leq t$,

$$\begin{aligned} &f(t, s, x, dy) \\ &= \Big(\varpi_{1,1}(t, s, x_1)\mathbb{1}_{E_1 \times \Delta}(x) + \varpi_{1,2}(t, s, x_2)\mathbb{1}_{\Delta \times E_2}(x)\Big)\phi_1(x, dy_1) \otimes \delta_\Delta(dy_2) \\ &\quad + \Big(\varpi_{2,1}(t, s, x_1)\mathbb{1}_{E_1 \times \Delta}(x) + \varpi_{2,2}(t, s, x_2)\mathbb{1}_{\Delta \times E_2}(x)\Big)\delta_\Delta(dy_1) \otimes \phi_2(x, dy_2). \end{aligned}$$

In the above, $\varpi_{i,j}$, $i,j = 1,2$, are appropriately regular positive functions modeling the impact of the marks of the jth coordinate N^j of N on the intensity of the marks of the ith coordinate N^i of N,[3] ϕ_i is a probability kernel from $(E^\Delta, \mathcal{E}^\Delta)$ to $(E_i, \mathcal{E}_i)$, and $x = (x_1, x_2)$. We see that with these definitions the measure $\kappa(t,dy)$ does not charge any measurable set $A \subset E_1 \times E_2$. Consequently, at any event time T_n only one coordinate of N exhibits an event. More precisely, $\kappa(t,A) = 0$ for $t \geq 0$ and any measurable set $A \subset E_1 \times E_2$. This implies that $\nu(t,A) = 0$ for $t \geq 0$, and thus

$$N((0,t],A) = 0$$

for $t \geq 0$. In particular, this implies that there are no common event times for N, and thus the coordinates of N can be considered as marked point processes given as

$$\begin{aligned}N^1((0,t],A_1) &= N((0,t],A_1 \times \{E_2,\Delta\}) = N((0,t],A_1 \times \Delta)\\ &= N^{idio,1}((0,t],A_1) = \int_{(0,t]\times E^\Delta} \mathbb{1}_{A_1\times\Delta}(x)N(ds,dx),\end{aligned}$$

for $t \geq 0$ and $A_1 \in \mathcal{E}_1$, and

$$\begin{aligned}N^2((0,t],A_2) &= N((0,t],\{E_1,\Delta\} \times A_2) = N((0,t],\Delta \times A_2)\\ &= N^{idio,2}((0,t],A_2) = \int_{(0,t]\times E^\Delta} \mathbb{1}_{\Delta\times A_2}(x)N(ds,dx),\end{aligned}$$

for $t \geq 0$ and $A_2 \in \mathcal{E}_2$.

After integrating $\kappa(t,dy)$ over $A_1 \times \{\Delta, E_2\}$ we get

$$\begin{aligned}\kappa^1(t,A_1) &= \kappa(t,A_1 \times \{\Delta,E_2\})\\ &= \eta_1(t,A_1) + \int_{(0,t)\times E^\Delta} \varpi_{1,2}(t,s,x_2)\mathbb{1}_{\Delta\times E_2}(x)\phi_1(x,A_1)N(ds,dx)\\ &\quad + \int_{(0,t)\times E^\Delta} \varpi_{1,1}(t,s,x_1)\mathbb{1}_{E_1\times\Delta}(x)\phi_1(x,A_1)N(ds,dx)\\ &= \eta_1(t,A_1) + \int_{(0,t)\times E_2} \varpi_{1,2}(t,s,x_2)\phi_1((\Delta,x_2),A_1)N^2(ds,dx_2)\\ &\quad + \int_{(0,t)\times E_1} \varpi_{1,1}(t,s,x_1)\phi_1((x_1,\Delta),A_1)N^1(ds,dx_1)\\ &= \eta_1(t,A_1) + \int_{(0,t)\times E_2} \varpi_{1,2}(t,s,x_2)\phi_1((\Delta,x_2),A_1)N^{idio,2}(ds,dx_2)\\ &\quad + \int_{(0,t)\times E_1} \varpi_{1,1}(t,s,x_1)\phi_1((x_1,\Delta),A_1)N^{idio,1}(ds,dx_1).\end{aligned} \tag{11.22}$$

Since $\phi_1(x,E_1) = 1$, we note, upon taking $A_1 = E_1$, that $\kappa(t,E_1 \times \{\Delta,E_2\})$ agrees with the intensity $\lambda_1(t)$ appearing in equation (1) in Embrechts et al. (2011), with the caveat that we are integrating over $(0,t) \times E^\Delta$ rather than over $(-\infty,t) \times E^\Delta$.

[3] The intensity of the marks of N^i is a concept analogous to the intensity of the marks of N. See Remark 11.3(iii) for the latter.

We end this example by noticing that upon taking $E_1 = E_2 = \{1\}$ we obtain the bivariate version of the multivariate Hawkes process studied in Bacry et al. (2013b). Analogous remarks apply to κ^2.

Each deterministic function $\varpi_{i,j}$ appearing in the above example is called a *decay and impact function*. Frequently, $\varpi_{i,j}$ is of the form

$$\varpi_{i,j}(t,s,x_j) = w_{i,j}(t,s)g_{i,j}(x_j).$$

The function $w_{i,j}$ is called the *decay function*, and the function $g_{i,j}$ is called the *impact function*. Typically, the function $g_{i,j}$ is a constant function.

Example 11.19 Generalized bivariate Hawkes process with no common event times and exponential decay function This is a special case of Example 11.18 with mark space

$$E^\Delta = \Big\{(1,\Delta),(\Delta,1),(1,1)\Big\}$$

and with an additional structure imposed on $\eta(t,dy)$ and $f(t,s,x,dy)$.

Specifically, we let

$$\eta(t,dy) = \eta_1(t)\delta_{(1,\Delta)}(dy) + \eta_2(t)\delta_{(\Delta,1)}(dy),$$

where $\eta_i(t) = \alpha_i + (\eta_i(0) - \alpha_i)e^{-\beta_i t}$, and for $0 \le s \le t$ we let

$$\begin{aligned} f(t,s,x,dy) = {} & e^{-\beta_1(t-s)}\left(\vartheta_{1,2}\mathbb{1}_{\Delta\times\{1\}}(x) + \vartheta_{1,1}\mathbb{1}_{\{1\}\times\Delta}(x)\right)\delta_{(1,\Delta)}(dy) \\ & + e^{-\beta_2(t-s)}\left(\vartheta_{2,1}\mathbb{1}_{\{1\}\times\Delta}(x) + \vartheta_{2,2}\mathbb{1}_{\Delta\times\{1\}}(x)\right)\delta_{(\Delta,1)}(dy). \end{aligned}$$

In particular, here we have $\varpi_{i,j}(t,s,x_j) = w_{i,j}(t,s)g_{i,j}(x_j)$ with $w_{i,j}(t,s) = e^{-\beta_i(t-s)}$ and $g_{i,j}(x_j) = \vartheta_{i,j}$.

This implies that $\nu(t,\{(1,1)\}) = 0$ for $t \ge 0$, and thus

$$N((0,t],\{(1,1)\}) = 0$$

for $t \ge 0$. In particular, the latter result implies that there are no common event times for N and thus the coordinates of N can be considered as counting processes given as follows:

$$\begin{aligned} N_t^1 &= N^1((0,t],\{1\}) = N((0,t],\{1\}\times\{1,\Delta\}) = N((0,t],\{1\}\times\Delta) \\ &= N^{idio,1}((0,t],\{1\}) = \int_{(0,t]\times E^\Delta} \mathbb{1}_{\{1\}\times\Delta}(x)N(ds,dx), \quad t \ge 0, \end{aligned}$$

and

$$\begin{aligned} N_t^2 &= N^2((0,t],\{1\}) = N((0,t],\{1,\Delta\}\times\{1\}) = N((0,t],\Delta\times\{1\}) \\ &= N^{idio,2}((0,t],\{1\}) = \int_{(0,t]\times E^\Delta} \mathbb{1}_{\Delta\times\{1\}}(x)N(ds,dx), \quad t \ge 0. \end{aligned}$$

As in Example 11.18 we have that

$$\begin{aligned} N_t^1 &= N^1((0,t],\{1\}) = N^{idio,1}((0,t],\{1\}) \\ &= \int_{(0,t]\times E^\Delta} \mathbb{1}_{\{1\}\times\Delta}(x)N(ds,dx), \quad t \ge 0, \end{aligned}$$

and

$$N_t^2 = N^2((0,t],\{1\}) = N^{idio,2}((0,t],\{1\})$$
$$= \int_{(0,t]\times E^\Delta} \mathbb{1}_{\Delta\times\{1\}}(x) N(ds,dx), \quad t \geq 0$$

are point processes. Their jump intensities, say λ^1 and λ^2, are given as, for $t \geq 0$,

$$\begin{aligned}\lambda^1(t) &:= \kappa^1(t,\{1\})\\ &= \eta_1(t) + \int_{(0,t)\times E^\Delta} e^{-\beta_1(t-s)}\Big(\vartheta_{1,2}\mathbb{1}_{\Delta\times\{1\}}(x) + \vartheta_{1,1}\mathbb{1}_{\{1\}\times\Delta}(x)\Big)N(ds,dx)\\ &= \alpha_1 + (\eta_1(0)-\alpha_1)e^{-\beta_1 t}\\ &\quad + e^{-\beta_1(t-s)}\Big(\int_{(0,t)\times\{1\}} \vartheta_{1,2}N^{idio,2}(ds,dx_2) + \int_{(0,t)\times\{1\}} \vartheta_{1,1}N^{idio,1}(ds,dx_1)\Big)\\ &= \alpha_1 + (\eta_1(0)-\alpha_1)e^{-\beta_1 t} + \int_{(0,t)} e^{-\beta_1(t-s)}\Big(\vartheta_{1,1}dN_s^1 + \vartheta_{1,2}dN_s^2\Big)\end{aligned}$$

and, similarly,

$$\lambda^2(t) = \alpha_2 + (\eta_2(0)-\alpha_2)e^{-\beta_2 t} + \int_{(0,t)} e^{-\beta_2(t-s)}\Big(\vartheta_{2,2}dN_s^2 + \vartheta_{2,1}dN_s^1\Big).$$

Example 11.20 Bivariate Poisson process This example illustrates that the bivariate Poisson process (cf. Example 5.4) is a special instance of a multivariate generalized Hawkes process. Indeed, taking $E_1 = E_2 = \{1\}$ we get the mark space

$$E^\Delta = \Big\{(1,\Delta),(\Delta,1),(1,1)\Big\}.$$

Here, the event times coincide with the jumps of the coordinates of N. In particular, the mark $(1,\Delta)$ indicates that the first coordinate N^1 jumps at a given event time but the second coordinate N^2 does not. The mark $(1,1)$ indicates that both coordinates jump at a given event time.

Next, letting $f \equiv 0$ and

$$\eta(t,dy) = \lambda_{(1,0)}\delta_{(1,\Delta)}(dy) + \lambda_{(0,1)}\delta_{(\Delta,1)}(dy) + \lambda_{(1,1)}\delta_{(1,1)}(dy),$$

where $\lambda_{(i,j)}$, $i,j \in \{0,1\}$, are positive constants, we recover the jump intensities of the bivariate Poisson process given in Bielecki et al. (2008b). In particular, the jump intensities of the coordinate N^1 and the coordinate N^2, considered as point processes, are given as

$$\kappa(t,\{1\}\times\{\Delta,1\}) = \lambda_{(1,0)} + \lambda_{(1,1)}$$

and

$$\kappa(t,\{\Delta,1\}\times\{1\}) = \lambda_{(0,1)} + \lambda_{(1,1)},$$

respectively.

Example 11.21 Generalized bivariate Hawkes process with common event times The mark space is the same as that given in Example 11.18. We will generalize Example 11.18 by allowing common event times. To this end, in order to define the kernel κ we take the kernel η to be in the form

$$\eta(t,dy) = \eta_1(t,dy_1)\otimes\delta_\Delta(dy_2) + \delta_\Delta(dy_1)\otimes\eta_2(t,dy_2) + \eta_c(t,dy_1,dy_2),$$

where η_i for $i=1,2$ are probability kernels from $(\mathbb{R}_+,\mathcal{B}(\mathbb{R}_+))$ to $(E_i,\mathcal{E}_i)$ and η_c is a probability kernel from $(\mathbb{R}_+,\mathcal{B}(\mathbb{R}_+))$ to $(E^\Delta,\mathcal{E}^\Delta)$ satisfying

$$\eta_c(t,E_1\times\Delta) = \eta_c(t,\Delta\times E_2) = 0.$$

The kernel f is given, for $0\le s\le t$, by

$$\begin{aligned} f(t,s,x,dy) = \Big(& w_{1,1}(t,s)g_{1,1}(x_1)\mathbb{1}_{E_1\times\Delta}(x) + w_{1,2}(t,s)g_{1,2}(x_2)\mathbb{1}_{\Delta\times E_2}(x) \\ &+ w_{1,c}(t,s)g_{1,c}(x)\mathbb{1}_{E_1\times E_2}(x)\Big)\phi_1(x,dy_1)\otimes\delta_\Delta(dy_2) \\ &+ \Big(w_{2,1}(t,s)g_{2,1}(x_1)\mathbb{1}_{E_1\times\Delta}(x) + w_{2,2}(t,s)g_{2,2}(x_2)\mathbb{1}_{\Delta\times E_2}(x) \\ &+ w_{2,c}(t,s)g_{2,c}(x)\mathbb{1}_{E_1\times E_2}(x)\Big)\delta_\Delta(dy_1)\otimes\phi_2(x,dy_2) \qquad (11.23) \\ &+ \Big(w_{c,1}(t,s)g_{c,1}(x_1)\mathbb{1}_{E_1\times\Delta}(x) + w_{c,2}(t,s)g_{c,2}(x_2)\mathbb{1}_{\Delta\times E_2}(x) \\ &+ w_{c,c}(t,s)g_{c,c}(x_1,x_2)\mathbb{1}_{E_1\times E_2}(x)\Big)\phi_c(x,dy_1,dy_2), \end{aligned}$$

where ϕ_i is a probability kernel from $(E^\Delta,\mathcal{E}^\Delta)$ to $(E_i,\mathcal{E}_i)$ for $i=1,2$ and ϕ_c is a probability kernel from $(E^\Delta,\mathcal{E}^\Delta)$ to $(E^\Delta,\mathcal{E}^\Delta)$ satisfying

$$\phi_c(x,E_1\times\Delta) = \phi_c(x,\Delta\times E_2) = 0.$$

The *decay functions* $w_{i,j}$ and the *impact functions* $g_{i,j}$, $i,j=1,2,c$, are appropriately regular and deterministic. Moreover, the decay functions are positive and the impact functions are nonnegative. In particular, this implies that the kernel f is deterministic and nonnegative.

We will now interpret the various terms that appear in the expressions for η and f above:

- $\eta_1(t,dy_1)\otimes\delta_\Delta(dy_2)$ represents the autonomous portion of the intensity, at time t, of the event that the marks of the coordinate N^1 take values in the set $dy_1\subset E_1$, with no marks occurring for N^2;
- $\eta_c(t,dy_1,dy_2)$ represents the autonomous portion of the intensity, at time t, of the event that the marks of both the coordinates N^1 and N^2 take values in the set $dy_1dy_2\subset E_1\times E_2$;
- the term

$$\begin{aligned} &\int_{(0,t)\times E^\Delta} w_{1,1}(t,s)g_{1,1}(x_1)\mathbb{1}_{E_1\times\Delta}(x)\phi_1(x,dy_1)\otimes\delta_\Delta(dy_2)N(ds,dx) \\ &= \int_{(0,t)\times E_1} w_{1,1}(t,s)g_{1,1}(x_1)\phi_1((x_1,\Delta),dy_1)\otimes\delta_\Delta(dy_2)N^{idio,1}(ds,dx_1) \end{aligned}$$

represents the idiosyncratic impact of the coordinate N^1 alone on the intensity, at time t, of the event that the marks of the coordinate N^1 take values in the set $dy_1 \subset E_1$, with no marks occurring for N^2;

- the term

$$\int_{(0,t)\times E^\Delta} w_{1,2}(t,s)g_{1,2}(x_2)\mathbb{1}_{\Delta\times E_2}(x)\phi_1(x,dy_1)\otimes\delta_\Delta(dy_2)N(ds,dx)$$
$$=\int_{(0,t)\times E_2} w_{1,2}(t,s)g_{1,2}(x_2)\phi_1((\Delta,x_2),dy_1)\otimes\delta_\Delta(dy_2)N^{idio,2}(ds,dx_2)$$

represents the idiosyncratic impact of the coordinate N^2 alone on the intensity, at time t, of the event that the marks of the coordinate N^1 take values in the set $dy_1 \subset E_1$, with no marks occurring for N^2;

- the term $\int_{(0,t)\times E^\Delta} w_{1,c}(t,s)g_{1,c}(x)\mathbb{1}_{E_1\times E_2}(x)\phi_1(x,dy_1)\otimes\delta_\Delta(dy_2)N(ds,dx)$ represents the joint impact of the coordinates N^1 and N^2 on the intensity, at time t, of the event that the marks of the coordinate N^1 take values in the set $dy_1 \subset E_1$, with no marks occurring for N^2;
- the term

$$\int_{(0,t)\times E^\Delta} w_{c,1}(t,s)g_{c,1}(x_1)\mathbb{1}_{E_1\times\Delta}(x)\phi_c(x,dy_1,dy_2)N(ds,dx)$$
$$=\int_{(0,t)\times E_1} w_{c,1}(t,s)g_{c,1}(x_1)\mathbb{1}_{E_1\times\Delta}(x)\phi_c((x_1,\Delta),dy_1,dy_2)N^{idio,1}(ds,dx_1)$$

represents the idiosyncratic impact of the coordinate N^1 alone on the intensity, at time t, of the event that the marks of both coordinates N^1 and N^2 take values in the set $dy_1dy_2 \subset E_1\times E_2$;

- the term $\int_{(0,t)\times E^\Delta} w_{c,c}(t,s)g_{c,c}(x_1)\mathbb{1}_{E_1\times\Delta}(x)\phi_c(x,dy_1,dy_2)N(ds,dx)$ represents the joint impact of the coordinates N^1 and N^2 on the intensity, at time t, of the event that the marks of both the coordinates N^1 and N^2 take values in the set $dy_1dy_2 \subset E_1\times E_2$.

In particular, it will be seen that the terms contributing to the occurrence of common events are $\eta_c(t,dy_1,dy_2)$ and

$$\Big(g_{c,1}(x_1)\mathbb{1}_{E_1\times\Delta}(x)+g_{c,2}(x_2)\mathbb{1}_{\Delta\times E_2}(x)+g_{c,c}(x_1,x_2)\mathbb{1}_{E_1\times E_2}(x)\Big)\phi_c(x,dy_1,dy_2).$$

Upon integrating $\kappa(t,dy)$ over $A_1\times\{\Delta,E_2\}$ we get

$$\begin{aligned}
&\kappa^1(t,A_1)=\kappa(t,A_1\times\{\Delta,E_2\})\\
&=\eta_1(t,A_1)+\eta_c(t,A_1\times E_2)\\
&\quad+\int_{(0,t)\times E^\Delta}\Big(w_{1,1}(t,s)g_{1,1}(x_1)\mathbb{1}_{E_1\times\Delta}(x)+w_{1,2}(t,s)g_{1,2}(x_2)\mathbb{1}_{\Delta\times E_2}(x)\\
&\qquad\qquad+w_{1,c}(t,s)g_{1,c}(x)\mathbb{1}_{E_1\times E_2}(x)\Big)\phi_1(x,A_1)N(ds,dx)\\
&\quad+\int_{(0,t)\times E^\Delta}\Big(w_{c,1}(t,s)g_{c,1}(x_1)\mathbb{1}_{E_1\times\Delta}(x)+w_{c,2}(t,s)g_{c,2}(x_2)\mathbb{1}_{\Delta\times E_2}(x)\\
&\qquad\qquad+w_{c,c}(t,s)g_{c,c}(x_1,x_2)\mathbb{1}_{E_1\times E_2}(x)\Big)\phi_c(x,A_1\times E_2)N(ds,dx)
\end{aligned}$$

$$
\begin{aligned}
&= \eta_1(t,A_1) + \eta_c(t,A_1\times E_2) \\
&\quad + \int_{(0,t)\times E^\Delta} w_{1,1}(t,s)g_{1,1}(x_1)\phi_1((x_1,\Delta),A_1)\mathbb{1}_{E_1\times\Delta}(x)N(ds,dx) \\
&\quad + \int_{(0,t)\times E^\Delta} w_{1,2}(t,s)g_{1,2}(x_2)\phi_1((\Delta,x_2),A_1)\mathbb{1}_{\Delta\times E_2}(x)N(ds,dx) \\
&\quad + \int_{(0,t)\times E^\Delta} w_{1,c}(t,s)g_{1,c}(x)\phi_1((x_1,x_2),A_1)\mathbb{1}_{E_1\times E_2}(x)N(ds,dx) \\
&\quad + \int_{(0,t)\times E^\Delta} w_{c,1}(t,s)g_{c,1}(x_1)\phi_c((x_1,\Delta),A_1\times E_2)\mathbb{1}_{E_1\times\Delta}(x)N(ds,dx) \\
&\quad + \int_{(0,t)\times E^\Delta} w_{c,2}(t,s)g_{c,2}(x_2)\phi_c(((\Delta,x_2),A_1\times E_2)\mathbb{1}_{\Delta\times E_2}(x)N(ds,dx) \\
&\quad + \int_{(0,t)\times E^\Delta} w_{c,c}(t,s)g_{c,c}(x_1,x_2)\phi_c((x_1,x_2),A_1\times E_2)\mathbb{1}_{E_1\times E_2}(x)N(ds,dx).
\end{aligned}
$$

After rearranging terms in the expression above we obtain

$$
\begin{aligned}
\kappa^1(t,A_1) &= \eta_1(t,A_1) + \eta_c(t,A_1\times E_2) \\
&\quad + \int_{(0,t)\times E_1} w_{1,1}(t,s)g_{1,1}(x_1)\phi_1((x_1,\Delta),A_1)N^{idio,1}(ds,dx_1) \\
&\quad + \int_{(0,t)\times E_2} w_{1,2}(t,s)g_{1,2}(x_2)\phi_1((\Delta,x_2),A_1)N^{idio,2}(ds,dx_2) \\
&\quad + \int_{(0,t)\times E} w_{1,c}(t,s)g_{1,c}(x)\phi_1((x_1,x_2),A_1)\mathbb{1}_{E_1\times E_2}(x)N(ds,dx) \\
&\quad + \int_{(0,t)\times E_1} w_{c,1}(t,s)g_{c,1}(x_1)\phi_c((x_1,\Delta),A_1\times E_2)N^{idio,1}(ds,dx_1) \\
&\quad + \int_{(0,t)\times E_2} w_{c,2}(t,s)g_{c,2}(x_2)\phi_c(((\Delta,x_2),A_1\times E_2)N^{idio,2}(ds,dx_2) \\
&\quad + \int_{(0,t)\times E} w_{c,c}(t,s)g_{c,c}(x_1,x_2)\phi_c((x_1,x_2),A_1\times E_2)\mathbb{1}_{E_1\times E_2}(x)N(ds,dx).
\end{aligned}
$$

Finally, we have

$$
\begin{aligned}
\kappa^1(t,A_1) &= \eta_1(t,A_1) + \eta_c(t,A_1\times E_2) \\
&\quad + \int_{(0,t)\times E_1} \Big(w_{1,1}(t,s)g_{1,1}(x_1)\phi_1((x_1,\Delta),A_1) \\
&\qquad\qquad + w_{c,1}(t,s)g_{c,1}(x_1)\phi_c((x_1,\Delta),A_1\times E_2)\Big) N^{idio,1}(ds,dx_1) \\
&\quad + \int_{(0,t)\times E_2} \Big(w_{1,2}(t,s)g_{1,2}(x_2)\phi_1((\Delta,x_2),A_1) \\
&\qquad\qquad + w_{c,2}(t,s)g_{c,2}(x_2)\phi_c((\Delta,x_2),A_1\times E_2)\Big) N^{idio,2}(ds,dx_2) \\
&\quad + \int_{(0,t)\times E} \Big(w_{1,c}(t,s)g_{1,c}(x)\phi_1((x_1,x_2),A_1) \\
&\qquad\qquad + w_{c,c}(t,s)g_{c,c}(x_1,x_2)\phi_c((x_1,x_2),A_1\times E_2)\Big) \mathbb{1}_{E_1\times E_2}(x)N(ds,dx).
\end{aligned}
\tag{11.24}
$$

The above formula should be compared with formula (11.22) in order to appreciate the impact of the common events on the excitation of the coordinate N^1.

Note that, upon setting $\eta_c = 0$ and $\phi_c = 0$, we may reduce the present example to Example 11.18. In particular, the absence of common event times implies that

$$\int_{(0,t)\times E} \Big(w_{1,c}(t,s) g_{1,c}(x) \phi_1((x_1,x_2),A_1) \\ + w_{c,c}(t,s) g_{c,c}(x_1,x_2) \phi_c((x_1,x_2),A_1 \times E_2) \Big) \mathbb{1}_{E_1 \times E_2}(x) N(ds,dx) = 0.$$

Thus, formula (11.24) reduces to formula (11.22).

Example 11.22 Generalized trivariate Hawkes process with common event times The mark space is

$$E^\Delta = E_1^\Delta \times E_2^\Delta \times E_3^\Delta \setminus \{(\Delta,\Delta,\Delta)\} \\ = \{(y_1,y_2,\Delta),(y_1,\Delta,y_3),(\Delta,y_2,y_3),(y_1,y_2,y_3) : y_1 \in E_1, y_2 \in E_2, y_3 \in E_3\},$$

so it is similar to that given in Example 11.18. Here we generalize Example 11.18 by allowing for common event times. To this end, in order to define the kernel κ we take the kernel η to be in the form

$$\eta(t,dy) = \eta_{\{1\}}(t,dy_1) \otimes \delta_{(\Delta,\Delta)}(dy_2,dy_3) + \eta_{\{2\}}(t,dy_2) \otimes \delta_{(\Delta,\Delta)}(dy_1,dy_3) \\ + \eta_{\{3\}}(t,dy_3) \otimes \delta_{(\Delta,\Delta)}(dy_1,dy_2) + \eta_{\{1,2\}}(t,dy_1,dy_2) \otimes \delta_\Delta(dy_3) \\ + \eta_{\{1,3\}}(t,dy_1,dy_3) \otimes \delta_\Delta(dy_2) + \eta_{\{2,3\}}(t,dy_2,dy_3) \otimes \delta_\Delta(dy_1) \\ + \eta_{\{1,2,3\}}(t,dy_1,dy_2,dy_3),$$

where $\eta_{\{i\}}$ for $i = 1,2,3$ is a probability kernel from $(\mathbb{R}_+,\mathcal{B}(\mathbb{R}_+))$ to $(E_i,\mathcal{E}_i)$, $\eta_{\{i,j\}}$ for $i,j = 1,2,3$, $i \neq j$, is a probability kernel from $(\mathbb{R}_+,\mathcal{B}(\mathbb{R}_+))$ to $((E_i \times E_j)^\Delta,(\mathcal{E}_i \otimes \mathcal{E}_j)^\Delta)$ satisfying

$$\eta_{\{i,j\}}(t,E_i \times \Delta) = \eta_{\{i,j\}}(t,\Delta \times E_j) = 0,$$

and $\eta_{\{1,2,3\}}$ is a probability kernel from $(\mathbb{R}_+,\mathcal{B}(\mathbb{R}_+))$ to $(E^\Delta,\mathcal{E}^\Delta)$ satisfying

$$\eta_{\{1,2,3\}}(t,E_1 \times \Delta \times \Delta) = \eta_{\{1,2,3\}}(t,\Delta \times E_2 \times \Delta) = \eta_{\{1,2,3\}}(t,\Delta \times \Delta \times E_3) \\ = \eta_{\{1,2,3\}}(t,E_1 \times E_2 \times \Delta) = \eta_{\{1,2,3\}}(t,\Delta \times E_2 \times E_3) \\ = \eta_{\{1,2,3\}}(t,E_1 \times \Delta \times E_3) = 0.$$

The kernel f is given, for $0 \le s \le t$, by

$$
\begin{aligned}
& f(t,s,x,dy) \\
&= \Big(\varpi_{\{1\},\{1\}}(t,s,x_1)\mathbb{1}_{E_1\times\Delta\times\Delta}(x) + \varpi_{\{1\},\{2\}}(t,s,x_2)\mathbb{1}_{\Delta\times E_2\times\Delta}(x) \\
&\quad + \varpi_{\{1\},\{3\}}(t,s,x_3)\mathbb{1}_{\Delta\times\Delta\times E_3}(x) + \varpi_{\{1\},\{1,2\}}(t,s,x_1,x_2)\mathbb{1}_{E_1\times E_2\times\Delta}(x) \\
&\quad + \varpi_{\{1\},\{1,3\}}(t,s,x_1,x_3)\mathbb{1}_{E_1\times\Delta\times E_3}(x) + \varpi_{\{1\},\{2,3\}}(t,s,x_2,x_3)\mathbb{1}_{\Delta\times E_2\times E_3}(x) \\
&\quad + \varpi_{\{1\},\{1,2,3\}}(t,s,x_1,x_2,x_3)\mathbb{1}_{E_1\times E_2\times E_3}(x)\Big)\phi_{\{1\}}(x,dy_1)\otimes\delta_{(\Delta,\Delta)}(dy_2,dy_3) \\
&\quad + \Big(\varpi_{\{2\},\{1\}}(t,s,x_1)\mathbb{1}_{E_1\times\Delta\times\Delta}(x) + \varpi_{\{2\},\{2\}}(t,s,x_2)\mathbb{1}_{\Delta\times E_2\times\Delta}(x) \\
&\quad\quad + \varpi_{\{2\},\{3\}}(t,s,x_3)\mathbb{1}_{\Delta\times\Delta\times E_3}(x) + \varpi_{\{2\},\{1,2\}}(t,s,x_1,x_2)\mathbb{1}_{E_1\times E_2\times\Delta}(x) \\
&\quad\quad + \varpi_{\{2\},\{1,3\}}(t,s,x_1,x_3)\mathbb{1}_{E_1\times\Delta\times E_3}(x) + \varpi_{\{2\},\{2,3\}}(t,s,x_2,x_3)\mathbb{1}_{\Delta\times E_2\times E_3}(x) \\
&\quad\quad + \varpi_{\{2\},\{1,2,3\}}(t,s,x_1,x_2,x_3)\mathbb{1}_{E_1\times E_2\times E_3}(x)\Big)\phi_{\{2\}}(x,dy_2)\otimes\delta_{(\Delta,\Delta)}(dy_1,dy_3) \\
&\quad + \Big(\varpi_{\{3\},\{1\}}(t,s,x_1)\mathbb{1}_{E_1\times\Delta\times\Delta}(x) + \varpi_{\{3\},\{2\}}(t,s,x_2)\mathbb{1}_{\Delta\times E_2\times\Delta}(x) \\
&\quad\quad + \varpi_{\{3\},\{3\}}(t,s,x_3)\mathbb{1}_{\Delta\times\Delta\times E_3}(x) + \varpi_{\{3\},\{1,2\}}(t,s,x_1,x_2)\mathbb{1}_{E_1\times E_2\times\Delta}(x) \\
&\quad\quad + \varpi_{\{3\},\{1,3\}}(t,s,x_1,x_3)\mathbb{1}_{E_1\times\Delta\times E_3}(x) + \varpi_{\{3\},\{2,3\}}(t,s,x_2,x_3)\mathbb{1}_{\Delta\times E_2\times E_3}(x) \\
&\quad\quad + \varpi_{\{3\},\{1,2,3\}}(t,s,x_1,x_2,x_3)\mathbb{1}_{E_1\times E_2\times E_3}(x)\Big)\phi_{\{3\}}(x,dy_3)\otimes\delta_{(\Delta,\Delta)}(dy_1,dy_2) \\
&\quad + \Big(\varpi_{\{1,2\},\{1\}}(t,s,x_1)\mathbb{1}_{E_1\times\Delta\times\Delta}(x) + \varpi_{\{1,2\},\{2\}}(t,s,x_2)\mathbb{1}_{\Delta\times E_2\times\Delta}(x) \\
&\quad\quad + \varpi_{\{1,2\},\{3\}}(t,s,x_3)\mathbb{1}_{\Delta\times\Delta\times E_3}(x) + \varpi_{\{1,2\},\{1,2\}}(t,s,x_1,x_2)\mathbb{1}_{E_1\times E_2\times\Delta}(x) \\
&\quad\quad + \varpi_{\{1,2\},\{1,3\}}(t,s,x_1,x_3)\mathbb{1}_{E_1\times\Delta\times E_3}(x) + \varpi_{\{1,2\},\{2,3\}}(t,s,x_2,x_3)\mathbb{1}_{\Delta\times E_2\times E_3}(x) \\
&\quad\quad + \varpi_{\{1,2\},\{1,2,3\}}(t,s,x_1,x_2,x_3)\mathbb{1}_{E_1\times E_2\times E_3}(x)\Big)\phi_{\{1,2\}}(x,dy_1,dy_2)\otimes\delta_{\Delta}(dy_3) \\
&\quad + \Big(\varpi_{\{1,3\},\{1\}}(t,s,x_1)\mathbb{1}_{E_1\times\Delta\times\Delta}(x) + \varpi_{\{1,3\},\{2\}}(t,s,x_2)\mathbb{1}_{\Delta\times E_2\times\Delta}(x) \\
&\quad\quad + \varpi_{\{1,3\},\{3\}}(t,s,x_3)\mathbb{1}_{\Delta\times\Delta\times E_3}(x) + \varpi_{\{1,3\},\{1,2\}}(t,s,x_1,x_2)\mathbb{1}_{E_1\times E_2\times\Delta}(x) \\
&\quad\quad + \varpi_{\{1,3\},\{1,3\}}(t,s,x_1,x_3)\mathbb{1}_{E_1\times\Delta\times E_3}(x) + \varpi_{\{1,3\},\{2,3\}}(t,s,x_2,x_3)\mathbb{1}_{\Delta\times E_2\times E_3}(x) \\
&\quad\quad + \varpi_{\{1,3\},\{1,2,3\}}(t,s,x_1,x_2,x_3)\mathbb{1}_{E_1\times E_2\times E_3}(x)\Big)\phi_{\{1,3\}}(x,dy_1,dy_3)\otimes\delta_{\Delta}(dy_2) \\
&\quad + \Big(\varpi_{\{2,3\},\{1\}}(t,s,x_1)\mathbb{1}_{E_1\times\Delta\times\Delta}(x) + \varpi_{\{2,3\},\{2\}}(t,s,x_2)\mathbb{1}_{\Delta\times E_2\times\Delta}(x) \\
&\quad\quad + \varpi_{\{2,3\},\{3\}}(t,s,x_3)\mathbb{1}_{\Delta\times\Delta\times E_3}(x) + \varpi_{\{2,3\},\{1,2\}}(t,s,x_1,x_2)\mathbb{1}_{E_1\times E_2\times\Delta}(x) \\
&\quad\quad + \varpi_{\{2,3\},\{1,3\}}(t,s,x_1,x_3)\mathbb{1}_{E_1\times\Delta\times E_3}(x) + \varpi_{\{2,3\},\{2,3\}}(t,s,x_2,x_3)\mathbb{1}_{\Delta\times E_2\times E_3}(x) \\
&\quad\quad + \varpi_{\{2,3\},\{1,2,3\}}(t,s,x_1,x_2,x_3)\mathbb{1}_{E_1\times E_2\times E_3}(x)\Big)\phi_{\{2,3\}}(x,dy_2,dy_3)\otimes\delta_{\Delta}(dy_1) \\
&\quad + \Big(\varpi_{\{1,2,3\},\{1\}}(t,s,x_1)\mathbb{1}_{E_1\times\Delta\times\Delta}(x) + \varpi_{\{1,2,3\},\{2\}}(t,s,x_2)\mathbb{1}_{\Delta\times E_2\times\Delta}(x) \\
&\quad\quad + \varpi_{\{1,2,3\},\{3\}}(t,s,x_3)\mathbb{1}_{\Delta\times\Delta\times E_3}(x) + \varpi_{\{1,2,3\},\{1,2\}}(t,s,x_1,x_2)\mathbb{1}_{E_1\times E_2\times\Delta}(x) \\
&\quad\quad + \varpi_{\{1,2,3\},\{1,3\}}(t,s,x_1,x_3)\mathbb{1}_{E_1\times\Delta\times E_3}(x) + \varpi_{\{1,2,3\},\{2,3\}}(t,s,x_2,x_3)\mathbb{1}_{\Delta\times E_2\times E_3}(x) \\
&\quad\quad + \varpi_{\{1,2,3\},\{1,2,3\}}(t,s,x_1,x_2,x_3)\mathbb{1}_{E_1\times E_2\times E_3}(x)\Big)\phi_{\{1,2,3\}}(x,dy_1,dy_2,dy_3),
\end{aligned}
$$

where $\phi_{\{i\}}$ is a probability kernel from $(E^\Delta, \mathcal{E}^\Delta)$ to $(E_i, \mathcal{E}_i)$ for $i = 1,2,3$, $\phi_{\{i,j\}}$ is a probability kernel from $(E^\Delta, \mathcal{E}^\Delta)$ to $((E_i \times E_j)^\Delta, (\mathcal{E}_i \times \mathcal{E}_j)^\Delta)$ for $i, j = 1,2,3$, $i \neq j$, satisfying

$$\phi_{\{i,j\}}(x, E_i \times \Delta) = \phi_{\{i,j\}}(x, \Delta \times E_j) = 0,$$

and $\phi_{\{1,2,3\}}$ is a probability kernel from $(E^\Delta, \mathcal{E}^\Delta)$ to $(E^\Delta, \mathcal{E}^\Delta)$ satisfying

$$\phi_{\{1,2,3\}}(x, E_1 \times E_2 \times E_3) = 1$$

or, equivalently,

$$\begin{aligned}\phi_{\{1,2,3\}}(x, E_1 \times \Delta \times \Delta) &= \phi_{\{1,2,3\}}(x, \Delta \times E_2 \times \Delta) = \phi_{\{1,2,3\}}(x, \Delta \times \Delta \times E_3) \\ &= \phi_{\{1,2,3\}}(x, E_1 \times E_2 \times \Delta) = \phi_{\{1,2,3\}}(x, \Delta \times E_2 \times E_3) \\ &= \phi_{\{1,2,3\}}(x, E_1 \times \Delta \times E_3) = 0.\end{aligned}$$

The decay and impact functions $\varpi_{\mathcal{I},\mathcal{J}}$, $\mathcal{I}, \mathcal{J} \subset \{1, \ldots, d\}$, are appropriately regular, positive, and deterministic. This implies in particular that the kernel f is deterministic and nonnegative.

Next, we will interpret the various terms that appear in the expressions for η and f above. To this end we first introduce some notation. For $\mathcal{J} = \{d_1, \ldots, d_J\} \subset \{1, \ldots, d\}$, using the same Cartesian product convention as in formula (11.16), we let

$$\begin{aligned}E^{\mathcal{J}} = &\Big(\mathop{\times}_{j=1}^{d_1-1} \{\Delta\}\Big) \times E_{d_1} \times \Big(\mathop{\times}_{j=d_1+1}^{d_2-1} \{\Delta\}\Big) \times E_{d_2} \times \Big(\mathop{\times}_{j=d_2+1}^{d_3-1} \{\Delta\}\Big) \\ &\times \cdots \times \Big(\mathop{\times}_{j=d_{J-1}+1}^{d_J-1} \{\Delta\}\Big) \times E_{d_J} \times \Big(\mathop{\times}_{j=d_J+1}^{d} \{\Delta\}\Big) \qquad (11.25)\end{aligned}$$

and

$$E_{\mathcal{J}} := \mathop{\times}_{i \in \mathcal{J}} E_i.$$

By $y_{\mathcal{J}}$ (or $x_{\mathcal{J}}$) we denote the elements of $E_{\mathcal{J}}$.

Using the above we now interpret the various terms that appear in the expressions for η and f.

- $\eta_{\mathcal{I}}(t, dy_{\mathcal{I}})$ for $\mathcal{I} \subset \{1, \ldots, d\}$ such that $|\mathcal{I}| \geq 1$ represents the autonomous portion of the intensity at time t of the event that the $\mathcal{I}$th idiosyncratic group of coordinates $N^{idio,\mathcal{I}}$ take values in the set $dy_{\mathcal{I}} \subset E_{\mathcal{I}}$;
- the term

$$\int_{(0,t)\times E^\Delta} \varpi_{\mathcal{I},\mathcal{J}}(t, s, x_{\mathcal{J}}) \mathbb{1}_{E^{\mathcal{J}}}(x) \phi_{\mathcal{I}}(x, dy_{\mathcal{I}}) \otimes \Big(\otimes_{j \in \mathcal{I}^c} \delta_\Delta(dy_j)\Big) N(ds, dx)$$

represents the idiosyncratic impact of the $\mathcal{J}$th idiosyncratic group of coordinates $N^{idio,\mathcal{J}}$ on the intensity at time t of the event that only the $\mathcal{I}$th idiosyncratic coordinate $N^{idio,\mathcal{I}}$ takes a value in the set $dy_{\mathcal{I}} \subset E_{\mathcal{I}}$.

Upon integrating $\kappa(t,dy)$ over $A_1 \times \{\Delta, E_2\} \times \{\Delta, E_3\}$ we get

$$
\begin{aligned}
&\kappa^1(t,A_1) = \kappa(t,A_1 \times \{\Delta,E_2\} \times \{\Delta,E_3\}) \\
&= \eta_{\{1\}}(t,A_1) + \eta_{\{1,2\}}(t,A_1 \times E_2) + \eta_{\{1,3\}}(t,A_1 \times E_3) + \eta_{\{1,2,3\}}(t,A_1 \times E_2 \times E_3) \\
&\quad + \int_{(0,t)\times E^\Delta} \Big[\Big(\varpi_{\{1\},\{1\}}(t,s,x_1) \mathbb{1}_{E_1\times\Delta\times\Delta}(x) + \varpi_{\{1\},\{2\}}(t,s,x_2) \mathbb{1}_{\Delta\times E_2\times\Delta}(x) \\
&\qquad\qquad + \varpi_{\{1\},\{3\}}(t,s,x_3) \mathbb{1}_{\Delta\times\Delta\times E_3}(x) + \varpi_{\{1\},\{1,2\}}(t,s,x_1,x_2) \mathbb{1}_{E_1\times E_2\times\Delta}(x) \\
&\qquad\qquad + \varpi_{\{1\},\{1,3\}}(t,s,x_1,x_3) \mathbb{1}_{E_1\times\Delta\times E_3}(x) + \varpi_{\{1\},\{2,3\}}(t,s,x_2,x_3) \mathbb{1}_{\Delta\times E_2\times E_3}(x) \\
&\qquad\qquad + \varpi_{\{1\},\{1,2,3\}}(t,s,x_1,x_2,x_3) \mathbb{1}_{E_1\times E_2\times E_3}(x) \Big) \phi_{\{1\}}(x,A_1) \\
&\quad + \Big(\varpi_{\{1,2\},\{1\}}(t,s,x_1) \mathbb{1}_{E_1\times\Delta\times\Delta}(x) + \varpi_{\{1,2\},\{2\}}(t,s,x_2) \mathbb{1}_{\Delta\times E_2\times\Delta}(x) \\
&\qquad\qquad + \varpi_{\{1,2\},\{3\}}(t,s,x_3) \mathbb{1}_{\Delta\times\Delta\times E_3}(x) + \varpi_{\{1,2\},\{1,2\}}(t,s,x_1,x_2) \mathbb{1}_{E_1\times E_2\times\Delta}(x) \\
&\qquad\qquad + \varpi_{\{1,2\},\{1,3\}}(t,s,x_1,x_3) \mathbb{1}_{E_1\times\Delta\times E_3}(x) + \varpi_{\{1,2\},\{2,3\}}(t,s,x_2,x_3) \mathbb{1}_{\Delta\times E_2\times E_3}(x) \\
&\qquad\qquad + \varpi_{\{1,2\},\{1,2,3\}}(t,s,x_1,x_2,x_3) \mathbb{1}_{E_1\times E_2\times E_3}(x) \Big) \phi_{\{1,2\}}(x,A_1 \times E_2) \\
&\quad + \Big(\varpi_{\{1,3\},\{1\}}(t,s,x_1) \mathbb{1}_{E_1\times\Delta\times\Delta}(x) + \varpi_{\{1,3\},\{2\}}(t,s,x_2) \mathbb{1}_{\Delta\times E_2\times\Delta}(x) \\
&\qquad\qquad + \varpi_{\{1,3\},\{3\}}(t,s,x_3) \mathbb{1}_{\Delta\times\Delta\times E_3}(x) + \varpi_{\{1,3\},\{1,2\}}(t,s,x_1,x_2) \mathbb{1}_{E_1\times E_2\times\Delta}(x) \\
&\qquad\qquad + \varpi_{\{1,3\},\{1,3\}}(t,s,x_1,x_3) \mathbb{1}_{E_1\times\Delta\times E_3}(x) + \varpi_{\{1,3\},\{2,3\}}(t,s,x_2,x_3) \mathbb{1}_{\Delta\times E_2\times E_3}(x) \\
&\qquad\qquad + \varpi_{\{1,3\},\{1,2,3\}}(t,s,x_1,x_2,x_3) \mathbb{1}_{E_1\times E_2\times E_3}(x) \Big) \phi_{\{1,3\}}(x,A_1 \times E_3) \\
&\quad + \Big(\varpi_{\{1,2,3\},\{1\}}(t,s,x_1) \mathbb{1}_{E_1\times\Delta\times\Delta}(x) + \varpi_{\{1,2,3\},\{2\}}(t,s,x_2) \mathbb{1}_{\Delta\times E_2\times\Delta}(x) \\
&\qquad\qquad + \varpi_{\{1,2,3\},\{3\}}(t,s,x_3) \mathbb{1}_{\Delta\times\Delta\times E_3}(x) + \varpi_{\{1,2,3\},\{1,2\}}(t,s,x_1,x_2) \mathbb{1}_{E_1\times E_2\times\Delta}(x) \\
&\qquad\qquad + \varpi_{\{1,2,3\},\{1,3\}}(t,s,x_1,x_3) \mathbb{1}_{E_1\times\Delta\times E_3}(x) + \varpi_{\{1,2,3\},\{2,3\}}(t,s,x_2,x_3) \mathbb{1}_{\Delta\times E_2\times E_3}(x) \\
&\qquad\qquad + \varpi_{\{1,2,3\},\{1,2,3\}}(t,s,x_1,x_2,x_3) \mathbb{1}_{E_1\times E_2\times E_3}(x) \Big) \phi_{\{1,2,3\}}(x,A_1 \times E_2 \times E_3) \Big] N(ds,dx).
\end{aligned}
$$

Rearranging the terms in the above equation we obtain

$$
\begin{aligned}
\kappa^1(t,A_1) &= \eta_{\{1\}}(t,A_1) + \eta_{\{1,2\}}(t,A_1 \times E_2) + \eta_{\{1,3\}}(t,A_1 \times E_3) \\
&\quad + \eta_{\{1,2,3\}}(t,A_1 \times E_2 \times E_3) \\
&\quad + \int_{(0,t)\times E^\Delta} \Big(\varpi_{\{1\},\{1\}}(t,s,x_1) \phi_{\{1\}}(x,A_1) \\
&\qquad\qquad + \varpi_{\{1,2\},\{1\}}(t,s,x_1) \phi_{\{1,2\}}(x,A_1 \times E_2) \\
&\qquad\qquad + \varpi_{\{1,3\},\{1\}}(t,s,x_1) \phi_{\{1,3\}}(x,A_1 \times E_3) \\
&\qquad\qquad + \varpi_{\{1,2,3\},\{1\}}(t,s,x_1) \phi_{\{1,2,3\}}(x,A_1 \times E_2 \times E_3) \Big) \\
&\qquad\qquad \times \mathbb{1}_{E_1\times\Delta\times\Delta}(x) N(ds,dx) \\
&\quad + \int_{(0,t)\times E^\Delta} \Big(\varpi_{\{1\},\{2\}}(t,s,x_2) \phi_{\{1\}}(x,A_1) \\
&\qquad\qquad + \varpi_{\{1,2\},\{2\}}(t,s,x_2) \phi_{\{1,2\}}(x,A_1 \times E_2) \\
&\qquad\qquad + \varpi_{\{1,3\},\{2\}}(t,s,x_2) \phi_{\{1,3\}}(x,A_1 \times E_3)
\end{aligned}
$$

$$
\begin{aligned}
&\quad + \varpi_{\{1.2.3\}.\{2\}}(t,s,x_2)\phi_{\{1.2.3\}}(x,A_1 \times E_2 \times E_3)\Big) \\
&\quad \times \mathbb{1}_{\Delta \times E_2 \times \Delta}(x) N(ds,dx) \\
&+ \int_{(0.t)\times E^\Delta} \Big(\varpi_{\{1\}.\{3\}}(t,s,x_3)\phi_{\{1\}}(x,A_1) \\
&\quad + \varpi_{\{1.2\}.\{3\}}(t,s,x_3)\phi_{\{1.2\}}(x,A_1 \times E_2) \\
&\quad + \varpi_{\{1.3\}.\{3\}}(t,s,x_3)\phi_{\{1.3\}}(x,A_1 \times E_3) \\
&\quad + \varpi_{\{1.2.3\}.\{3\}}(t,s,x_3)\phi_{\{1.2.3\}}(x,A_1 \times E_2 \times E_3)\Big) \\
&\quad \times \mathbb{1}_{\Delta \times \Delta \times E_3}(x) N(ds,dx) \\
&+ \int_{(0.t)\times E^\Delta} \Big(\varpi_{\{1\}.\{1.2\}}(t,s,x_1,x_2)\phi_{\{1\}}(x,A_1) \\
&\quad + \varpi_{\{1.2\}.\{1.2\}}(t,s,x_1,x_2)\phi_{\{1.2\}}(x,A_1 \times E_2) \\
&\quad + \varpi_{\{1.3\}.\{1.2\}}(t,s,x_1,x_2)\phi_{\{1.3\}}(x,A_1 \times E_3) \\
&\quad + \varpi_{\{1.2.3\}.\{1.2\}}(t,s,x_1,x_2)\phi_{\{1.2.3\}}(x,A_1 \times E_2 \times E_3)\Big) \\
&\quad \times \mathbb{1}_{E_1 \times E_2 \times \Delta}(x) N(ds,dx) \\
&+ \int_{(0.t)\times E^\Delta} \Big(\varpi_{\{1\}.\{1.3\}}(t,s,x_1,x_3)\phi_{\{1\}}(x,A_1) \\
&\quad + \varpi_{\{1.2\}.\{1.3\}}(t,s,x_1,x_3)\phi_{\{1.2\}}(x,A_1 \times E_2) \\
&\quad + \varpi_{\{1.3\}.\{1.3\}}(t,s,x_1,x_3)\phi_{\{1.3\}}(x,A_1 \times E_3) \\
&\quad + \varpi_{\{1.2.3\}.\{1.3\}}(t,s,x_1,x_3)\phi_{\{1.2.3\}}(x,A_1 \times E_2 \times E_3)\Big) \\
&\quad \times \mathbb{1}_{E_1 \times \Delta \times E_3}(x) N(ds,dx) \\
&+ \int_{(0.t)\times E^\Delta} \Big(\varpi_{\{1\}.\{2.3\}}(t,s,x_2,x_3)\phi_{\{1\}}(x,A_1) \\
&\quad + \varpi_{\{1.2\}.\{2.3\}}(t,s,x_2,x_3)\phi_{\{1.2\}}(x,A_1 \times E_2) \\
&\quad + \varpi_{\{1.3\}.\{2.3\}}(t,s,x_2,x_3)\phi_{\{1.3\}}(x,A_1 \times E_3) \\
&\quad + \varpi_{\{1.2.3\}.\{2.3\}}(t,s,x_2,x_3)\phi_{\{1.2.3\}}(x,A_1 \times E_2 \times E_3)\Big) \\
&\quad \times \mathbb{1}_{\Delta \times E_2 \times E_3}(x) N(ds,dx) \\
&+ \int_{(0.t)\times E^\Delta} \Big(\varpi_{\{1\}.\{1.2.3\}}(t,s,x_1,x_2,x_3)\phi_{\{1\}}(x,A_1) \\
&\quad + \varpi_{\{1.2\}.\{1.2.3\}}(t,s,x_1,x_2,x_3)\phi_{\{1.2\}}(x,A_1 \times E_2) \\
&\quad + \varpi_{\{1.3\}.\{1.2.3\}}(t,s,x_1,x_2,x_3)\phi_{\{1.3\}}(x,A_1 \times E_3) \\
&\quad + \varpi_{\{1.2.3\}.\{1.2.3\}}(t,s,x_1,x_2,x_3)\phi_{\{1.2.3\}}(x,A_1 \times E_2 \times E_3)\Big) \\
&\quad \times \mathbb{1}_{E_1 \times E_2 \times E_3}(x) N(ds,dx).
\end{aligned}
$$

We can rewrite this equation as follows:

$$
\begin{aligned}
\kappa^1(t,A_1) &= \eta_{\{1\}}(t,A_1) + \eta_{\{1,2\}}(t,A_1 \times E_2) + \eta_{\{1,3\}}(t,A_1 \times E_3) \\
&\quad + \eta_{\{1,2,3\}}(t,A_1 \times E_2 \times E_3) \\
&\quad + \int_{(0.t)\times E_1} h^1_{\{1\}}(t,s,x_1,A_1) N^{idio.\{1\}}(ds,dx)
\end{aligned}
$$

$$
\begin{aligned}
&+\int_{(0,t)\times E_2} h^1_{\{2\}}(t,s,x_2,A_1)N^{idio,\{2\}}(ds,dx)\\
&+\int_{(0,t)\times E_3} h^1_{\{3\}}(t,s,x_3,A_1)N^{idio,\{3\}}(ds,dx)\\
&+\int_{(0,t)\times E_1\times E_2} h^1_{\{1,2\}}(t,s,x_1,x_2,A_1)N^{idio,\{1,2\}}(ds,dx)\\
&+\int_{(0,t)\times E_1\times E_3} h^1_{\{1,3\}}(t,s,x_1,x_3,A_1)N^{idio,\{1,3\}}(ds,dx)\\
&+\int_{(0,t)\times E_2\times E_3} h^1_{\{2,3\}}(t,s,x_2,x_3,A_1)N^{idio,\{2,3\}}(ds,dx)\\
&+\int_{(0,t)\times E_1\times E_2\times E_3} h^1_{\{1,2,3\}}(t,s,x_1,x_2,x_3,A_1)N^{idio,\{1,2,3\}}(ds,dx),
\end{aligned}
$$

where

$$
\begin{aligned}
h^1_{\{1\}}(t,s,x_1,A_1) &= \varpi_{\{1\},\{1\}}(t,s,x_1)\phi_{\{1\}}(x,A_1)\\
&\quad+\varpi_{\{1,2\},\{1\}}(t,s,x_1)\phi_{\{1,2\}}(x,A_1\times E_2)\\
&\quad+\varpi_{\{1,3\},\{1\}}(t,s,x_1)\phi_{\{1,3\}}(x,A_1\times E_3)\\
&\quad+\varpi_{\{1,2,3\},\{1\}}(t,s,x_1)\phi_{\{1,2,3\}}(x,A_1\times E_2\times E_3),\\
h^1_{\{2\}}(t,s,x_2,A_1) &= \varpi_{\{1\},\{2\}}(t,s,x_2)\phi_{\{1\}}(x,A_1)\\
&\quad+\varpi_{\{1,2\},\{2\}}(t,s,x_2)\phi_{\{1,2\}}(x,A_1\times E_2)\\
&\quad+\varpi_{\{1,3\},\{2\}}(t,s,x_2)\phi_{\{1,3\}}(x,A_1\times E_3)\\
&\quad+\varpi_{\{1,2,3\},\{2\}}(t,s,x_2)\phi_{\{1,2,3\}}(x,A_1\times E_2\times E_3),\\
h^1_{\{3\}}(t,s,x_3,A_1) &= \varpi_{\{1\},\{3\}}(t,s,x_3)\phi_{\{1\}}(x,A_1)\\
&\quad+\varpi_{\{1,2\},\{3\}}(t,s,x_3)\phi_{\{1,2\}}(x,A_1\times E_2)\\
&\quad+\varpi_{\{1,3\},\{3\}}(t,s,x_3)\phi_{\{1,3\}}(x,A_1\times E_3)\\
&\quad+\varpi_{\{1,2,3\},\{3\}}(t,s,x_3)\phi_{\{1,2,3\}}(x,A_1\times E_2\times E_3),\\
h^1_{\{1,2\}}(t,s,x_1,x_2,A_1) &= \varpi_{\{1\},\{1,2\}}(t,s,x_1,x_2)\phi_{\{1\}}(x,A_1)\\
&\quad+\varpi_{\{1,2\},\{1,2\}}(t,s,x_1,x_2)\phi_{\{1,2\}}(x,A_1\times E_2)\\
&\quad+\varpi_{\{1,3\},\{1,2\}}(t,s,x_1,x_2)\phi_{\{1,3\}}(x,A_1\times E_3)\\
&\quad+\varpi_{\{1,2,3\},\{1,2\}}(t,s,x_1,x_2)\phi_{\{1,2,3\}}(x,A_1\times E_2\times E_3),\\
h^1_{\{1,3\}}(t,s,x_1,x_3,A_1) &= \varpi_{\{1\},\{1,3\}}(t,s,x_1,x_3)\phi_{\{1\}}(x,A_1)\\
&\quad+\varpi_{\{1,2\},\{1,3\}}(t,s,x_1,x_3)\phi_{\{1,2\}}(x,A_1\times E_2)\\
&\quad+\varpi_{\{1,3\},\{1,3\}}(t,s,x_1,x_3)\phi_{\{1,3\}}(x,A_1\times E_3)\\
&\quad+\varpi_{\{1,2,3\},\{1,3\}}(t,s,x_1,x_3)\phi_{\{1,2,3\}}(x,A_1\times E_2\times E_3),
\end{aligned}
$$

$$
\begin{aligned}
h^1_{\{2,3\}}(t,s,x_2,x_3,A_1) &= \varpi_{\{1\},\{2,3\}}(t,s,x_2,x_3)\phi_{\{1\}}(x,A_1) \\
&\quad + \varpi_{\{1,2\},\{2,3\}}(t,s,x_2,x_3)\phi_{\{1,2\}}(x,A_1 \times E_2) \\
&\quad + \varpi_{\{1,3\},\{2,3\}}(t,s,x_2,x_3)\phi_{\{1,3\}}(x,A_1 \times E_3) \\
&\quad + \varpi_{\{1,2,3\},\{2,3\}}(t,s,x_2,x_3)\phi_{\{1,2,3\}}(x,A_1 \times E_2 \times E_3), \\
h^1_{\{1,2,3\}}(t,s,x_1,x_2,x_3,A_1) &= \varpi_{\{1\},\{1,2,3\}}(t,s,x_1,x_2,x_3)\phi_{\{1\}}(x,A_1) \\
&\quad + \varpi_{\{1,2\},\{1,2,3\}}(t,s,x_1,x_2,x_3)\phi_{\{1,2\}}(x,A_1 \times E_2) \\
&\quad + \varpi_{\{1,3\},\{1,2,3\}}(t,s,x_1,x_2,x_3)\phi_{\{1,3\}}(x,A_1 \times E_3) \\
&\quad + \varpi_{\{1,2,3\},\{1,2,3\}}(t,s,x_1,x_2,x_3)\phi_{\{1,2,3\}}(x,A_1 \times E_2 \times E_3).
\end{aligned}
$$

11.2.4 Construction of a Generalized Multivariate Hawkes Process via Poisson Thinning

Earlier in this chapter, in Theorem 11.13, we proved the existence of a generalized multivariate Hawkes process N with a given compensator. This result referred to the canonical space. Here, we provide a construction of the generalized multivariate Hawkes process N (with deterministic kernels η and f) which does not refer to our canonical space. In particular, formula (11.21) is not used to construct the measure N.

The construction provided here adapts the thinning method (see e.g. Chapter 9 in Last and Brandt (1995), Brémaud and Massoulié (1996), Massoulié (1998), Liniger (2009), or Chapter 6 in Çınlar (2011)).

We will construct a generalized multivariate Hawkes process admitting the Hawkes kernel given as

$$
\kappa(t,dy) = \bar{\eta}(t,y)Q_1(dy) + \left(\int_{(0,t)\times E^\Delta} \bar{f}(t,s,x,y)N(ds,dx)\right) Q_2(dy), \tag{11.26}
$$

where $\bar{\eta} : \mathbb{R}_+ \times E^\Delta \to \mathbb{R}_+$, $\bar{f} : \mathbb{R}_+ \times \mathbb{R}_+ \times E^\Delta \times E^\Delta \to \mathbb{R}_+$ are deterministic functions, Q_1 and Q_2 are finite measures on E^Δ such that for every $T > 0$, $s \geq 0$, and $x \in E^\Delta$ it holds that

$$
\int_0^T \int_{E^\Delta} \bar{\eta}(t,y)Q_1(dy)dt < \infty
$$

and

$$
\int_s^{s+T} \int_{E^\Delta} \bar{f}(t,s,x,y)Q_2(dy)dt < \infty.
$$

Without loss of generality, we will take Q_1 and Q_2 to be probability measures.

Let us take a probability space $(\Omega, \mathcal{F}, \mathbb{P})$ that is rich enough to support two independent Poisson random measures M_1 and M_2 on $\mathbb{R}_+ \times \mathbb{R}_+ \times E^\Delta$ such that, for $i = 1,2$, the measure M_i admits an intensity measure, say μ_i, of the form

$$
\mu_i(dt,da,dy) = dt \otimes da \otimes Q_i(dy). \tag{11.27}
$$

That is, for $B \in \mathcal{B}(\mathbb{R}_+ \times \mathbb{R}_+) \otimes \mathcal{E}^\Delta$ we have

$$\mu_i(B) = \mathbb{E}(M_i(B)), \quad i = 1,2.$$

We endow the space $(\Omega, \mathcal{F}, \mathbb{P})$ with the filtration $\mathbb{F} = (\mathcal{F}_t)_{t \geq 0}$ generated by M_1 and M_2. That is,

$$\mathbb{F} = \mathbb{F}^{M_1} \vee \mathbb{F}^{M_2},$$

where $\mathbb{F}^{M_i} = (\mathcal{F}_t^{M_i})_{t \geq 0}$, $i = 1,2$, with

$$\mathcal{F}_t^{M_i} = \sigma\Big(M_i((a,b] \times B) : a < b \leq t,\ B \in \mathcal{B}(\mathbb{R}_+) \otimes \mathcal{B}(\mathbb{R}_+) \otimes \mathcal{E}^\Delta\Big), \quad t \geq 0$$

(see (1.6.1) in Last and Brandt (1995)). As usual we denote by $\mathcal{P}$ the predictable σ-field on $\Omega \times \mathbb{R}_+$ and we let $\widetilde{\mathcal{P}} = \mathcal{P} \otimes \mathcal{E}^\Delta$.

Let us recall that, for $k = 1,2$, the Poisson measure M_k with intensity measure μ_k can be constructed as

$$M_k(ds, da, dy) = \sum_{i \geq 1} \delta_{(S_i^k, A_i^k, Y_i^k)}(ds, da, dy), \tag{11.28}$$

where $(S_i^k, A_i^k, Y_i^k)_{i \geq 1}$ is an appropriately defined sequence of random variables (see e.g. Theorem 6.4 in Peszat and Zabczyk (2007)).

We note that for any $T > 0$ the random number, say $\mathfrak{I}(T)$, of indices i for which $0 < A_i^1 \leq \bar{\eta}(S_i^1, Y_i^1)$ and $S_i^1 \leq T$, is a random variable and is finite with probability 1. Indeed, the expected value of $\mathfrak{I}(T)$ is given by

$$\mathbb{E}\mathfrak{I}(T) = \mathbb{E}\int_0^T \int_{\mathbb{R}_+ \times E^\Delta} \mathbb{1}_{(0, \bar{\eta}(s,y)]}(a) M_1(ds, da, dy).$$

Since

$$\begin{aligned}
&\mathbb{E}\int_0^T \int_{\mathbb{R}_+ \times E^\Delta} \mathbb{1}_{(0, \bar{\eta}(s,y)]}(a) M_1(ds, da, dy) \\
&\quad = \mathbb{E}\int_0^T \int_{E^\Delta} \int_{\mathbb{R}_+} \mathbb{1}_{(0, \bar{\eta}(s,y)]}(a) da Q_1(dy) ds \\
&\quad = \int_0^T \int_{E^\Delta} \bar{\eta}(s,y) Q_1(dy) ds < \infty,
\end{aligned}$$

we see that $\mathbb{E}\mathfrak{I}(T)$ is finite and thus $\mathfrak{I}(T)$ is finite with probability 1. We can use this observation to construct a random field $\bar{\lambda}$ and a sequence $(T_n, X_n)_{n \geq 1}$, which are essential ingredients of our construction of a generalized Hawkes process.

First, let us define

$$i_0 = 0, \quad i_m = \inf\{i > i_{m-1} : 0 < A_i^1 \leq \bar{\eta}(S_i^1, Y_i^1)\}, \quad m \geq 1.$$

For $m \geq 1$ we define a sequence of random variables

$$(T_m^1, Z_m^1) = \begin{cases} (S_{i_m}^1, Y_{i_m}^1) & \text{if } i_m < \infty, \\ (\infty, \partial) & \text{if } i_m = \infty. \end{cases} \tag{11.29}$$

We will show now that the sequence $(T_m^1, Z_m^1)_{m=1}^\infty$ can be ordered with respect to the first, i.e. the time, coordinate. Observe that with probability 1 we have

$$\begin{aligned} &\operatorname{card}\{m : T_m^1 \in [0,T]\} \\ &\quad = \operatorname{card}\{i : S_i^1 \in [0,T] \text{ and } 0 < A \leq \bar{\eta}(S_i^1, Y_i^1)\} = \mathfrak{I}(T) < \infty. \end{aligned}$$

In effect, for almost all ω, the set $\{T_m^1(\omega) : T_m^1(\omega) \in [0,T]\}$ does not have any finite accumulation point, so it can be ordered. Thus, one can construct order statistics for the sequence of random variables T_m^1.

So we may, and we do, assume that the sequence $(T_i^1, Z_i^1)_{i=1}^\infty$ is ordered with respect to the first coordinate, i.e.

$$T_1^1 < T_2^1 < \cdots < T_n^1 < \cdots$$

We will now define a process $\bar{\lambda}$ that will play an important role in our construction; to this end we will proceed inductively.

Step 1: We put

$$T_1 = T_1^1, \quad X_1 = Z_1^1, \quad \bar{\lambda}(t,y) = \lambda_1(t,y) \quad \text{and} \quad \lambda_1(t,y) = 0$$

for $(t,y) \in]\!]0, T_1]\!] \times E^\Delta$.

Step 2: Let λ_2 be a random field on $\mathbb{R}_+ \times E^\Delta$ defined by

$$\lambda_2(t,y) := \bar{f}(t, T_1, X_1, y) \mathbb{1}_{\{t > T_1\}}.$$

Consider points (S_l^2, A_l^2, Y_l^2) determining the Poisson random measure M_2, which additionally satisfy

$$0 < A_l^2 \leq \lambda_2(S_l^2, Y_l^2) \quad \text{and} \quad S_l^2 > T_1. \tag{11.30}$$

Since for every T we have

$$\begin{aligned} &\mathbb{E}\left(\int_{T_1}^{T_1+T} \int_{\mathbb{R}_+ \times E^\Delta} \mathbb{1}_{\{(0,\lambda_2(s,y)]\}}(a) M_2(ds,da,dy) \Big| \mathcal{F}_{T_1} \right) \\ &\quad = \mathbb{E}\left(\int_{T_1}^{T_1+T} \int_{\mathbb{R}_+ \times E^\Delta} \mathbb{1}_{\{(0,\lambda_2(s,y)]\}}(a) da Q_2(dy) ds \Big| \mathcal{F}_{T_1} \right) \\ &\quad = \int_{T_1}^{T_1+T} \int_{E^\Delta} \lambda_2(s,y) Q_2(dy) ds \\ &\quad = \int_{T_1}^{T_1+T} \int_{E^\Delta} \bar{f}(s, T_1, X_1, y) Q_2(dy) ds < \infty, \end{aligned}$$

we may order the sequence $(S_l^2, A_l^2, Y_l^2)_{l \geq 1}$ with respect to S_l^2 in a way analogous to the way in which we ordered the sequence $(T_i^1, Z_i^1)_{i \geq 1}$ with respect to T_i^1.

Let us take the first triple (S_l^2, A_l^2, Y_l^2) in this ordered sequence that satisfies (11.30), i.e. the triple with minimal S_l^2, and denote it by $(S_{l_1}^2, A_{l_1}^2, Y_{l_1}^2)$. Now, if $S_{l_1}^2 < T_2^1$ we take

$$T_2 = S_{l_1}^2, \quad X_2 = Y_{l_1}^2,$$

otherwise, we take

$$T_2 = T_2^1, \quad X_2 = Z_2^1.$$

Having these relations we define $\bar{\lambda}$ on $]\!]T_1, T_2]\!] \times E^\Delta$ by

$$\bar{\lambda}(t,y) = \lambda_2(t,y).$$

Step 3: Suppose that we have a sequence $(T_i, X_i)_{i=1}^n$ and $\bar{\lambda}$ given on $]\!]0, T_n]\!] \times E^\Delta$ by

$$\bar{\lambda}(t,y) = \lambda_n(t,y),$$

where

$$\lambda_n(t,y) = \sum_{i=1}^{n-1} \bar{f}(t, T_i, X_i, y) \mathbb{1}_{\{t > T_i\}}.$$

Now we find the point (T_{n+1}, X_{n+1}) and extend $\bar{\lambda}$ to $]\!]0, T_{n+1}]\!] \times E^\Delta$. Given λ^n we consider points (S_l^2, A_l^2, Y_l^2) which satisfy

$$A_l^2 \leq \lambda_n(S_l^2, Y_l^2) \quad \text{and} \quad S_l^2 > T_n.$$

By analogous arguments to those used before, these points can be ordered with respect to the "time" component S_l^2. Let us take the first such triple, (S_l^2, A_l^2, Y_l^2), i.e. the triple with minimal S_l^2, and denote it by $(S_{l_n}^2, A_{l_n}^2, Y_{l_n}^2)$. Let us take

$$j(T_n) = \min\{j : T_j^1 > T_n\}.$$

If $S_{l_n}^2 < T_{j(T_n)}^1$, we put

$$T_{n+1} = S_{l_n}^2, \quad X_{n+1} = Y_{l_n}^2;$$

otherwise, we take

$$T_{n+1} = T_{j(T_n)}^1, \quad X_{n+1} = Z_{j(T_n)}^1.$$

Having T_{n+1}, we define $\bar{\lambda}$ on $]\!]T_n, T_{n+1}]\!] \times E^\Delta$ by

$$\bar{\lambda}(t,y) = \lambda_{n+1}(t,y) = \lambda_n(t,y) + \bar{f}(t, T_n, X_n, y) \mathbb{1}_{\{t > T_n\}}.$$

Step 4: Let $T_\infty = \lim_{n \to \infty} T_n$ and on $]\!]T_\infty, \infty[\![$ put

$$\bar{\lambda}(t,y) = 0.$$

Thus in the above construction we have defined a random field $\bar{\lambda}$ on $\mathbb{R}_+ \times E^\Delta$ such that

$$\bar{\lambda}(t,y) = \mathbb{1}_{[\![0,T_\infty[\![}(t) \sum_{n:T_n<t} \bar{f}(t,T_n,X_n,y)$$

and a sequence $(T_n,X_n)_{n\geq 1}$ giving a random measure N on $(\mathbb{R}_+ \times E^\Delta, \mathcal{B}(\mathbb{R}_+)\otimes\mathcal{E}^\Delta)$ of the form

$$N(dt,dx) = \sum_{n\geq 1} \delta_{(T_n,X_n)}(dt,dx)\mathbb{1}_{\{T_n<\infty\}}.$$

Since T_n is an increasing sequence of times the sequence $(T_n,X_n)_{n\geq 1}$ is a marked point process. Observe that for $t < T_\infty$ there exists n such that $T_n < t \leq T_{n+1}$, and the sequence $(T_i)_{i=1}^n$ consists of the first n elements of the monotone sequence constructed out of the set $\left\{T_1^1,\ldots,T_n^1,S_{j_1}^2,\ldots S_{j_n}^2\right\}$, i.e. it consists of all T_m^1 such that $T_m^1 \leq t$ and all $S_{j_k}^2$ such that $S_{j_k}^2 < t$. Therefore $\bar{\lambda}$ constructed in such a way satisfies

$$\bar{\lambda}(t,y) = \mathbb{1}_{[\![0,T_\infty[\![}(t)\left(\sum_{m:T_m^1<t} \bar{f}(t,T_m^1,Z_m^1,y) + \sum_{k:S_{j_k}^2<t} \bar{f}(t,S_{j_k}^2,Y_{j_k}^2,y)\right).$$

We may write this equation in the form

$$\bar{\lambda}(t,y) = \mathbb{1}_{[\![0,T_\infty[\![}(t)\Bigg(\sum_{i:S_i^1<t} \bar{f}(t,S_i^1,Y_i^1,y)\mathbb{1}_{\{0<A_i^1\leq\bar{\eta}(S_i^1,Y_i^1)\}}$$
$$+ \sum_{j:S_j^2<t} \bar{f}(t,S_j^2,Y_j^2,y)\mathbb{1}_{\{0<A_j^2\leq\bar{\lambda}(S_j^2,Y_j^2)\}}\Bigg).$$

Hence $\bar{\lambda}$ is a solution of

$$\bar{\lambda}(t,y) = \mathbb{1}_{[\![0,T_\infty[\![}(t)\int_{(0,t)\times\mathbb{R}_+\times E^\Delta} \bar{f}(t,s,x,y)\mathbb{1}_{[\![0,T_\infty[\![}(s)$$
$$\times\left[\mathbb{1}_{(0,\bar{\eta}(s,x)]}(a)M_1(ds,da,dx)+\mathbb{1}_{(0,\bar{\lambda}(s,x)]}(a)M_2(ds,da,dx)\right],\ y\in E^\Delta,\ t\geq 0. \tag{11.31}$$

Using this $\bar{\lambda}$ we now define a random measure $\bar{N}$ on $(\mathbb{R}_+ \times E^\Delta, \mathcal{B}(\mathbb{R}_+)\otimes\mathcal{E}^\Delta)$ by

$$\bar{N}(dt,dy) = \mathbb{1}_{[\![0,T_\infty[\![}(t)\int_{\mathbb{R}_+} \mathbb{1}_{(0,\bar{\eta}(t,y)]}(a)M_1(dt,da,dy)$$
$$+\mathbb{1}_{[\![0,T_\infty[\![}(t)\int_{\mathbb{R}_+} \mathbb{1}_{(0,\bar{\lambda}(t,y)]}(a)M_2(dt,da,dy). \tag{11.32}$$

Then, for any $\omega \in \Omega$, $s > 0$, and $B \in \mathcal{E}^\Delta$, we have

$$\begin{aligned}
&\bar{N}(\omega, [0,s] \times B) \\
&\quad := \int_0^s \mathbb{1}_{[\![0,T_\infty(\omega)[\![}(t) \int_{\mathbb{R}_+ \times B} \mathbb{1}_{(0,\bar{\eta}(t,y)]}(a) M_1(\omega, dt, da, dy) \\
&\qquad + \int_0^s \mathbb{1}_{[\![0,T_\infty(\omega)[\![}(t) \int_{\mathbb{R}_+ \times B} \mathbb{1}_{(0,\bar{\lambda}(t,y)]}(a) M_2(\omega, dt, da, dy) \\
&\quad = \sum_{l \geq 1} \mathbb{1}_{\{S_l^1(\omega) < T_\infty(\omega),\ S_l^1(\omega) \leq s,\ 0 < A_l^1(\omega) \leq \bar{\eta}(S_l^1(\omega), Y_l^1(\omega)),\ Y_l^1(\omega) \in B\}} \\
&\qquad + \sum_{l \geq 1} \mathbb{1}_{\{S_{j_l}^2(\omega) < T_\infty(\omega),\ S_{j_l}^2(\omega) \leq s,\ 0 < A_{j_l}^2(\omega) \leq \bar{\lambda}(S_{j_l}^2(\omega), Y_{j_l}^1(\omega)),\ Y_{j_l}^1(\omega) \in B\}}.
\end{aligned}$$

Now note that for a fixed ω the sequence $(T_i(\omega), X_i(\omega))_{i \geq 1}$ is constructed by choosing points $(S_l^1(\omega), Y_l^1(\omega))_{l \geq 1}$ and $(S_{j_l}^2(\omega), Y_{j_l}^2(\omega))_{l \geq 1}$ out of the sequences $(S_m^1(\omega), A_m^1(\omega), Y_m^1(\omega))_{m \geq 1}$ and $(S_{j_k}^2(\omega), A_{j_k}^2(\omega), Y_{j_k}^2(\omega))_{k \geq 1}$ satisfying

$$0 < A_l^1(\omega) \leq \bar{\eta}(S_l^1(\omega), Y_l^1(\omega))$$

and

$$0 < A_{j_l}^2(\omega) \leq \bar{\lambda}(S_{j_l}^2(\omega), Y_{j_l}^2(\omega)).$$

Hence, by the construction of the sequence $(T_i, X_i)_{i \geq 1}$ and the definition of N, we see that

$$\bar{N}(\omega, [0,s] \times B) = \sum_{i \geq 1} \mathbb{1}_{\{T_i(\omega) \leq s, T_i(\omega) < T_\infty(\omega), X_i(\omega) \in B\}} = N(\omega, [0,s] \times B),$$

so that $N = \bar{N}$. Thus in view of (11.32) we can write (11.31) in the form

$$\bar{\lambda}(t,y) = \mathbb{1}_{[\![0,T_\infty[\![}(t) \int_{(0,t) \times E^\Delta} \bar{f}(t,s,x,y) N(ds, dx). \tag{11.33}$$

Since $N = \bar{N}$, using (11.32) and the fact that M_i admits the intensity measure μ_i given by (11.27) we conclude that, for any nonnegative $\widetilde{\mathcal{P}}$-measurable function h on $\Omega \times \mathbb{R}_+ \times E^\Delta$,

$$\begin{aligned}
&\mathbb{E} \int_{\mathbb{R}_+ \times E^\Delta} h(t,y) N(dt, dy) \\
&\quad = \mathbb{E} \int_{\mathbb{R}_+ \times \mathbb{R}_+ \times E^\Delta} \mathbb{1}_{[\![0,T_\infty[\![}(t) h(t,y) \mathbb{1}_{(0,\bar{\eta}(t,y)]}(a) M_1(dt, da, dy) \\
&\qquad + \mathbb{E} \int_{\mathbb{R}_+ \times \mathbb{R}_+ \times E^\Delta} \mathbb{1}_{[\![0,T_\infty[\![}(t) h(t,y) \mathbb{1}_{(0,\bar{\lambda}(t,y)]}(a) M_2(dt, da, dy) \\
&\quad = \mathbb{E} \int_{\mathbb{R}_+ \times E^\Delta} h(t,y) \mathbb{1}_{[\![0,T_\infty[\![}(t) [\bar{\eta}(t,y) Q_1(dy) + \bar{\lambda}(t,y) Q_2(dy)] dt.
\end{aligned}$$

In view of Theorem II.1.8 in Jacod and Shiryaev (2003) this shows that the $\mathbb{F}$-compensator of N is given by

$$\bar{\nu}(t, dy) = \mathbb{1}_{[\![0,T_\infty[\![}(t)(\bar{\eta}(t,y) Q_1(dy) + \bar{\lambda}(t,y) Q_2(dy)) = \mathbb{1}_{[\![0,T_\infty[\![} \kappa(t, dy),$$

with κ as given in (11.26); the second equality comes from (11.33). Since $\mathbb{F}^N \subset \mathbb{F}$, from (11.26) and (11.5) we see that κ is $\mathbb{F}^N$-predictable and thus $\bar{\nu}$ is an $\mathbb{F}^N$-compensator of N.

Consequently, N is a generalized multivariate Hawkes process with Hawkes kernel defined in (11.26).

Remark 11.23 Our goal was to construct a generalized multivariate Hawkes process, so we took E^Δ as a mark space; however, in the construction above we could take arbitrary $\mathcal{X}$ as a mark space to construct a generalized Hawkes process.

11.3 Markovian Aspects of a Generalized Multivariate Hawkes Process

An important class of Hawkes processes considered in the literature is that of Hawkes processes for which the Hawkes kernel is given in terms of exponential decay functions. See e.g. Çınlar (2011), Oakes (1975), and Zhu (2013). One interesting and useful aspect of such processes is that they can be extended to Markov processes, a feature that we term the Markovian aspects of Hawkes processes.

In this section we will discuss the Markovian aspects of generalized multivariate Hawkes processes with Hawkes kernel given in terms of exponential decay functions.

To simplify the presentation we consider the bivariate case only and, for concreteness, we let $E_1 = E_2 = \{1\}$ so that

$$E^\Delta = \{(1,\Delta),(\Delta,1),(1,1)\}.$$

Next, we take

$$\eta(t,dy) := \eta_1(t)\delta_{(1,\Delta)}(dy) + \eta_2(t)\delta_{(\Delta,1)}(dy) + \eta_c(t)\delta_{(1,1)}(dy),$$

where

$$\eta_i(t) := \alpha_i + (\eta_i(0) - \alpha_i)e^{-\beta_i t}, \quad i \in \{1,2,c\},$$

and $\alpha_i, \eta_i(0), \beta_i$ are nonnegative constants.

Finally, we assume that, for $0 \le s \le t$, the kernel f is as in Example 11.21, with decay functions $w_{i,j}$ in the exponential form

$$w_{i,j}(t,s) = e^{-\beta_i(t-s)}, \quad i,j \in \{1,2,c\},$$

with constant (nonnegative) impact functions

$$\begin{aligned}
g_{1,1}(x_1) &= \vartheta_{1,1}, \quad g_{1,2}(x_2) = \vartheta_{1,2}, \quad g_{1,c}(x) = \vartheta_{1,c},\\
g_{2,1}(x_1) &= \vartheta_{2,1}, \quad g_{2,2}(x_2) = \vartheta_{2,2}, \quad g_{2,c}(x) = \vartheta_{2,c},\\
g_{c,1}(x_1) &= \vartheta_{c,1}, \quad g_{c,2}(x_2) = \vartheta_{c,2}, \quad g_{c,c}(x) = \vartheta_{c,c},
\end{aligned}$$

and with Dirac kernels

$$\phi_1(x,dy_1) = \delta_1(dy_1), \quad \phi_2(x,dy_2) = \delta_1(dy_2), \quad \phi_c(x,dy_1,dy_2) = \delta_{(1,1)}(dy_1,dy_2).$$

Thus, the kernel f is of the form

$$
\begin{aligned}
&f(t,s,x,dy) \\
&= e^{-\beta_1(t-s)}\Big(\vartheta_{1,1}\mathbb{1}_{\{1\}\times\Delta}(x)+\vartheta_{1,2}\mathbb{1}_{\Delta\times\{1\}}(x)+\vartheta_{1,c}\mathbb{1}_{\{1\}\times\{1\}}(x)\Big)\delta_{(1,\Delta)}(dy) \\
&\quad + e^{-\beta_2(t-s)}\Big(\vartheta_{2,1}\mathbb{1}_{\{1\}\times\Delta}(x)+\vartheta_{2,2}\mathbb{1}_{\Delta\times\{1\}}(x)+\vartheta_{2,c}\mathbb{1}_{\{1\}\times\{1\}}(x)\Big)\delta_{(\Delta,1)}(dy) \\
&\quad + e^{-\beta_c(t-s)}\Big(\vartheta_{c,1}\mathbb{1}_{\{1\}\times\Delta}(x)+\vartheta_{c,2}\mathbb{1}_{\Delta\times\{1\}}(x)+\vartheta_{c,c}\mathbb{1}_{\{1\}\times\{1\}}(x)\Big)\delta_{(1,1)}(dy).
\end{aligned}
\tag{11.34}
$$

Using the above specifications, by Theorem 11.13 we end up with a Hawkes process N whose Hawkes kernel κ is of the form

$$
\kappa(t,dy) = \lambda_t^1\delta_{(1,\Delta)}(dy) + \lambda_t^2\delta_{(\Delta,1)}(dy) + \lambda_t^c\delta_{(1,1)}(dy), \tag{11.35}
$$

where, for $i = 1,2,c$, we have $\lambda_0^i := \eta_i(0)$ and

$$
\begin{aligned}
\lambda_t^i &= \alpha_i + (\lambda_0^i - \alpha_i)e^{-\beta_i t} \\
&\quad + \int_{(0,t)\times E^\Delta} e^{-\beta_i(t-u)}\Big(\vartheta_{i,1}\mathbb{1}_{\{1\}\times\Delta}(x) \\
&\qquad\qquad + \vartheta_{i,2}\mathbb{1}_{\Delta\times\{1\}}(x) + \vartheta_{i,c}\mathbb{1}_{\{1\}\times\{1\}}(x)\Big)N(du,dx).
\end{aligned}
\tag{11.36}
$$

The coordinates of N (see (11.8)) reduce here to counting (point) processes

$$
N_t^1 = N^1((0,t],\{1\}) = N((0,t],\{1\}\times\{1,\Delta\}) \tag{11.37}
$$

and

$$
N_t^2 = N^2((0,t],\{1\}) = N((0,t],\{1,\Delta\}\times\{1\}). \tag{11.38}
$$

It is straightforward to verify (upon appropriate integration of the kernel κ) that the $\mathbb{F}^N$-intensity of the process N^i, say $\widehat{\lambda}^i$, is given as

$$
\widehat{\lambda}_t^i = \lambda_t^i + \lambda_t^c, \quad t \geq 0, i = 1,2. \tag{11.39}
$$

Now let us consider a bivariate counting process $\widetilde{N} := (N^1, N^2)$. Note that we may, and we will, identify the process $\widetilde{N}$ with our bivariate generalized Hawkes process N in the following way:

$$
T_0 = 0, \quad T_n = \inf\Big\{t > T_{n-1} : \Delta\widetilde{N}_t \neq (0,0)\Big\},
$$

and, for $i = 1,2$,

$$
X_n^i = \begin{cases} 1 & \text{if } \Delta N_{T_n}^i = 1, \\ \Delta & \text{if } \Delta N_{T_n}^i = 0. \end{cases}
$$

Also, note that we may, and we will, identify the process $\widetilde{N}$ with a random measure on $\mathbb{R}_+ \times \widetilde{E}$, with $\widetilde{E} = \{(1,0),(0,1),(1,1)\}$, whose $\mathbb{F}^{\widetilde{N}}$–compensator is given by

$$
\widetilde{\nu}(dt,dy) = \mathbb{1}_{]\!]0,T_\infty[\![}\widetilde{\kappa}(t,dy)dt, \tag{11.40}
$$

where

$$\widetilde{\kappa}(t,dy) = \lambda_t^1 \delta_{(1,0)}(dy) + \lambda_t^2 \delta_{(0,1)}(dy) + \lambda_t^c \delta_{(1,1)}(dy). \tag{11.41}$$

Thus, we may slightly abuse the terminology and call $\widetilde{N}$ a generalized bivariate Hawkes process.

Let

$$\bar{N}_t^c = [N^1, N^2]_u, \quad \bar{N}_t^1 = N_u^1 - \bar{N}_u^c, \quad \bar{N}_t^2 = N_u^2 - \bar{N}_u^c,$$

where $[N^1, N^2]$ is the square bracket of N^1, N^2 (see Appendix A). Then, for $i = 1, 2, c$, the equality (11.36) can be written as

$$\begin{aligned} \lambda_t^i &= \alpha_i + (\lambda_0^i - \alpha_i) e^{-\beta_i t} \\ &\quad + \int_{(0,t)} e^{-\beta_i (t-u)} \left(\vartheta_{i,1} d\bar{N}_u^1 + \vartheta_{i,2} d\bar{N}_u^2 + \vartheta_{i,c} d\bar{N}_u^c \right) \end{aligned} \tag{11.42}$$

for $t \geq 0$. This follows from the fact that $[N^1, N^2]$ counts common jumps of N^1 and N^2, so for $i = 1, 2$ the process $\bar{N}^i$ counts the idiosyncratic jumps of N^i, i.e. the jumps that do not occur simultaneously with the jumps of N^j, $j \neq i$. In particular, expression (11.42) allows us to give an interpretation of the parameters $\vartheta_{i,j}$, $i, j \in \{1, 2, c\}$: the parameter $\vartheta_{i,j}$ describes the impact of jump of the process $\bar{N}^j$ on the intensity of $\bar{N}^i$.

From (11.42) and (11.41) we see that for any $t > 0$ the quantity $\widetilde{\nu}(dt, dy)$ given in (11.40) depends on the entire path of $\widetilde{N}$ until time t. Thus, by Theorem 4 in He and Wang (1984), the generalized bivariate Hawkes process $\widetilde{N}$ is not a Markov process.

However, as we will show, the process

$$Z = (\lambda_t^1, \lambda_t^2, \lambda_t^c, N_t^1, N_t^2)_{t \geq 0}$$

is a Markov process, with generator A acting on $C_c^\infty(\mathbb{R}_+^5)$ given by

$$\begin{aligned} &Av(\lambda^1, \lambda^2, \lambda^c, n^1, n^2) \\ &= \beta_1(\alpha_1 - \lambda^1) \frac{\partial}{\partial \lambda^1} v(\lambda^1, \lambda^2, \lambda^c, n^1, n^2) + \beta_2(\alpha_2 - \lambda^2) \frac{\partial}{\partial \lambda^2} v(\lambda^1, \lambda^2, \lambda^c, n^1, n^2) \\ &\quad + \beta_c(\alpha_c - \lambda^c) \frac{\partial}{\partial \lambda^c} v(\lambda^1, \lambda^2, \lambda^c, n^1, n^2) \\ &\quad + (v(\lambda^1 + \vartheta_{1,1}, \lambda^2 + \vartheta_{2,1}, \lambda^c + \vartheta_{c,1}, n^1 + 1, n^2) - v(\lambda^1, \lambda^2, \lambda^c, n^1, n^2))\lambda^1 \\ &\quad + (v(\lambda^1 + \vartheta_{1,2}, \lambda^2 + \vartheta_{2,2}, \lambda^c + \vartheta_{c,2}, n^1, n^2 + 1) - v(\lambda^1, \lambda^2, \lambda^c, n^1, n^2))\lambda^2 \\ &\quad + (v(\lambda^1 + \vartheta_{1,c}, \lambda^2 + \vartheta_{2,c}, \lambda^c + \vartheta_{c,c}, n^1 + 1, n^2 + 1) - v(\lambda^1, \lambda^2, \lambda^c, n^1, n^2))\lambda^c. \end{aligned} \tag{11.43}$$

In order to briefly sketch the proof of the Markov property of Z we proceed as follows.

First note that (11.42) can be written as

$$\lambda_t^i - \alpha_i = e^{-\beta_i t} \left(\lambda_0^i - \alpha_i + \int_{(0,t)} e^{\beta_i u} \left(\vartheta_{i,1} d\bar{N}_u^1 + \vartheta_{i,2} d\bar{N}_u^2 + \vartheta_{i,c} d\bar{N}_u^c \right) \right).$$

Hence, using stochastic integration by parts, one can show that λ^i can be represented as

$$\lambda_t^i = \lambda_0^i + \int_0^t \beta_i(\alpha_i - \lambda_u^i)du + \int_{(0,t)} \left(\vartheta_{i,1}d\bar{N}_u^1 + \vartheta_{i,2}d\bar{N}_u^2 + \vartheta_{i,c}d\bar{N}_u^c\right).$$

This and (11.41) imply that the process Z is an $\mathbb{F}^Z$-semimartingale with characteristics (with respect to the cut-off function $h(x) = x\mathbb{1}_{|x|<1}$)

$$B_t = \int_0^t b_u du, \quad C_t = 0_{4\times 4},$$

$$\nu(dt, dy_1, dy_2, dy_c, dz_1, dz_2) = \nu_t(dy_1, dy_2, dy_c, dz_1, dz_2)dt,$$

where

$$b_t := (\beta_1(\alpha_1 - \lambda_{t-}^1),\ \beta_2(\alpha_2 - \lambda_{t-}^2),\ \beta_c(\alpha_c - \lambda_{t-}^c),\ 0,\ 0)'$$

and

$$\begin{aligned}
&\nu_t(dy_1, dy_2, dy_c, dz_1, dz_2) \\
&\quad := \lambda_{u-}^1 \delta_{(\vartheta_{1.1}, \vartheta_{2.1}, \vartheta_{c.1}, 1, 0)}(dy_1, dy_2, dy_c, dz_1, dz_2) \\
&\qquad + \lambda_{u-}^2 \delta_{(\vartheta_{1.2}, \vartheta_{2.2}, \vartheta_{c.2}, 0, 1)}(dy_1, dy_2, dy_c, dz_1, dz_2) \\
&\qquad + \lambda_{u-}^c \delta_{(\vartheta_{1.c}, \vartheta_{2.c}, \vartheta_{c.c}, 1, 1)}(dy_1, dy_2, dy_c, dz_1, dz_2).
\end{aligned} \tag{11.44}$$

This, by Theorem II.2.42 of Jacod and Shiryaev (2003), implies that for any function $v \in C_b^2(\mathbb{R}^5)$ the process M^v given as

$$\begin{aligned}
M_t^v &= v(Z_t) - \int_0^t Av(Z_u)du \\
&= v(\lambda_t^1, \lambda_t^2, \lambda_t^c, N_t^1, N_t^2) - \int_0^t Av(\lambda_u^1, \lambda_u^2, \lambda_u^c, N_u^1, N_u^2)du, \quad t \geq 0
\end{aligned}$$

is an $\mathbb{F}^Z$-local martingale. Hence, for any $v \in C_c^\infty(\mathbb{R}^5)$, the process defined above is a martingale under $\mathbb{P}$ since v and Av are bounded, which follows from the fact that $v \in C_c^\infty(\mathbb{R}^5)$ has compact support and thus the local martingale M^v is a martingale for such a v. Consequently, the process Z solves the martingale problem for (A, ρ), where ρ is the deterministic initial distribution of Z , i.e. $\rho(dz) = \delta_{Z_0}(dz)$.

We will now verify that Z is a Markov process, with generator A given in (11.43), using Theorem 4.4.1 in Ethier and Kurtz (1986).

For this, we first observe that the parameters determining A, i.e.

$$\mathcal{I} = \{1, \ldots, 5\}, \quad \mathcal{J} = \emptyset, \quad a = 0, \quad \alpha = 0, \quad c = 0, \quad \gamma = 0, \quad m = 0,$$

$$b = (\alpha_1\beta_1, \alpha_2\beta_2, \alpha_3\beta_3, 0, 0)', \quad \beta = \text{diag}(-\beta_1, -\beta_2, -\beta_c, 0, 0),$$

$$\mu_1 = \delta_{(\vartheta_{1.1}, \vartheta_{2.1}, \vartheta_{c.1}, 1, 0)}, \quad \mu_2 = \delta_{(\vartheta_{1.2}, \vartheta_{2.2}, \vartheta_{c.2}, 0, 1)},$$

$$\mu_3 = \delta_{(\vartheta_{1.c}, \vartheta_{2.c}, \vartheta_{c.c}, 1, 1)}, \quad \mu_4 = \mu_5 = 0,$$

are admissible in the sense of Definition 2.6 in Duffie et al. (2003).

Thus, invoking Theorem 2.7 in Duffie et al. (2003) we conclude that there exists a unique regular affine semigroup $(P_t)_{t\geq 0}$ with infinitesimal generator A given by (11.43). Hence, there exists a unique regular affine process with generator A and with transition function P defined by $(P_t)_{t\geq 0}$. Since A is a generator of a regular affine process, it satisfies the Hille–Yosida conditions (see Theorem 1.2.6 in Ethier and Kurtz (1986)) relative to the Banach space $B(\mathbb{R}^5)$ – real-valued, bounded, and measurable functions on $\mathbb{R}^5$. Moreover, from Corollary 1.1.6 in Ethier and Kurtz (1986) it follows that A is a closed operator. Now, using Theorem 4.4.1 in Ethier and Kurtz (1986) we obtain that Z is a Markov process with generator A. Moreover, P is the transition function of Z.

We end this section by pointing to an additional feature of the process Z, which presents itself under some aditional assumptions. To this end we fix $i \in \{1,2\}$ and assume that the parameters of λ^k, $k = 1,2,c$, satisfy

$$\vartheta_{i.j} = \vartheta_{c.j} = 0, \quad j \neq i,\ j \in \{1,2\},$$

$$\beta_i = \beta_c \quad \text{and} \quad \vartheta_{i.c} + \vartheta_{c.c} = \vartheta_{i.i} + \vartheta_{c.i}.$$

Then, using analogous arguments to those employed earlier in this section (also compare Example 11.32, Section 6.6 in Çınlar (2011)), we conclude that for $i = 1,2$ the process $Y^i := (\lambda^i_t + \lambda^c_t, N^i_t)_{t\geq 0}$ is a Markov process in the filtration $\mathbb{F}^Z$. We term this feature of Z the Markovian consistency of Z related to the coordinate N^i.

11.4 Hawkes Structures

In this section we define and discuss Hawkes structures in the context of canonical generalized multivariate Hawkes processes. The applications of Hawkes structures will be presented in Section 12.4.

We proceed by stating an important definition that underlies the concept of Hawkes structures.

Definition 11.24 We say that a marked point process M is a *quasi-Hawkes process* if it is a coordinate of some canonical generalized multivariate Hawkes process, say K, defined on a probability space $(\Omega, \mathcal{F}, \mathbb{P}_\nu)$. We call K a ν-constructor of M.

Clearly, each coordinate N^i, $i = 1, \ldots, d$, of a canonical generalized multivariate Hawkes process N is a quasi-Hawkes process.

Let M^i, $i = 1, \ldots, d$, be quasi-Hawkes processes with common d-variate ν-constructor, say K, with Hawkes kernel $\bar{\kappa}$ on $(\Omega, \mathcal{F}, \mathbb{P}_\nu)$. Let the mark space of M^i be $(E_i, \mathcal{E}_i)$. Therefore, M^i has kernel κ^i such that (see (11.13))

$$\kappa^i(t, A) = \bar{\kappa}(t, A^i), \quad t \geq 0,\ A \in \mathcal{E}_i.$$

Hence by (11.4) we have

$$\begin{aligned}\kappa^i(t,A) &= \bar{\eta}(t,A^i) + \int_{(0,t)\times\mathcal{X}} \bar{f}(t,s,x,A^i)K(ds,dx) \\ &= \widehat{\kappa}^i(K,t,A), \quad t\geq 0,\ A\in\mathcal{E}_i, \qquad (11.45)\end{aligned}$$

for some appropriate mapping $\widehat{\kappa}^i$, $i=1,2,\dots,d$.

Therefore M^i admits the Hawkes kernel uniquely determined by the function $\widehat{\kappa}^i$. These considerations lead to

Definition 11.25 Let N be a canonical generalized multivariate Hawkes process defined on a probability space $(\Omega,\mathcal{F},\mathbb{P}_\gamma)$ with mark space E^Δ given by (11.7), with Hawkes kernel κ, and with $(\mathbb{F}^N,\mathbb{P}_\gamma)$-compensator γ. We say that the process N is a *Hawkes structure* for the family $\{M^i,\ i=1,\dots,d\}$ with kernels determined by the mappings $\left\{\widehat{\kappa}^i,\ i=1,\dots,d\right\}$, if the kernel κ of N satisfies

$$\kappa(t,A^i) = \widehat{\kappa}^i(N,t,A), \quad t\geq 0,\ A\in\mathcal{E}_i. \qquad (11.46)$$

Clearly, the common ν-constructor N on $(\Omega,\mathcal{F},\mathbb{P}_\nu)$ for M^i, $i=1,\dots,d$, is an example of the Hawkes structure for the family M^i, $i=1,\dots,d$.

The construction of a Hawkes structure N for M^i, $i=1,\dots,d$, boils down to solving the system of equations (11.46) for the Hawkes kernel κ. Once κ is obtained, one constructs a kernel γ as follows (see (11.3)):

$$\gamma(dt,dy) := \mathbb{1}_{]\!]0,T_\infty[\![}(t)\kappa(t,dy)dt; \qquad (11.47)$$

then one can use Theorem 11.13 to obtain the probability space $(\Omega,\mathcal{F},\mathbb{P}_\gamma)$, so that N is a canonical generalized multivariate Hawkes process with $(\mathbb{F}^N,\mathbb{P}_\gamma)$-compensator γ; thus N is a Hawkes structure for M^i, $i=1,\dots,d$.

We will now present two relevant examples.

Example 11.26 We take $E_1=E_2=\{1\}$, so that

$$E^\Delta = \{(1,\Delta),(\Delta,1),(1,1)\}.$$

Now let M^i, $i=1,2$, be quasi-Hawkes processes with common constructor K, such that the Hawkes kernel κ^i of M^i is given as

$$\begin{aligned}\kappa^i(t,\{1\}) = \bar{\alpha}_i + (\bar{\alpha}-\bar{\lambda}_i)^{-\beta t} &+ \int_{(0,t)\times\{1\}} e^{-\beta(t-s)}\bar{\vartheta}_{i,1}K^{idio,1}(ds,dx_1) \\ &+ \int_{(0,t)\times\{1\}} e^{-\beta(t-s)}\bar{\vartheta}_{i,2}K^{idio,2}(ds,dx_2) \qquad (11.48)\\ &+ \int_{(0,t)\times\{1\}\times\{1\}} e^{-\beta(t-s)}\bar{\vartheta}_{i,c}K(ds,dx),\end{aligned}$$

where $\bar{\alpha}_i$, $\bar{\lambda}_i$, $\bar{\vartheta}_{i,1}$, $\bar{\vartheta}_{i,2}$, $\bar{\vartheta}_{i,c}$, $\bar{\beta}$ are positive constants. From the definition of $K^{idio,i}$ it follows that for $t\geq 0$ we have $\kappa^i(t,\{1\}) = \widehat{\kappa}^i(K,t,\{1\})$, where $\widehat{\kappa}^i(K,t,\{1\})$ is given by

$$\widehat{\kappa}^i(K,t,\{1\}) = \bar{\alpha}_i + (\bar{\alpha} - \bar{\lambda}_i)^{-\beta t} + \int_{(0,t)\times\{1\}\times\{\Delta\}} e^{-\beta(t-s)}\bar{\vartheta}_{i,1}K(ds,dx)$$
$$+ \int_{(0,t)\times\{\Delta\}\times\{1\}} e^{-\beta(t-s)}\bar{\vartheta}_{i,2}K(ds,dx)$$
$$+ \int_{(0,t)\times\{1\}\times\{1\}} e^{-\beta(t-s)}\bar{\vartheta}_{i,c}K(ds,dx).$$

We will now show how to construct multiple Hawkes structures for the family $\{M^1,M^2\}$. To this end let us consider a Hawkes process N whose Hawkes kernel κ is given by (11.35) and (11.36) with $\beta_1 = \beta_2 = \beta_c = \beta$. Thus for $\{1\}^i$(see (11.9) for the definition of A^i for a set A) we have

$$\begin{aligned}\kappa(t,\{1\}^i) &= \lambda_t^i + \lambda_t^c \\ &= (\alpha_i + \alpha_c) + (\lambda_0^i - \alpha_i)e^{-\beta t} + (\lambda_0^c - \alpha_c)e^{-\beta t} \\ &\quad + \int_{(0,t)\times E^\Delta} e^{-\beta(t-u)}\Big(\vartheta_{i,1}\mathbb{1}_{\{1\}\times\Delta}(x) \\ &\qquad + \vartheta_{i,2}\mathbb{1}_{\Delta\times\{1\}}(x) + \vartheta_{i,c}\mathbb{1}_{\{1\}\times\{1\}}(x)\Big)N(du,dx). \\ &\quad + \int_{(0,t)\times E^\Delta} e^{-\beta(t-u)}\Big(\vartheta_{c,1}\mathbb{1}_{\{1\}\times\Delta}(x) \\ &\qquad + \vartheta_{c,2}\mathbb{1}_{\Delta\times\{1\}}(x) + \vartheta_{c,c}\mathbb{1}_{\{1\}\times\{1\}}(x)\Big)N(du,dx),\end{aligned}$$

which may be also written in the form

$$\begin{aligned}\kappa(t,\{1\}^i) &= (\alpha_i + \alpha_c) + (\lambda_0^i + \lambda_0^c - \alpha_i - \alpha_c)e^{-\beta t} \\ &\quad + \int_{(0,t)\times E^\Delta} e^{-\beta(t-u)}\Big((\vartheta_{i,1} + \vartheta_{c,1})\mathbb{1}_{\{1\}\times\Delta}(x) + (\vartheta_{i,2} + \vartheta_{c,2})\mathbb{1}_{\Delta\times\{1\}}(x) \\ &\qquad + (\vartheta_{i,c} + \vartheta_{c,c})\mathbb{1}_{\{1\}\times\{1\}}(x)\Big)N(du,dx) \\ &= (\alpha_i + \alpha_c) + (\lambda_0^i + \lambda_0^c - \alpha_i - \alpha_c)e^{-\beta t} \\ &\quad + \int_{(0,t)\times\{1\}} e^{-\beta(t-u)}(\vartheta_{i,1} + \vartheta_{c,1})N^{idio,1}(du,dx) \\ &\quad + \int_{(0,t)\times\{1\}} e^{-\beta(t-u)}(\vartheta_{i,2} + \vartheta_{c,2})N^{idio,2}(du,dx) \\ &\quad + \int_{(0,t)\times\{1\}\times\{1\}} e^{-\beta(t-u)}(\vartheta_{i,c} + \vartheta_{c,c})\mathbb{1}_{\{1\}\times\{1\}}(x)N(du,dx).\end{aligned}$$

Note that in this case, i.e. for, $A = \{1\}$, (11.46) takes the form

$$\kappa(t,\{1\}^i) = \widehat{\kappa}^i(N,t,\{1\}^i), \quad t \geq 0. \tag{11.49}$$

Invoking (11.48) we see that for $i=1,2$ the above condition is equivalent to the following system of equalities:

$$\begin{aligned} \vartheta_{i,1}+\vartheta_{c,1} &= \bar{\vartheta}_{i,1}, & \alpha_i+\alpha_c &= \bar{\alpha}_i, \\ \vartheta_{i,2}+\vartheta_{c,2} &= \bar{\vartheta}_{i,2}, & \lambda_0^i+\lambda_0^c &= \bar{\lambda}_i, \\ \vartheta_{i,c}+\vartheta_{c,c} &= \bar{\vartheta}_{i,c}, & \beta_i &= \bar{\beta}. \end{aligned} \tag{11.50}$$

In order to determine a Hawkes structure for M^i, $i=1,2$, one needs to consider the above two (for $i=1,2$) systems of equalities as two systems of linear equations with unknowns $\{\alpha_i,\lambda_0^i,\beta_i,\vartheta_{i,j}: i,j\in\{1,2,c\}\}$. This system admits multiple nonnegative solutions and, with every such solution of the above system, there is an associated Hawkes structure for the family $\{M^1,M^2\}$. Any solution

$$\{\alpha_i,\lambda_0^i,\beta_i,\vartheta_{i,j}: i,j\in\{1,2,c\}\}$$

of (11.50) determines a kernel κ satisfying (11.49) for $i=1,2$. Finally, any such kernel gives a Hawkes structure for the family $\{M^1,M^2\}$.

The next example generalizes the previous one.

Example 11.27 Consider the bivariate mark space

$$E^\Delta = E_1^\Delta\times E_2^\Delta\setminus\{(\Delta,\Delta)\} = \{(\Delta,y_2),(y_1,\Delta),(y_1,y_2): y_1\in E_1, y_2\in E_2\}.$$

Let M^i, $i=1,2$, be quasi-Hawkes processes with ν-constructor K which is a canonical multivariate Hawkes process on $(\mathbb{F}^K,\mathbb{P}_\nu)$ with the above mark space. Let $\bar{h}_{i,j}$, for $i=1,2$, $j\in\{1,2,c\}$, be nonnegative functions satisfying

$$\bar{h}_{i,j}(t,s)=0 \quad \text{for } s>t$$

and $\widehat{h}_{i,j}$, for $i=1,2$, $j\in\{1,2,c\}$, be kernels which do not depend on $\omega\in\Omega$. Suppose that, for $i=1,2$,

$$\begin{aligned} \kappa^i(t,A) = \chi_i(t,A) &+ \int_{(0,t)\times E_1} \bar{h}_{i,1}(t,s)\widehat{h}_{i,1}(x_1,A)K^{idio,1}(ds,dx_1) \\ &+ \int_{(0,t)\times E_2} \bar{h}_{i,2}(t,s)\widehat{h}_{i,2}(x_2,A)K^{idio,2}(ds,dx_2) \\ &+ \int_{(0,t)\times E} \bar{h}_{i,c}(t,s)\widehat{h}_{i,c}(x,A)K(ds,dx), \quad A\in\mathcal{E}_i. \end{aligned}$$

By the definitions of $K^{idio,1}(ds,dx_1)$ and $K^{idio,2}(ds,dx_2)$ we have

$$\begin{aligned} \kappa^i(t,A) = \chi_i(t,A) &+ \int_{(0,t)\times E_1\times\{\Delta\}} \bar{h}_{i,1}(t,s)\widehat{h}_{i,1}(x_1,A)K(ds,dx) \\ &+ \int_{(0,t)\times\{\Delta\}\times E_2} \bar{h}_{i,2}(t,s)\widehat{h}_{i,2}(x_2,A)K(ds,dx) \\ &+ \int_{(0,t)\times E} \bar{h}_{i,c}(t,s)\widehat{h}_{i,c}(x,A)K(ds,dx), \quad i=1,2. \end{aligned}$$

Thus, we find that $\widehat{\kappa}^i$ satisfying (11.45) has the form

$$\begin{aligned}\widehat{\kappa}^i(K,t,A) &= \chi_i(t,A)\\ &\quad+\int_{(0,t)\times E_1\times\{\Delta\}} \bar{h}_{i,1}(t,s)\widehat{h}_{i,1}(x_1,A)K(ds,dx)\\ &\quad+\int_{(0,t)\times\{\Delta\}\times E_2} \bar{h}_{i,2}(t,s)\widehat{h}_{i,2}(x_2,A)K(ds,dx)\\ &\quad+\int_{(0,t)\times E} \bar{h}_{i,c}(t,s)\widehat{h}_{i,c}(x,A)K(ds,dx), \quad i=1,2. \end{aligned} \tag{11.51}$$

Now, we look for Hawkes structures N for $\{M^1,M^2\}$ in the form presented in Example 11.21. So, we suppose that the Hawkes kernel κ of N is given in terms of η and f as in Example 11.21. We show that there are multiple specifications of η and f defining a κ which satisfies (11.46). Of course each specification gives the Hawkes structure N for $\{M^1,M^2\}$.

Since $\widehat{\kappa}^i$ as given by (11.51) and $\kappa(t,A^i) = \kappa^i(t,A)$ as given by (11.24) should both satisfy (11.46) they have to be equal. Comparing (11.24) and (11.51) we see that our task amounts to showing that there are multiple specifications of η and f for which the following equations are satisfied for $i=1$, $t\geq s\geq 0$, $A\in\mathcal{E}_1$, $x_j\in E_j$, and $j=1,2$:

$$\eta_1(t,A)+\eta_c(t,A\times E_2)=\chi_1(t,A),$$

$$\begin{aligned}w_{1,1}(t,s)g_{1,1}(x_1)\phi_1((x_1,\Delta),A)+w_{c,1}(t,s)g_{c,1}(x_1)\phi_c((x_1,\Delta),A\times E_2)&\\ =\bar{h}_{1,1}(t,s)\widehat{h}_{1,1}(x_1,A)&,\\ w_{1,2}(t,s)g_{1,2}(x_2)\phi_1((\Delta,x_2),A)+w_{c,2}(t,s)g_{c,2}(x_2)\phi_c((\Delta,x_2),A\times E_2)&\\ =\bar{h}_{1,2}(t,s)\widehat{h}_{1,2}(x_2,A)&,\\ w_{1,c}(t,s)g_{1,c}(x)\phi_1((x_1,x_2),A)+w_{c,c}(t,s)g_{c,c}(x_1,x_2)\phi_c((x_1,x_2),A\times E_2)&\\ =\bar{h}_{1,c}(t,s)\widehat{h}_{1,c}(x,A)&,\end{aligned}$$

and that analogous equations are satisfied for $i=2$, $t\geq s\geq 0$, $A\in\mathcal{E}_2$, $x_j\in E_j$, and $j=1,2$.

It is clear that such multiple specifications of η and f exist, which speaks in favor of the multiplicity of Hawkes structures for the family $\{M^1,M^2\}$.

11.5 Hawkes Consistency of Generalized Multivariate Hawkes Processes

Recall that for a generalized multivariate Hawkes process N (see Definition 11.5) we denote by N^i its ith coordinate, defined in (11.8), which is the marked point process with mark space E_i.

As we know, any coordinate N^i of a generalized multivariate Hawkes process N is a quasi-Hawkes process. Nevertheless, N^i is not necessarily a generalized $\mathbb{F}^{N^i}$-Hawkes process. The reason is that its $(\mathbb{F}^{N^i},\mathbb{P})$-compensator may not be representable in terms of a Hawkes kernel. If, however, N^i is a generalized $\mathbb{F}^{N^i}$-Hawkes process then we will be dealing with the Hawkes consistency property of N with respect to N^i. Formally, we have

Definition 11.28 We say that a generalized multivariate Hawkes process N is *Hawkes consistent with respect to its coordinate* N^i if the marked point process N^i is a generalized $\mathbb{F}^{N^i}$-Hawkes process. The process N is said to have the *Hawkes consistency property* if it is Hawkes consistent with respect to all of its coordinates.

We should note that the concept of Hawkes consistency does not appear in the definition of the Hawkes structure. One reason for this is that, as it appears, Hawkes consistency is quite a constraining property. So, imposing the Hawkes consistency requirement on a Hawkes structure would considerably limit the applicability and appeal of the Hawkes structure. Nevertheless, we will discuss Hawkes consistency in some detail in order to provide more insight into the nature of generalized multivariate Hawkes processes.

Even though the following proposition provides only sufficient conditions for Hawkes consistency, which may not be necessary, it gives a sense of why the Hawkes consistency property is a constraining property.

Proposition 11.29 *Let N be a generalized multivariate Hawkes process with kernel*

$$\kappa(t,dy) = \eta(t,dy) + \int_{(0,t)\times E^\Delta} f(t,s,x,dy)N(ds,dx).$$

Fix $i \in \{1,\dots,d\}$ and suppose that the following conditions are satisfied for every $A \in \mathcal{E}_i$ and for all $t,s \geq 0$:

$$f(t,s,x,A^i) = 0 \quad \text{for all } x \in E^\Delta \text{ such that } x_i = \Delta, \tag{11.52}$$

$$f(t,s,x,A^i) = f(t,s,y,A^i) \quad \text{for all } x,y \in E^\Delta \text{ such that } x_i = y_i \in E_i, \tag{11.53}$$

where A^i is defined in (11.9). *Then*

(i) *The kernel κ^i from $(\Omega \times \mathbb{R}_+, \mathcal{F} \otimes \mathcal{B}(\mathbb{R}_+))$ to $(E_i, \mathcal{E}_i)$ defined in* (11.13) *is $\mathbb{F}^N$-predictable and takes the form*

$$\kappa^i(t,dz) = \eta^i(t,dz) + \int_{(0,t)\times E_i} f^i(t,s,x_i,dz)N^i(ds,dx_i), \tag{11.54}$$

where, for all $t,s \geq 0$, $x_i \in E_i$,

$$\eta^i(t,dz) = \eta(t, E_1^\Delta \times \cdots \times E_{i-1}^\Delta \times dz \times E_{i+1}^\Delta \times \cdots \times E_d^\Delta), \tag{11.55}$$

$$f^i(t,s,x_i,dz) = f(t,s,(\Delta,\dots,\Delta,x_i,\Delta,\dots,\Delta), E_1^\Delta \times \cdots \times E_{i-1}^\Delta \times dz \times E_{i+1}^\Delta \times \ldots \times E_d^\Delta). \tag{11.56}$$

Moreover, the coordinate N^i is a generalized $\mathbb{F}^N$-Hawkes process with Hawkes kernel κ^i given by (11.54).

(ii) *If additionally η and f are deterministic then N is Hawkes consistent with respect to its coordinate N^i, i.e. the coordinate N^i is a generalized $\mathbb{F}^{N^i}$-Hawkes process with Hawkes kernel κ^i given by* (11.54).

Proof Fix $i \in \{1,\dots,d\}$, $t \geq s$, $A \in \mathcal{E}_i$. By assumption (11.52) we have, for every $x \in E^\Delta$,

$$f(t,s,x,A^i) = f(t,s,x,A^i)\mathbb{1}_{E_i}(x_i). \tag{11.57}$$

Condition (11.53) yields the following equalities for every $x \in E^\Delta$ and $(y^j)_{j\neq i} \in \times_{j\neq i} E_j^\Delta$:

$$\begin{aligned} f(t,s,x,A^i)\mathbb{1}_{E_i}(x_i) &= f(t,s,(y_1,\dots,y^{i-1},x_i,y^{i+1},\dots,y^d),A^i)\mathbb{1}_{E_i}(x_i) \\ &= f(t,s,(\Delta,\dots,\Delta,x_i,\Delta,\dots,\Delta),A^i)\mathbb{1}_{E_i}(x_i). \end{aligned} \tag{11.58}$$

Let us take the kernel κ^i, defined in (11.13), from $(\Omega \times \mathbb{R}_+, \mathcal{F} \otimes \mathcal{B}(\mathbb{R}_+))$ to $(E_i, \mathcal{E}_i)$. Taking into account the representations (11.10) and (11.11) we obtain from (11.57) and (11.58)

$$\begin{aligned} \kappa^i(t,A) &= \kappa(t,A^i) \\ &= \eta(t,A^i) + \int_{(0,t)\times E^\Delta} f(t,s,x,A^i)N(ds \times dx) \\ &= \eta(t,A^i) + \sum_{T_n<t} f(t,T_n,(\Delta,\dots,\Delta,X_n^i,\Delta,\dots,\Delta),A^i)\mathbb{1}_{\{X_n^i\in E_i\}} \\ &= \eta(t,A^i) + \sum_{T_m^i<t} f(t,T_m^i,(\Delta,\dots,\Delta,Y_m^i,\Delta,\dots,\Delta),A^i)\mathbb{1}_{\{Y_m^i\in E_i\}} \\ &= \eta(t,A^i) + \int_{(0,t)\times E_i} f(t,s,(\Delta,\dots,\Delta,x_i,\Delta,\dots,\Delta),A^i)N^i(ds,dx_i) \\ &= \eta^i(t,A) + \int_{(0,t)\times E_i} f^i(t,s,x_i,A^i)N^i(ds,dx_i), \end{aligned}$$

where the third equality follows from (11.10). Consequently, invoking Proposition 11.8 we see that N^i is a generalized $\mathbb{F}^N$-Hawkes process with Hawkes kernel given by (11.54). This completes the proof of (i).

Now we prove (ii). We first observe that

$$\kappa^i(t,A) = \eta^i(t,A) + \sum_{m:T_m^i<t} f^i(t,T_m^i,Y_m^i,A^i). \tag{11.59}$$

Combining (11.12) and (11.59) and invoking the assumption that η and f are deterministic, we see that the measure ν^i is $\mathbb{F}^{N^i}$-predictable. Consequently, using Proposition 11.10 and Remark 11.11(ii) we conclude that ν^i is the $\mathbb{F}^{N^i}$-compensator of N^i. So N^i is a generalized Hawkes process with Hawkes kernel given by (11.54). This completes the proof of (ii). □

Note that the postulate (11.52) for fixed i implies that the marks corresponding to events associated only with the coordinates $N^j, j \neq i$, do not affect the occurrence of events related to the ith coordinate. However, the postulate (11.53) implies that whenever the occurrence of events related to the ith coordinate is also affected by an event associated with the other coordinates, the magnitude of such excitation does not depend on the part of the mark corresponding to coordinates $N^j, j \neq i$.

Example 11.30 We place ourselves here in the set-up of Section 11.3. Specifically, we consider here the generalized bivariate Hawkes process discussed in that section.

Using Proposition 11.29 we will now study the question of what conditions imposed on the parameters of this Hawkes kernel are sufficient for:

(a) the Hawkes consistency of N with respect to the coordinate N^1 and with respect to the coordinate N^2;
(b) the Hawkes consistency property of N.

We first consider the case of N^1. For $i = 1$ the conditions (11.52) and (11.53) take the form

$$f(t,s,(\Delta,1),\{1\} \times E_2^\Delta) = 0,$$
$$f(t,s,(1,\Delta),\{1\} \times E_2^\Delta) = f(t,s,(1,1),\{1\} \times E_2^\Delta), \quad \text{for } 0 \leq s \leq t.$$

Using (11.34) we obtain the following equivalent system of equations:

$$e^{-\beta_1(t-s)}\vartheta_{1,2} + e^{-\beta_c(t-s)}\vartheta_{c,2} = 0,$$
$$e^{-\beta_1(t-s)}\vartheta_{1,1} + e^{-\beta_c(t-s)}\vartheta_{c,1} = e^{-\beta_1(t-s)}\vartheta_{1,c} + e^{-\beta_c(t-s)}\vartheta_{c,c}, \quad \text{for } 0 \leq s \leq t.$$

Thus, since all the coefficients $\vartheta_{i,j}$ are nonnegative, the above system is equivalent to

$$\vartheta_{1,2} = \vartheta_{c,2} = 0, \quad \beta_1 = \beta_c, \quad \vartheta_{1,1} + \vartheta_{c,1} = \vartheta_{1,c} + \vartheta_{c,c} \tag{11.60}$$

or

$$\vartheta_{1,2} = \vartheta_{c,2} = 0, \quad \beta_1 \neq \beta_c, \quad \vartheta_{1,1} = \vartheta_{1,c}, \quad \vartheta_{c,1} = \vartheta_{c,c}. \tag{11.61}$$

Consequently, in view of Proposition 11.29, if either (11.60) or (11.61) holds then N has the Hawkes consistency property with respect to the coordinate N^1. The Hawkes kernel κ^1 is determined by η^1 and f^1, which are given as

$$\eta^1(t,dy_1) = (\eta_1(t) + \eta_c(t))\delta_1(dy_1)$$
$$= \left(\alpha_1 + \alpha_c + (\eta_1(0) - \alpha_1)e^{-\beta_1 t} + (\eta_c(0) - \alpha_c)e^{-\beta_c t}\right)\delta_1(dy_1),$$
$$f^1(t,s,dy_1,x_1) = (e^{-\beta_1(t-s)}\vartheta_{1,1} + e^{-\beta_c(t-s)}\vartheta_{c,1})\delta_1(dy_1).$$

Note that if (11.60) holds then the above formulae simplify to

$$\eta^1(t,dy_1) = \left(\alpha_1 + \alpha_c + (\eta_1(0) - \alpha_1 + \eta_c(0) - \alpha_c)e^{-\beta_1 t}\right)\delta_1(dy_1),$$
$$f^1(t,s,dy_1,x_1) = e^{-\beta_1(t-s)}(\vartheta_{1,1} + \vartheta_{c,1})\delta_1(dy_1).$$

In the case of N^2 the conditions (11.52) and (11.53) take the form

$$\begin{aligned} &f(t,s,(1,\Delta),E_1^\Delta \times \{1\}) = 0,\\ &f(t,s,(\Delta,1),E_1^\Delta \times \{1\}) = f(t,s,(1,1),E_1^\Delta \times \{1\}), \quad \text{for } 0 \le s \le t. \end{aligned}$$

Using (11.34) again we see that these conditions take the form

$$\begin{aligned} &e^{-\beta_2(t-s)}\vartheta_{2,1} + e^{-\beta_c(t-s)}\vartheta_{c,1} = 0,\\ &e^{-\beta_2(t-s)}\vartheta_{2,2} + e^{-\beta_c(t-s)}\vartheta_{c,2} = e^{-\beta_2(t-s)}\vartheta_{2,c} + e^{-\beta_c(t-s)}\vartheta_{c,c}, \quad \text{for } 0 \le s \le t. \end{aligned}$$

The above is equivalent to

$$\vartheta_{2,1} = \vartheta_{c,1} = 0, \quad \beta_2 = \beta_c, \quad \vartheta_{2,2} + \vartheta_{c,2} = \vartheta_{2,c} + \vartheta_{c,c} \tag{11.62}$$

or

$$\vartheta_{2,1} = \vartheta_{c,1} = 0, \qquad \beta_2 \neq \beta_c, \quad \vartheta_{2,2} = \vartheta_{2,c}, \quad \vartheta_{c,2} = \vartheta_{c,c}. \tag{11.63}$$

Invoking Proposition 11.29 again we conclude that if either (11.62) or (11.63) holds then N has the Hawkes consistency property with respect to the coordinate N^2. The Hawkes kernel κ^2 of N^2 is determined by η^2 and f^2, which are given as

$$\begin{aligned} \eta^2(t,dy_2) &= (\eta_2(t) + \eta_c(t))\delta_1(dy_2)\\ &= \left(\alpha_2 + \alpha_c + (\eta_2(0) - \alpha_2)e^{-\beta_2 t} + (\eta_c(0) - \alpha_c)e^{-\beta_c t}\right)\delta_1(dy_2),\\ f^2(t,s,dy_2,x_2) &= (e^{-\beta_2(t-s)}\vartheta_{2,2} + e^{-\beta_c(t-s)}\vartheta_{c,2})\delta_1(dy_2). \end{aligned}$$

If (11.62) holds then the above simplify to

$$\begin{aligned} \eta^2(t,dy_2) &= (\alpha_2 + \alpha_c + (\eta_2(0) - \alpha_2 + \eta_c(0) - \alpha_c)e^{-\beta_2 t})\delta_1(dy_2),\\ f^2(t,s,dy_2,x_2) &= (\vartheta_{2,2} + \vartheta_{c,2})e^{-\beta_2(t-s)}\delta_1(dy_2). \end{aligned}$$

(b) Now we examine the Hawkes consistency of N. We will make use of Proposition 11.29 again. Recalling that the Hawkes consistency of N occurs if N is Hawkes consistent with respect to both N^1 and N^2, and taking into account considerations from item (a) above we conclude that the Hawkes consistency of N holds if:

(1) conditions (11.60) and (11.62) hold;

$$\begin{aligned} &\vartheta_{1,2} = \vartheta_{c,2} = \vartheta_{2,1} = \vartheta_{c,1} = 0, \quad \vartheta_{1,1} = \vartheta_{1,c} + \vartheta_{c,c},\\ &\beta_1 = \beta_2 = \beta_c, \quad \vartheta_{2,2} = \vartheta_{2,c} + \vartheta_{c,c}; \end{aligned}$$

(2) or conditions (11.60) and (11.63) hold;

$$\begin{aligned} &\vartheta_{1,2} = \vartheta_{c,2} = \vartheta_{2,1} = \vartheta_{c,1} = \vartheta_{c,c} = 0, \quad \vartheta_{1,1} = \vartheta_{1,c},\\ &\beta_1 = \beta_c \neq \beta_2, \quad \vartheta_{2,2} = \vartheta_{2,c}; \end{aligned}$$

(3) or conditions (11.61) and (11.62) hold;

$$\vartheta_{1,2} = \vartheta_{c,2} = \vartheta_{2,1} = \vartheta_{c,1} = \vartheta_{c,c} = 0, \quad \vartheta_{1,1} = \vartheta_{1,c},$$
$$\beta_2 = \beta_c \neq \beta_1, \quad \vartheta_{2,2} = \vartheta_{2,c};$$

(4) or conditions (11.61) and (11.63) hold;

$$\vartheta_{1,2} = \vartheta_{c,2} = \vartheta_{2,1} = \vartheta_{c,1} = \vartheta_{c,c} = 0, \quad \vartheta_{1,1} = \vartheta_{1,c},$$
$$\beta_1 \neq \beta_c \neq \beta_2, \quad \vartheta_{2,2} = \vartheta_{2,c}.$$

In case (1) we have $\beta_1 = \beta_2 = \beta_c$, and thus the kernel f takes the form

$$\begin{aligned} f(t,s,x,dy) \\ = e^{-\beta_1(t-s)} \Bigg(& \Big((\vartheta_{1,c} + \vartheta_{c,c}) \mathbb{1}_{\{1\}\times\Delta}(x) + \vartheta_{1,c} \mathbb{1}_{\{1\}\times\{1\}}(x) \Big) \delta_{(1,\Delta)}(dy) \\ & + \Big((\vartheta_{2,c} + \vartheta_{c,c}) \mathbb{1}_{\Delta\times\{1\}}(x) + \vartheta_{2,c} \mathbb{1}_{\{1\}\times\{1\}}(x) \Big) \delta_{(\Delta,1)}(dy) \\ & + \vartheta_{c,c} \mathbb{1}_{\{1\}\times\{1\}}(x) \delta_{(1,1)}(dy) \Bigg). \end{aligned}$$

In cases (2), (3), and (4) we have $\vartheta_{c,i} = 0$ for $i = 1,2,c$, and thus the kernel f takes the form

$$\begin{aligned} f(t,s,x,dy) = {} & e^{-\beta_1(t-s)} \vartheta_{1,c} \mathbb{1}_{\{1\}}(x_1) \delta_{(1,\Delta)}(dy) \\ & + e^{-\beta_2(t-s)} \vartheta_{2,c} \mathbb{1}_{\{1\}}(x_2) \delta_{(\Delta,1)}(dy). \end{aligned}$$

It is important to note that case (1) exhibits both self- and mutual excitations, and common event times. Contrariwise, in cases (2), (3), and (4) no common event times are present and only self-excitation takes place.

Example 11.31 We now place ourselves in the framework of Example 11.21. Following in the footsteps of the preceding example, we will employ Proposition 11.29 to study the question of what conditions imposed on the decay and impact functions of this Hawkes kernel are sufficient for:

(a) the Hawkes consistency of N with respect to the coordinate N^1 and with respect to the coordinate N^2;
(b) the Hawkes consistency property of N.

We begin by analyzing the consistency of N with respect to N^1. In this case conditions (11.52) and (11.53) take the following form. For every $A_1 \in \mathcal{E}_1, t \geq s$,

$$f(t,s,(\Delta,x_2),A_1 \times E_2^{\Delta}) = 0, \quad x_2 \in E_2, \tag{11.64}$$
$$f(t,s,(x_1,x_2),A_1 \times E_2^{\Delta}) = f(t,s,(x_1,\Delta),A_1 \times E_2^{\Delta}), \quad x_1 \in E_1, \quad x_2 \in E_2. \tag{11.65}$$

In order to ease the following presentation we fix $x_1 \in E_1$, $x_2 \in E_2$ in the rest of the example. We assume additionally that the decay functions satisfy

$$w_{1.2}(t,s) = w_{1.1}(t,s) = w_{1.c}(t,s) = w_1(t-s),$$

$$w_{2.2}(t,s) = w_{2.1}(t,s) = w_{2.c}(t,s) = w_2(t-s),$$

$$w_{c.2}(t,s) = w_{c.1}(t,s) = w_{c.c}(t,s) = w_c(t-s).$$

Thus, with f given in (11.23), the conditions (11.64) and (11.65) are equivalent to the following system of equalities:

$$\begin{aligned} &w_1(t-s)g_{1.2}(x_2)\phi_1((\Delta,x_2),A_1) \\ &\quad + w_c(t-s)g_{c.2}(x_2)\phi_c((\Delta,x_2),A_1 \times E_2) = 0, \end{aligned} \tag{11.66}$$

$$\begin{aligned} &w_1(t-s)g_{1.1}(x_1)\phi_1((x_1,\Delta),A_1) \\ &\quad + w_c(t-s)g_{c.1}(x_1)\phi_c((x_1,\Delta),A_1 \times E_2) \\ &\quad = w_1(t-s)g_{1.c}(x_1,x_2)\phi_1((x_1,x_2),A_1) \\ &\qquad + w_c(t-s)g_{c.c}(x_1,x_2)\phi_c((x_1,x_2),A_1 \times E_2), \end{aligned} \tag{11.67}$$

for every $A_1 \in \mathcal{E}_1$, $x_1 \in E_1$, $x_2 \in E_2$, and $t \geq s$.

It is straightforward to verify, via inspection of (11.66) and (11.67), that

(i) if the functions w_1 and w_c are linearly independent and if the following equalities hold for every $A_1 \in \mathcal{E}_1$, $x_1 \in E_1$, and $x_2 \in E_2$:

$$g_{1.2}(x_2) = g_{c.2}(x_2) = 0, \tag{11.68}$$

$$g_{1.1}(x_1)\phi_1((x_1,\Delta),A_1) = g_{1.c}(x_1,x_2)\phi_1((x_1,x_2),A_1), \tag{11.69}$$

$$g_{c.1}(x_1)\phi_c((x_1,\Delta),A_1 \times E_2) = g_{c.c}(x_1,x_2)\phi_c((x_1,x_2),A_1 \times E_2);$$

(ii) or, if the functions w_1 and w_c are linearly dependent (which implies that $w_1(t-s) = a_1 w_c(t-s)$ for some constant $a_1 \neq 0$), if (11.68) holds, and if for every $A_1 \in \mathcal{E}_1$, $x_1 \in E_1$, and $x_2 \in E_2$ the following equality is satisfied,

$$\begin{aligned} &g_{1.1}(x_1)\phi_1((x_1,\Delta),A_1) + a_1 g_{c.1}(x_1)\phi_c((x_1,\Delta),A_1 \times E_2) \\ &\quad = g_{1.c}(x_1,x_2)\phi_1((x_1,x_2),A_1) + a_1 g_{c.c}(x_1,x_2)\phi_c((x_1,x_2),A_1 \times E_2), \end{aligned}$$

then, for f as given in (11.23), conditions (11.66) and (11.67) are obeyed, and thus conditions (11.52) and (11.53) are obeyed. Therefore, by Proposition 11.29, N^1 is an $\mathbb{F}^N$-Hawkes process. Moreover, since η and f are deterministic, N is Hawkes consistent with respect to N^1. The kernel f^1 takes the form

$$\begin{aligned} f^1(t,s,x_1,dy_1) &= w_1(t-s)g_{1.1}(x_1)\phi_1((x_1,\Delta),dy_1) \\ &\quad + w_c(t-s)g_{c.1}(x_1)\phi_c((x_1,\Delta),dy_1 \times E_2). \end{aligned}$$

A similar analysis can be made with regard to the Hawkes consistency of N with respect to N^2, of course. Specifically, in case of N^2, conditions (11.52) and (11.53) take the following form. For every $A_2 \in \mathcal{E}_2, t \geq s$,

$$f(t,s,(x_1,\Delta),E_1^\Delta \times A_2) = 0, \quad x_1 \in E_1,$$
$$f(t,s,(x_1,x_2),E_1^\Delta \times A_2) = f(t,s,(\Delta,x_2),E_1^\Delta \times A_2), \quad x_1 \in E_1, x_2 \in E_2.$$

By analogous reasoning to that in the case of N^1 we conclude that

(i) if w_2 and w_c are linearly independent functions, and if the following equalities hold for every $A_1 \in \mathcal{E}_1$, $x_1 \in E_1$, and $x_2 \in E_2$:

$$g_{2,1}(x_1) = g_{c,1}(x_1) = 0, \tag{11.70}$$
$$g_{2,2}(x_2)\phi_2((\Delta,x_2),A_2) = g_{2,c}(x_1,x_2)\phi_2((x_1,x_2),A_2), \tag{11.71}$$
$$g_{c,2}(x_2)\phi_c((\Delta,x_2),E_1 \times A_2) = g_{c,c}(x_1,x_2)\phi_c((x_1,x_2),E_1 \times A_2);$$

(ii) or, if w_2 and w_c are linearly dependent functions (which implies that $w_2(t-s) = a_2 w_c(t-s)$ for some constant $a_2 \neq 0$), and if (11.70) and the following equality hold for every $A_1 \in \mathcal{E}_1$, $x_1 \in E_1$, and $x_2 \in E_2$,

$$g_{2,2}(x_2)\phi_2((\Delta,x_2),A_2) + a_2 g_{c,2}(x_2)\phi_c((\Delta,x_2),E_1 \times A_2)$$
$$= g_{2,c}(x_1,x_2)\phi_2((x_1,x_2),A_2) + a_2 g_{c,c}(x_1,x_2)\phi_c((x_1,x_2),E_1 \times A_2),$$

then, for f as given by (11.23), conditions (11.52) and (11.53) hold. The kernel f^2 takes the form

$$f^2(t,s,x_2,dy_2) = w_2(t-s)g_{2,2}(x_2)\phi_2((\Delta,x_2),dy_2)$$
$$+ w_c(t-s)g_{c,2}(x_2)\phi_c((\Delta,x_2),dy_2 \times E_2).$$

As far as the Hawkes consistency of N is concerned, it is rather straightforward to verify that that N is Hawkes consistent if (11.68), (11.70) hold and if one of the following conditions is satisfied:

(i) the functions w_1 and w_c, as well as w_2 and w_c, are linearly dependent and the following equalities hold for every $A_1 \in \mathcal{E}_1$, $x_1 \in E_1$, and $x_2 \in E_2$,

$$g_{1,1}(x_1)\phi_1((x_1,\Delta),A_1)$$
$$= g_{1,c}(x_1,x_2)\phi_1((x_1,x_2),A_1) + a_1 g_{c,c}(x_1,x_2)\phi_c((x_1,x_2),A_1 \times E_2),$$
$$g_{2,2}(x_2)\phi_2((\Delta,x_2),A_2)$$
$$= g_{2,c}(x_1,x_2)\phi_2((x_1,x_2),A_2) + a_2 g_{c,c}(x_1,x_2)\phi_c((x_1,x_2),E_1 \times A_2);$$

(ii) either the linear independence of the functions w_1 and w_c or the linear independence of the functions w_2 and w_c does not hold, $g_{cc} = 0$ and (11.69) and (11.71) hold.

We complete the example by displaying the kernel f satisfying either of the above sets of conditions (i) or (ii).

Assuming (11.68), (11.70), and (i), we see that the kernel f takes the form

$$
\begin{aligned}
&f(t,s,x,dy)\\
&= w_1(t-s)\Big(g_{1.1}(x_1)\mathbb{1}_{E_1\times\Delta}(x)+g_{1.c}(x)\mathbb{1}_{E_1\times E_2}(x)\Big)\phi_1(x,dy_1)\otimes\delta_\Delta(dy_2)\\
&\quad + w_2(t-s)\Big(g_{2.2}(x_2)\mathbb{1}_{\Delta\times E_2}(x)+g_{2.c}(x)\mathbb{1}_{E_1\times E_2}(x)\Big)\delta_\Delta(dy_1)\otimes\phi_2(x,dy_2)\\
&\quad + w_c(t-s)\Big(g_{c.c}(x_1,x_2)\mathbb{1}_{E_1\times E_2}(x)\Big)\phi_c(x,dy_1,dy_2).
\end{aligned}
$$

Upon assuming (11.68), (11.70), and (ii), the kernel f takes the form

$$
\begin{aligned}
f(t,s,x,dy) &= w_1(t-s)g_{1.1}(x_1)\mathbb{1}_{E_1}(x_1)\phi_1((x_1,\Delta),dy_1)\otimes\delta_\Delta(dy_2)\\
&\quad + w_2(t-s)g_{2.2}(x_2)\mathbb{1}_{E_2}(x_2)\delta_\Delta(dy_1)\otimes\phi_2((\Delta,x_2),dy_2).
\end{aligned}
$$

Example 11.32 Let N be a generalized bivariate Hawkes process, as in Section 11.3, given on $(\Omega,\mathcal{F},\mathbb{P})$, with Hawkes kernel κ defined by (11.35) with λ^k for $k=1,2,c$ defined by (11.36). Suppose that the parameters of λ^k for $k=1,2,c$ satisfy

$$\vartheta_{i.j}=\vartheta_{c.j}=0, \quad j\neq i,\ j\in\{1,2\}, \tag{11.72}$$

and

$$\beta_i=\beta_c, \quad \vartheta_{i.c}+\vartheta_{c.c}=\vartheta_{i.i}+\vartheta_{c.i}. \tag{11.73}$$

Then, as we will verify below, for $i=1,2$, the coordinate N^i of N is a generalized $\mathbb{F}^N$-Hawkes process with Hawkes kernel κ^i given in terms of $\lambda^i+\lambda^c$ as

$$
\begin{aligned}
\kappa^i(t,dz) &= (\alpha_i+\alpha_c+(\lambda_0^i+\lambda_0^c-(\alpha_i+\alpha_c))e^{-\beta_i t})\delta_1(dz)\\
&\quad + \int_{(0,t)\times\{1\}}(\vartheta_{i.i}+\vartheta_{c.i})e^{-\beta_i(t-u)}\delta_1(dz)N^i(du,dx).
\end{aligned}\tag{11.74}
$$

To verify this claim, we first recall that N^i, a random measure on $(\mathbb{R}_+\times\{1\},\mathcal{B}(\mathbb{R}_+)\otimes\{\{1\},\emptyset\})$, can be interpreted as a counting point process and that the $\mathbb{F}^N$-intensity process of N^i is given as $\lambda^i+\lambda^c$ (see (11.39)). Thus, the $(\mathbb{F}^N,\mathbb{P})$-compensator of the random measure N^i is given by

$$\nu^i(dt,dx)=\mathbb{1}_{]\!]0,T^i_\infty[\![}(\lambda_t^i+\lambda_t^c)\delta_1(dz)dt. \tag{11.75}$$

Since (11.36) is equivalent to (11.42) we have

$$
\begin{aligned}
\lambda_t^i+\lambda_t^c &= (\alpha_i+\alpha_c)+(\lambda_0^i-\alpha_i)e^{-\beta_i t}+(\lambda_0^c-\alpha_c)e^{-\beta_c t}\\
&\quad + \int_{(0,t)}\Big((e^{-\beta_i(t-u)}\vartheta_{i.1}+e^{-\beta_c(t-u)}\vartheta_{c.1})d\bar{N}_u^1\\
&\qquad + (e^{-\beta_i(t-u)}\vartheta_{i.2}+e^{-\beta_c(t-u)}\vartheta_{c.2})d\bar{N}_u^2\\
&\qquad + (e^{-\beta_i(t-u)}\vartheta_{i.c}+e^{-\beta_c(t-u)}\vartheta_{c.c})d\bar{N}_u^c\Big), \quad t\geq 0.
\end{aligned}
$$

Hence, since (11.72) implies that for fixed $i \in \{1,2\}$ the term involving $d\bar{N}^j_u$, $j \neq i$, vanishes, we obtain

$$\begin{aligned}\lambda^i_t + \lambda^c_t = \alpha_i + \alpha_c + (\lambda^i_0 - \alpha_i)e^{-\beta_i t} + (\lambda^c_0 - \alpha_c)e^{-\beta_c t} \\ + \int_{(0,t)} (\vartheta_{i,i}e^{-\beta_i(t-u)} + \vartheta_{c,i}e^{-\beta_c(t-u)})d\bar{N}^i_u \\ + \int_{(0,t)} (\vartheta_{i,c}e^{-\beta_i(t-u)} + \vartheta_{c,c}e^{-\beta_c(t-u)})d\bar{N}^c_u, \quad t \geq 0.\end{aligned}$$

Next, using (11.73) we have

$$\vartheta_{i,i}e^{-\beta_i(t-u)} + \vartheta_{c,i}e^{-\beta_c(t-u)} = \vartheta_{i,c}e^{-\beta_i(t-u)} + \vartheta_{c,c}e^{-\beta_c(t-u)}$$

for $0 \leq u \leq t$, which together with $N^i_u = \bar{N}^i_u + \bar{N}^c$ implies that

$$\begin{aligned}\lambda^i_t + \lambda^c_t = {} & \alpha_i + \alpha_c + (\lambda^i_0 + \lambda^c_0 - (\alpha_i + \alpha_c))e^{-\beta_i t} \\ & + \int_{(0,t)} (\vartheta_{i,i} + \vartheta_{c,i})e^{-\beta_i(t-u)}dN^i_u, \quad t \geq 0.\end{aligned}$$

Consequently, we have

$$\begin{aligned}(\lambda^i_t + \lambda^c_t)\delta_1(dz) = \left(\alpha_i + \alpha_c + (\lambda^i_0 + \lambda^c_0 - (\alpha_i + \alpha_c))e^{-\beta_i t}\right)\delta_1(dz) \\ + \int_{(0,t)\times\{1\}} (\vartheta_{i,i} + \vartheta_{c,i})e^{-\beta_i(t-u)}\delta_1(dz)N^i(du,dx).\end{aligned}$$

Thus, using (11.75), we see that N^i is a generalized $\mathbb{F}^N$-Hawkes process with Hawkes kernel given by (11.74). In particular, N is Hawkes consistent with respect to N^i.

PART FOUR

APPLICATIONS OF STOCHASTIC STRUCTURES

12
Applications of Stochastic Structures

It will be demonstrated in this chapter that stochastic structures constitute a versatile tool that has many practical applications. Some applications have already been worked out, and some others are still to be addressed.

In this chapter we provide a survey of existing applications of stochastic structures, and we also suggest some new potential applications.

12.1 Markov and Conditionally Markov Structures

Several applications of Markov structures in finance and insurance have been studied.

Financial applications typically refer to the valuation and hedging of basket derivatives, i.e. financial products that derive their value from the values of a collection – a basket – of some underlying financial securities. Moreover, Markov structures have been applied with success in the valuation and hedging of counterparty risk. One of the key benefits of using Markov structures has been that they make it possible to separate the calibration of the pricing and hedging model for univariate data (credit default spreads, for example) from the calibration of the model for multivariate data (spreads on credit portfolio contracts, such as collateralized loan obligations or collateralized debt obligations). This aspect of Markov structures theory is of fundamental importance for the efficient calibration of a model to market data.

In insurance applications, conditional Markov structures may prove to be useful in premium evaluation for unemployment insurance products.

12.1.1 Portfolio Credit Risk Valuation and the Hedging and Valuation of Step-Up Bonds

In Bielecki et al. (2008a, 2014a,b) (see also the references therein), the theory of strong Markov structures was successfully applied to separate the calibration of stochastic dependence in the pool of 125 obligors (constituting an iTraxx index) from the calibration of the univariate characteristics of the individual obligors.

Additionally, in Bielecki et al. (2008a) this theory was applied to the valuation of so-called ratings-triggered step-up bonds and to the computation of step-up provision. In Bielecki et al. (2014a,b) the theory was additionally applied to the valuation of collateralized debt obligation (CDO) contracts and to the hedging of the these contracts using a pool of credit default swaps (CDS).

12.1.2 Counterparty Risk Valuation and Hedging

Counterparty risk in financial transactions is the risk that one counterparty in a given transaction will declare bankruptcy and consequently will not honor their contractual obligations. The first steps towards the application of Markov structures for the valuation of counterparty risk were made in Assefa et al. (2011). This was followed by applications discussed in several works. We refer to Crépey et al. (2014) and the references therein.

12.1.3 Premium Evaluation for Unemployment Insurance Products

In the paper Biagini et al. (2013), an interesting problem of the evaluation of premia for unemployment insurance products, for a pool of individuals, was considered.

In Bielecki et al. (2017a) we suggested a possible generalization of the model studied in Biagini et al. (2013); this generalization, we believe, may provide a more adequate way to deal with the computation of the premia.

Biagini et al. (2013) used the doubly stochastic Markov chain (DSMC) framework to model the dynamics of the employment status of an individual. The dynamics were modeled under a probability measure, say $\mathbb{P}$, called a real-world measure. Then, using these dynamics they aimed to compute for $t \in [0,T]$ an insurance premium, denoted as P_t. The evolution of the employment status of an individual k was given in terms of a Markov chain, say X^k, which takes values in the state space $S_k = \{1,2\}$, where the state "1" indicates that the individual is employed and "2" indicates that the individual is unemployed. It is assumed that the process X^k is an $(\mathbb{F}^Z, \mathbb{F}^{X^k})$-DSMC, where $\mathbb{F}^Z$ is a reference filtration generated by some factor process Z.

As stated earlier, the quantity to be computed for the individual k is the value of the insurance premium against unemployment. Roughly speaking, the premium P_t^k at time t is given as

$$P_t^k = \mathbb{E}_{\mathbb{P}}(\Phi_k(X^k)|\mathcal{G}_t^k),$$

where Φ_k is some random functional of the process X^k and where $\mathcal{G}_t^k = \mathcal{F}_t^Z \vee \mathcal{F}_t^{X^k}$. In particular, the premium at time $t = 0$ needs to be computed, i.e.

$$P_0^k = \mathbb{E}_{\mathbb{P}}(\Phi_k(X^k)|\mathcal{G}_0^k).$$

Note that we have written P_0^k as a conditional expectation, given $\mathcal{G}_0^k$, rather than an unconditional expectation as in formula (2) in Biagini et al. (2013).

Proposed Conditional Markov Chain Structures Approach

We think that, to evaluate premia for unemployment insurance products for a pool of individuals labeled as $k = 1, \ldots, n$, it is important to account for possible dependence between the processes X^k, $k = 1, \ldots, n$.

Thus, we think that it may be advantageous to enrich the model studied in Biagini et al. (2013) by considering a process $Y = (Y^1, \ldots, Y^n)$ which is a conditional Markov structure for the processes X^k, $k = 1, \ldots, n$.

Thanks to the structure property, the characteristics of the dependence between the processes X^k, $k = 1, \ldots, n$, can be estimated separately from the estimation of the distributional characteristics of each process X^k. The latter task can be efficiently executed using the methodology outlined in Biagini et al. (2013).

The premium P_t^k at time t is given in the CMC copula model as

$$P_t^k = \mathbb{E}(\Phi_k(Y^k)|\widehat{\mathcal{G}}_t^k),$$

where $\widehat{\mathcal{G}}_t^k = \mathcal{F}_t^Z \vee \mathcal{F}_t^{Y^k}$. If the process Y is constructed as a weak-only CMC copula between the processes X^k, $k = 1, \ldots, n$, then we have, with $\widehat{\mathcal{G}}_t = \mathcal{F}_t^Z \vee \mathcal{F}_t^Y$,

$$\mathbb{E}(\Phi_k(Y^k)|\widehat{\mathcal{G}}_t^k) \neq \mathbb{E}(\Phi_k(Y^k)|\widehat{\mathcal{G}}_t).$$

This, of course, means that the employment status of the entire pool influences the calculation of the individual premium, a feature which we think is important.

The theory of strong and weak CMC copulae can be extended in a straightforward manner to modeling the structured dependence between subgroups of processes X^k, $k = 1, \ldots, n$. This allows for the study of insurance premia modeling for relevant subgroups of employees.

12.1.4 Multiple Ion Channels

Using the theory of CMCs and the corresponding conditionally independent structure model (see Example 8.4), one can model the conditional independence between multiple ion channels that are otherwise linked via common stochastic factors, embedded in a filtration $\mathbb{F}$ which models a random environment.

Ball et al. (1994) considered a model that corresponds to a special case of CMCs. They assumed that single ion channels X^1, ..., X^n are independent conditionally on an environmental factor process, say Z^E, which is a Markov chain. Their model corresponds to setting the $\mathbb{F}^{Z^E}$-intensity of X^k, $k = 1, \ldots, n$, to

$$\Psi_t^k = \Psi(Z_t^E)$$

and the $\mathbb{F}^{Z^E}$-intensity of the joint process X to

$$\Lambda(Z_t^E) = \sum_{k=1}^{n} I_1 \otimes \cdots \otimes I_{k-1} \otimes \Psi(Z_t^E) \otimes I_{k+1} \otimes \cdots \otimes I_n, \quad t \in [0, T].$$

Using the theory of CMC structures presented in this volume, a random environmental factor process driving the intensities of channel switches between open and closed states can be made far more general.

12.1.5 Dynamic Risk Management in Central Clearing Parties

In Bielecki et al. (2018a) a dynamic model for risk management in a central clearing party (CCP) was proposed. The main goal of that paper was to develop a new methodology for computing some ingredients of a CCP's default waterfall, in a dynamic framework. In particular, the authors proposed a novel, risk sensitive method for computing the total default fund of a CCP. Risk sensitivity amounts to accounting for the credit migrations of the clearing members and the stochastic dependence between these migrations, which was modeled in terms of Markov structures. The proposed model of joint credit migrations needs to account for the marginal laws of the credit migrations of each member of a CCP, and that is precisely what Markov structures were used to achieve.

12.2 Special Semimartingale Structures

We will suggest here a possible application of the Poisson semimartingale structure discussed in Example 9.4.

12.2.1 Poisson Structure and Ruin Probabilities

We provide some actuarial background first.

Insuring the same risks using multiple insurers is a well-established practice in the insurance industry. A version of this practice is called reinsurance. Quite commonly, in the reinsurance situation, there are two insurers involved, one being the primary insurer and the other the secondary insurer (or reinsurer). If the random risk (loss) level at time $t \geq 0$ is $L_t > 0$, and if the ith insurer is responsible for providing insurance at level L^i_t, $i = 1,2$, then the indemnity clause requires that $L^1_t + L^2_t \leq L_t$. It is frequently assumed in actuarial modeling that $L^i = \delta^i L$, where $\delta^i \in (0,1)$ is a constant, $i = 1,2$, and $\delta^1 + \delta^2 = 1$.

We refer to, e.g., Avram et al. (2008), Asimit and Booneny (2017), Richmond (1995), or Section XIII.9 in Asmussen and Albrecher (2010) for a discussion of various aspects of the insuring of the same risk by multiple insurers.

Now, we consider two insurance companies, say I_1 and I_2, that serve multiple customers. We take process $N = (N^1, N^2)$ as in Example 9.4. The claims arrive at the insurers according to N, as follows:

- If at time t we have $\Delta N_t = (1,0)$ and $N^1_t = k$ then the kth claim, say X^1_k, distributed according to F^1, arrives at I^1; no claim arrives at I^2. In this case I^1 serves as the primary and the only insurer for the claim.
- If at time t we have $\Delta N_t = (0,1)$ and $N^2_t = k$ then the kth claim, say X^2_k, distributed according to F^2, arrives at I^2; no claim arrives at I^1. In this case I^2 serves as the primary and the only insurer for the claim.

- If at time t we have $\Delta N_t = (1,1)$, $N^1_t = k$ and $N^2_t = m$, then the kth claim, say X^1_k, distributed according to F^1 arrives at, I^1 and the mth claim, say X^2_m, distributed according to F^2, arrives at I^2. We have $L^1_t = X^1_k$, $L^2_t = X^2_m$, and $L_t = X^1_k + X^2_m$. In this case I^1 serves as the primary insurer for the claim L_t, and I^2 serves as the reinsurer for this claim.

In particular, suppose that $\lambda_1 = \lambda_{10} + \lambda_{11}$ is the intensity of arrival of claims to insurer I^1, and $\lambda_2 = \lambda_{01} + \lambda_{11}$ is the intensity of arrival of claims to insurer I^2. These intensities can be observed by each company. However, $\lambda_{10}, \lambda_{01}$, and λ_{11} typically cannot be observed by the individual companies. So, numerous Poisson structures can be fitted here.

In this toy model we will postulate that processes X^i_n, $i = 1,2$, $n = 1,2,\ldots$ are all independent, and that they are independent of N, under $\mathbb{P}$.

Next, for $i = 1,2$, we define a surplus process, say U^i, for insurer I^i, as

$$U^i_t = u^i + c^i t - \sum_{n=1}^{N^i_t} X^i_n, \quad t \geq 0,$$

where $u^i > 0$ is the (deterministic) initial capital, and c^i is the (deterministic) premium per unit of time.

One of the problems of interest is to compute the probabilities of ruin in the system

$$\pi(u^1,u^2) := \mathbb{P}(T < \infty | U^1_0 = u^1, U^2_0 = u^2), \tag{12.1}$$

where $T := T^1 \wedge T^2$ and

$$T^i := \inf\{t > 0 : U^i_t < 0\}$$

is the ruin time of the insurer I^i, $i = 1,2$. Clearly, the probabilities in (12.1) depend on the particular structure N, which must match the marginal intensities λ_1 and λ_2. In fact, one can prove (see Michalik (2015)) that the complementary probability function $\varphi(u^1,u^2) = 1 - \pi(u^1,u^2)$ satisfies the following integral equation:

$$\begin{aligned}
\varphi(u^1,u^2) &= \lambda_{10} \int_0^\infty \int_0^{u^1+c^1 t} \varphi(u^1 + c^1 t - x_1, u^2 + c^2 t) dF^1(x_1) e^{-\lambda_{10} t} dt \\
&+ \lambda_{01} \int_0^\infty \int_0^{u^2+c^2 t} \varphi(u^1 + c^1 t, u^2 + c^2 t - x_2) dF^2(x_2) e^{-\lambda_{01} t} dt \\
&+ \lambda_{11} \int_0^\infty \int_0^{u^1+c^1 t} \int_0^{u^2+c^2 t} \varphi(u^1 + c^1 t - x_1, u^2 + c^2 t - x_2) \\
&\qquad \times dF^2(x_2) dF^1(x_1) e^{-\lambda_{11} t} dt.
\end{aligned} \tag{12.2}$$

12.3 Archimedean Survival Process (ASP) Structures

In this section we present an application of ASP structures.

12.3.1 Decision Functionals

We consider an underlying probability space $(\Omega, \mathcal{G}, \mathbb{P})$, endowed with a filtration $\mathbb{G}$, and a real-valued, non-decreasing process X on this space. The probability measure $\mathbb{P}$, the filtration $\mathbb{G}$, and the process X may have various practical interpretations. For us, in this section, they will provide a basis for making either an acceptance or rejection decision, relevant to some practical interpretation. To this end we will introduce two types of decision functionals, say δ_i, $i = 1, 2$, taking two values: 1 or 0. The value 1 of the decision functional renders an acceptance decision, whereas the value 0 renders a rejection decision.

Now, let us fix two dates $t > s \geq 0$, the cut-off level $c > 0$, and the acceptance level $\alpha \in \mathbb{R}$. The acceptance/rejection decision involves the time t, but it is made at the time s and depends on the conditional distribution of the increment $X_t - X_s$, given the information $\mathcal{G}_s$.

The decision functional δ_1 is based on the conditional probability of exceeding the level c by the increment $X_t - X_s$ and is defined as

$$\delta_1(\alpha, c) := \begin{cases} 1 & \text{if } \mathbb{P}(X_t - X_s > c \mid \mathcal{G}_s) \geq \alpha, \\ 0 & \text{if } \mathbb{P}(X_t - X_s > c \mid \mathcal{G}_s) < \alpha. \end{cases} \tag{12.3}$$

The second decision functional is based on the restricted conditional expectation of the increment $X_t - X_s$ and is given by the formula

$$\delta_2(\alpha, c) = \begin{cases} 1 & \text{if } \mathbb{E}((X_t - X_s)\mathbb{1}_{(c,\infty)}(X_t - X_s) \mid \mathcal{G}_s) \geq \alpha, \\ 0 & \text{if } \mathbb{E}((X_t - X_s)\mathbb{1}_{(c,\infty)}(X_t - X_s) \mid \mathcal{G}_s) < \alpha. \end{cases} \tag{12.4}$$

Both these functionals are simple and intuitive and can be computed via Monte Carlo simulations or, as shown in the next subsection, via analytical formulae.

12.3.2 Explicit Formulae for Decision Functionals for ASPs

In this part we focus on decision functions corresponding to ASPs. As in Chapter 10 we consider an ASP X with $\alpha = \lambda = 1$ and with generating law ψ. Also, we take $T = 1$. The filtration of X is denoted by $\mathbb{F}$, and the filtration of X^i is denoted by $\mathbb{F}^i$.

Let us denote by $I_{a,b}(x)$ the regularized incomplete beta function defined, for $x \in [0, 1]$, by

$$I_{a,b}(x) = \frac{\int_0^x u^{a-1}(1-u)^{b-1} du}{\int_0^1 u^{a-1}(1-u)^{b-1} du}.$$

Also, let us denote by $\|\cdot\|_1$ the l_1-norm in $\mathbb{R}^n$.

The following two results are borrowed from Jakubowski and Pytel (2016).

Proposition 12.1 *Let X be an n-dimensional ASP with the generating law ψ. Fix $0 < s < t < 1$ and $c > 0$. Then we have*

(i)

$$\mathbb{P}\left(X_t^i - X_s^i > c \mid \mathcal{F}_s^i\right) = \frac{1}{\Psi_s(X_s^i)} \int_{z > X_s^i + c} \frac{f_{n-s}(z - X_s^i)}{f_n(z)} I_{n-t,t-s}\left(1 - \frac{c}{z - X_s^i}\right) \psi(dz). \tag{12.5}$$

(ii)

$$\mathbb{E}\left(\left(X_t^i - X_s^i\right) \mathbb{1}_{\{X_t^i - X_s^i > c\}} \mid \mathcal{F}_s^i\right) = \frac{t-s}{\Psi_s(X_s^i)} \int_{z > X_s^i + c} \frac{f_{n-s+1}(z - X_s^i)}{f_n(z)} I_{n-t,t-s+1}\left(1 - \frac{c}{z - X_s^i}\right) \psi(dz). \tag{12.6}$$

Proposition 12.2 *Let X be an n-dimensional ASP with generating law ψ. Fix $0 < s < t < 1$ and $c > 0$. Then we have*

(i)

$$\mathbb{P}\left(X_t^i - X_s^i > c \mid \mathcal{F}_s\right) = \frac{1}{\Psi_{ns}(\|X_s\|_1)} \int_{z > \|X_s\|_1 + c} \frac{f_{n-ns}(z - \|X_s\|_1)}{f_n(z)} I_{n-ns-t+s,t-s}\left(1 - \frac{c}{z - \|X_s\|_1}\right) \psi(dz), \tag{12.7}$$

(ii)

$$\mathbb{E}\left(\left(X_t^i - X_s^i\right) \mathbb{1}_{\{X_t^i - X_s^i > c\}} \mid \mathcal{F}_s\right) = \frac{t-s}{\Psi_{ns}(\|X_s\|_1)} \int_{z > \|X_s\|_1 + c} \frac{f_{n-ns+1}(z - \|X_s\|_1)}{f_n(z)} I_{n-ns-t+s,t-s+1}\left(1 - \frac{c}{z - \|X_s\|_1}\right) \psi(dz). \tag{12.8}$$

Remark 12.3 Note that, in the case of the strong Markov consistency property, the decision functionals δ_1 and δ_2 produce the same results regardless of whether at time s the idiosyncratic information $\mathcal{F}_s^i$ is used or the full information $\mathcal{F}_s$ is used. The opposite is seen, in general, in the case of the strictly weak Markov consistency property.

12.3.3 Application of ASP Structures in Maintenance Management

In maintenance management two general uncertainties are recognized: the time to failure and the rate of deterioration. Most mathematical models focus on the uncertainty about the time to failure, so they focus on describing the distributions of the lifetime of construction structures or machines.

However, it is believed to be generally more appropriate to base a failure model on the deterioration rate; we refer to the discussion in van Noortwijk (2007) in this regard. Using a gamma process to model a deterioration randomly occurring in time was first proposed in van Abdel-Hameed (1987). The deterioration of construction structures is a key issue in engineering. If the rate of deterioration exceeds a given level then a repair or a replacement of some elements is needed.

A simple model of deterioration assumes that it is described by an increasing continuous stochastic process D. It was assumed in van Noortwijk (2007) that D is a gamma process. A natural extension of a gamma process is a random gamma bridge process.

An interesting modeling problem arises if we consider a construction structure which contains several elements that influence the overall condition of the structure. For example, one can think of a bridge as the construction structure and the bridge's pillars as being the failure-prone elements. If one pillar is damaged then a repair of the structure is required even if the overall deterioration does not exceed a preventive maintenance level. Therefore, if we want to design a decision algorithm to generate maintenance decisions we need to model the deterioration with each component as a part of an interdependent structure. Naturally, the deterioration in time of similar components can be modeled in terms of random processes following the same laws and exhibiting some dependence between them. It appears that an ASP structure X fulfils these general requirements.

In this context, the coordinate X^i of X represents the deterioration process of the ith component. The decision, at time s, whether to repair the bridge can now be made by applying either δ_1 or δ_2 to all the components and to the time interval $[s,t]$, with use of either the idiosyncratic information or the full information.

12.4 Hawkes Structures

It will be argued in this section that Hawkes structures offer a tremendous potential for the consistent modeling of dynamic phenomena of diverse applications.

12.4.1 Seismology

In the Introduction to Ogata (1999) the author writes:

"Lists of earthquakes are published regularly by the seismological services of most countries in which earthquakes occur with frequency. These lists supply at least the epicenter of each shock, focal depth, origin time and instrumental magnitude.

Such records from a self-contained seismic region reveal time series of extremely complex structure. Large fluctuations in the numbers of shocks per time unit, complicated sequences of shocks related to each other, dependence on activity in other seismic regions, fluctuations of seismicity on a larger time scale, and changes in the detection level of shocks, all appear to be characteristic features of such records. In this manuscript the origin times are mainly considered to be modeled by point processes, other elements being largely ignored except that the ETAS model and its extensions use data of magnitudes and epicenters."

In particular, the dependence on (simultaneous) seismic activity in other seismic regions has been ignored in the classical univariate epidemic-type aftershock sequences (ETAS) model, and in all other models of which we are aware.

The ETAS model is a univariate self-exciting point process in which the shock intensity at time t, corresponding to a specific seismic location, is designated as (see equation (17) in Ogata (1999))

$$\lambda(t|H_t) = \mu + \sum_{t_m<t} \frac{K_m}{(t - t_m + c)^p}. \tag{12.9}$$

In the above formula, H_t stands for the history of aftershocks at the given location, μ represents the background occurrence rate of seismic activity at this location, the t_m-s are the times of occurrences of all the aftershocks that took place prior to time t at the specific seismic location, and

$$K_m = K_0 e^{\alpha(M_m - M_0)},$$

where M_m is the magnitude of the shock occurring at time t_m and M_0 is the cut-off magnitude of the data set; we refer to Ogata (1999) for details. As mentioned above, the dependence between (simultaneous) seismic activity in different seismic regions is ignored in the classical univariate ETAS model.

Hawkes structures may offer a good way of modeling joint seismic activities at various locations, as they can account for dependences between activities at different locations and for consistencies with local data.

We will now briefly describe such a model, which we name the *multivariate generalized ETAS model*. To this end we consider the generalized multivariate Hawkes process N (see Definition 11.5), where the index $i = 1, \ldots, d$ represents the ith seismic location and where the set $E_i = \mathfrak{M}_i := \{m_1, \ldots, m_n\}$ of marks is a discrete set whose elements represent the possible magnitudes of seismic shocks with epicenters at location i. In the corresponding Hawkes kernel κ the measure $\eta(t, dy)$ represents the time-t background distribution of shocks across all seismic regions and the measure $f(t, s, dy, x)$ represents the feedback effect.

For the purpose of illustration, let $d = 2$. Suppose that local seismic data are collected for each location, producing local kernels of the form

$$\begin{aligned}\kappa^i(t,\{y_i\}) &= \chi_i(t,\{y_i\}) \\ &\quad + \int_{(0,t)\times E_1} h_{i.1}(t,s,x_1,\{y_i\})N^{idio,1}(ds,dx_1) \\ &\quad + \int_{(0,t)\times E_2} h_{i.2}(t,s,x_2,\{y_i\})N^{idio,2}(ds,dx_2) \\ &\quad + \int_{(0,t)\times E_1\times E_2} h_{i.c}(t,s,x,\{y_i\})N^{idio,1,2}(ds,dx), \quad i=1,2. \end{aligned} \tag{12.10}$$

In particular, the quantity $\lambda^i(t) := \kappa^i(t,E_i) = \sum_{y_i\in\mathfrak{M}_i}\kappa^i(t,y_i)$ represents the time-t intensity of seismic activity at the ith location.

In order to produce an ETAS-type model, we postulate that for $x=(x_1,x_2)$

$$\sum_{y_i\in\mathfrak{M}_i} h_{i.j}(t,s,x_j,\{y_i\}) = \frac{K_{i,j,0}e^{\alpha_{i.j}(x_j-x_{j.0})}}{(t-s+c)^{p_{i.j}}}$$

for $j=1,2$ and

$$\sum_{y_i\in\mathfrak{M}_i} h_{i.c}(t,s,x,\{y_i\}) = \frac{K_{i,c,0}e^{\alpha_{i.c}[(x_1-x_{1.0})+(x_2-x_{2.0})]}}{(t-s+c)^{p_{i.c}}}.$$

Thus

$$\begin{aligned}\lambda^i(t) = \sum_{y_i\in\mathfrak{M}_i}\Bigg(&\chi_i(t,\{y_i\}) + \sum_{j=1}^{2}\sum_{t_{j.m}<t}\frac{K_{i,j,0}e^{\alpha_{i.j}(X_{j.t_{j.m}}-x_{j.0})}}{(t-s+c)^{p_{i.j}}} \\ &+ \sum_{t_{c.m}<t}\frac{K_{i,c,0}e^{\alpha_{i.c}[(X_{1.t_{c.m}}-x_{1.0})+(X_{2.t_{c.m}}-x_{2.0})]}}{(t-s+c)^{p_{i.c}}}\Bigg),\end{aligned} \tag{12.11}$$

where

- the $t_{j.m}$ are the times of occurrences of aftershocks that took place prior to time t only at the ith seismic location, and $X_{j.t_{j.m}}$ is the magnitude of the aftershock at location i that took place at time $t_{j.m}$;
- the $t_{c.m}$ are the times of occurrences of aftershocks that took place prior to time t at both seismic locations, and $X_{j.t_{c.m}}$ is the magnitude of the aftershock at location i that took place at time $t_{c.m}$.

Now, in order for N to be a corresponding Hawkes structure we need to construct kernels η and f in such way that the Hawkes kernel κ matches the marginals:

$$\kappa(t,\{y_i\}\times E_j) = \kappa^i(t,\{y_i\}), \quad y_i\in E_i,\ t\geq 0,\ i,j=1,2,\ i\neq j.$$

The classical univariate ETAS model was extended in Ogata (1998) to the (classical) univariate space–time ETAS model (see also Section 5 in Ogata (1999)). It is important to note that our Hawkes structure is a useful generalization of the space–time extension of the multivariate generalized ETAS model. In order to see this, let us consider the model (2.1) in Ogata (1998) with g as in Section 2.1 in Ogata (1998), i.e. in the original notation of Ogata (1998), which should not be confused with our notation,

$$\lambda(t,x,y|H_t) = \mu(x,y) + \int_0^t \int \int_A \int_{M_0}^{\infty} g(t-s,x-\xi,y-\eta;M)N(ds,d\xi,d\eta,dM). \tag{12.12}$$

Then, coming back to our generalized multivariate Hawkes process, let the seismic location $i = 1,2$ be identified with a point in the plane with coordinates $(a_i,b_i) \in \mathbb{R}^2$. Next, let the set of marks E_i be given as

$$E_i := D_i \times \mathfrak{M}_i, \tag{12.13}$$

where $D_i = [a_i - a_i', b_i - b_i'] \times [a_i + a_i'', b_i + b_i'']$ for some positive numbers a_i', a_i'', b_i', b_i''. Then, a Hawkes structure in the space–time ETAS framework can be constructed in analogy to the construction of the Hawkes structure in the classical ETAS framework, upon taking E_i as in (12.13) and choosing the counterparts of the functions $h_{i,j}$ in (12.10) in correspondence to various forms of the function g discussed in Section 2.2 in Ogata (1998).

12.4.2 Criminology and Terrorism

It was argued in Section 12.4.1 that Hawkes structures can generalize the space–time extension of the ETAS model used for describing seismic activity. In a similar fashion, one can provide a useful generalization of the space–time extension of the models proposed for describing criminal activity as well as models proposed for describing terrorist activity in given areas. For example, models discussed in Mohler et al. (2011) and Clark and Dixon (2017) can be generalized so account for possible simultaneous activities in various areas, using our generalized multivariate Hawkes processes.

12.4.3 Neurology, Neuroscience, and DNA Sequencing

Neurons are responsible for transmitting information in the brain. This process is controlled by changes in neuronal electrical activity. These changes can be detected and recorded by electrodes. Hawkes processes have been used to model neuronal electrical activity.

In Reynaud-Bouret et al. (2013) the authors proposed a Hawkes-type model for the joint activity of a group of neurons. The changes in activity of the ith individual neuron have been modeled by what is called a quasi-Hawkes process. More precisely, the instantaneous firing rates of d different neurons are modeled by a multivariate Hawkes process such that spike trains $N^1, \ldots, N^d$ are point processes, with the conditional intensity of the kth point process N^k defined by

$$\lambda_k(t) = \nu_k + \sum_{j=1}^{d} \int_{(0,t)} h_{j,k}(t-s)N^j(ds), \quad t \geq 0, \tag{12.14}$$

where the ν_k are positive parameters representing the spontaneous firing rates and where the $h_{j,k}$ are interaction functions; see formula (2) in Reynaud-Bouret et al. (2013).

The quasi-Hawkes-process model presented in Reynaud-Bouret et al. (2013), represented in part by formula (12.14), mixes the idiosyncratic excitation effects, that is, the effects of electrical changes in only one neuron, with possible joint excitation effects caused by simultaneous electrical changes in a group of neurons. Moreover, this model does not account for the possibly different strengths of the changes in neuronal electrical activity.

These shortcomings may be addressed using the theory of Hawkes structures. For simplicity of presentation, we consider a model with $d = 3$ spike trains. Specifically, the neuronal activity is modeled by a generalized Hawkes process N of dimension 3 with Hawkes kernel κ given by

$$\kappa(t, dy) = \eta(t, dy) + \int_{(0,t)\times E^{\Delta}} f(t,s,x,dy)N(ds,dx).$$

The state space E is given as $E = E_1 \times E_2 \times E_3$, where $\{E_i\}$ is a set of various possible strengths of changes in neuronal electrical activity in the ith spike train. The kernel η (the spontaneous rate) models the external excitation linked to all neurons, and the kernels f models the cross-excitation between the neurons.

In order for N to constitute a Hawkes structure for the marginal spike trains, we require that for each coordinate N^i the corresponding kernel κ^i satisfies, for $i = 1,2,3$,

$$\begin{aligned}
\kappa^i(t,A_i) = \chi_i(t,A_i) & \\
&+ \int_{(0,t)\times E_1} h_{i,1}(t,s,x_1,A_i)N^{idio,1}(ds,dx_1) \\
&+ \int_{(0,t)\times E_2} h_{i,2}(t,s,x_2,A_i)N^{idio,2}(ds,dx_2) \\
&+ \int_{(0,t)\times E_3} h_{i,3}(t,s,x_1,A_i)N^{idio,3}(ds,dx_3) \\
&+ \int_{(0,t)\times E_1\times E_2} h_{i,1,2}(t,s,x_1,x_2,A_i)N^{idio,1,2}(ds,dx_1,x_2) \\
&+ \int_{(0,t)\times E_1\times E_3} h_{i,1,3}(t,s,x_1,x_3,A_i)N^{idio,1,3}(ds,dx_1,x_3) \\
&+ \int_{(0,t)\times E_2\times E_3} h_{i,2,3}(t,s,x_2,x_3,A_i)N^{idio,2,3}(ds,dx_2,x_3) \qquad (12.15)\\
&+ \int_{(0,t)\times E} h_{i,1,2,3}(t,s,x,A_i)N^{idio,1,2,3}(ds,dx). \qquad (12.16)
\end{aligned}$$

In summary, Hawkes structures allow one to incorporate more information in a model of neuronal activity.

Remark 12.4 In the set-up of Reynaud-Bouret et al. (2013) the functions $h_{j,k}$ may take negative values. Accordingly, in Reynaud-Bouret et al. (2013), formula (12.14) takes the form

$$\lambda_k(t) = \max\left\{ \nu_k + \sum_{j=1}^{d} \int_{(0,t)} h_{j,k}(t-s)N^j(ds), t \geq 0, \quad 0 \right\}. \tag{12.17}$$

This generalization may easily be accounted for in the context of Hawkes structures.

The functions χ_i and functions $h_{i,1}, \ldots, h_{i,1,2,3}$ appearing in (12.15) need to be estimated, and the model needs to be simulated. This can perhaps be done by adapting and extending the estimation and simulation techniques presented in Iyengar (2002), Reynaud-Bouret et al. (2013), Reynaud-Bouret et al. (2014), and Hansen et al. (2015).

In a similar fashion to that discussed above, Hawkes structures may also find applications in biology, for modeling the process of occurrences of a particular event (e.g. DNA patterns or transcriptional regulatory elements) along a DNA sequence; see e.g. Reynaud-Bouret and Schbath (2010) or the Ph.D. thesis of Carstensen (2010) for the underlying research.

12.4.4 Finance

A series of papers (Bacry et al. (2013a, b) and Bacry and Muzy (2014)) introduced a multidimensional model for stock prices driven by (multivariate) Hawkes processes. This model for stock prices was formulated in Bacry et al. (2013a) via a marked point process $N = (T_n, Z_n)_{n \geq 1}$, where Z_n is a random variable taking values in $\{1, \ldots, 2d\}$, and the compensator ν of N has the following form (it is assumed that $T_\infty = \lim_{n\to\infty} T_n = \infty$):

$$\nu(dt, dy) = \sum_{i=1}^{2d} \delta_i(dy)\lambda_i(t)dt,$$

where

$$\lambda_i(t) = \mu_i + \sum_{j=1}^{2d} \int_{(0,t)} \phi_{i,j}(t-s)N(ds \times \{j\}), \quad t \geq 0,$$

with $\mu_i \in \mathbb{R}_+$ and functions $\phi_{i,j}$ from $\mathbb{R}_+$ to $\mathbb{R}_+$. Let us define the processes N^i, $i = 1, \ldots 2d$, by

$$N^i((0,t]) = \sum_{n \geq 1} \mathbb{1}_{\{T_n \leq t\} \cap \{Z_n = i\}}, \quad t \geq 0.$$

Note that the above implies that $N^1, \ldots, N^{2d}$ have no common jumps and that the $\mathbb{F}^N$-intensity of N^i is given by λ_i and can be written in the form

$$\lambda_i(t) = \mu_i + \sum_{j=1}^{2d} \int_{(0,t)} \phi_{i,j}(t-s)N^j(ds), \quad t \geq 0.$$

Bacry et al. (2013a) assumed that a d-dimensional vector of asset prices $S = (S^1, \ldots, S^d)$ is based on N via the representation

$$S_t^i = N^{2i-1}((0,t]) - N^{2i}((0,t]), \quad t \geq 0, i = 1, \ldots, d.$$

The obvious interpretation is that N^{2i-1} corresponds to an upward jump of the ith asset and N^{2i} to a downward jump of the ith asset. Bacry et al. (2013a) showed that within such framework some stylized facts about high frequency data, such as microstructure noise and the Epps effect, are reproduced.

Using generalized multivariate Hawkes processes we can easily generalize their model in several directions. In particular, a model of stock price movements driven by a generalized multivariate Hawkes process N allows for common jumps in the upward and/or downward direction. This can be done by setting the multivariate mark space of N to be

$$E^\Delta = \{e = (e_1, \ldots, e_{2d}) : e_i \in \{1, \Delta\}\} \setminus \{(\Delta, \ldots, \Delta)\}$$

and the $\mathbb{F}^N$-compensator of N to be

$$\nu(dt, dy) = \mathbb{1}_{]\!]0,T_\infty[\![}(t) \sum_{e \in E^\Delta} \delta_e(dy) \lambda_e(t) dt,$$

where

$$\lambda_e(t) = \mu_e + \int_{E^\Delta \times (0,t)} \phi_{e,x}(t-s) N(ds \times dx), \quad e \in E^\Delta, t \geq 0,$$

and where $\mu_e \in \mathbb{R}_+$ and $\phi_{e,x}$ is a function from $\mathbb{R}_+$ to $\mathbb{R}_+$.

Including the possibility of embedding co-jumps of the prices of various stocks in the common excitation mechanism, as we have done in our book, may turn out to be important in modeling evolution in general and in pricing basket options in particular.

APPENDICES

Appendix A Stochastic Analysis: Selected Concepts and Results Used in this Book

We gather here selected concepts and results from stochastic analysis that we use throughout the book. For a comprehensive study of these concepts and results we refer to e.g. He et al. (1992), Last and Brandt (1995), and Jacod and Shiryaev (2003).

Fix an underlying probability space $(\Omega, \mathcal{F}, \mathbb{P})$. All random objects discussed below, such as random variables, stochastic processes, or random measures, are defined on $(\Omega, \mathcal{F}, \mathbb{P})$.

We endow this space with a filtration $\mathbb{F} := (\mathcal{F}_t)_{t \geq 0}$ and we suppose that this filtration satisfies the *usual conditions*, i.e.

1. $\mathbb{F}$ is right-continuous: $\mathcal{F}_t = \bigcap_{s>t} \mathcal{F}_s$ for all $t \geq 0$.
2. If $A \subset B$ with $B \in \mathcal{F}$, and if $\mathbb{P}(B) = 0$ then $A \in \mathcal{F}_0$.

We also set $\mathcal{F}_\infty := \sigma(\bigcup_{t \geq 0} \mathcal{F}_t) \subseteq \mathcal{F}$.

For two filtrations $\mathbb{F}$ and $\mathbb{G}$ we denote by $\mathbb{F} \vee \mathbb{G}$ the fitration defined as $\{\mathcal{F}_t \vee \mathcal{G}_t\}_{t \geq 0}$, where $\mathcal{F}_t \vee \mathcal{G}_t$ denotes the smallest σ-algebra containing all sets from $\mathcal{F}_t$ and all sets from $\mathcal{G}_t$. Moreover, by $\mathbb{F} \triangledown \mathbb{G}$ we denote the right-continuous regularization of the filtration $\mathbb{F} \vee \mathbb{G}$, i.e. $\mathbb{F} \triangledown \mathbb{G} = \{\bigcap_{s>t} \mathcal{F}_s \vee \mathcal{G}_s\}_{t \geq 0}$.

Stochastic processes Let $(\mathcal{X}, \boldsymbol{\mathcal{X}})$ be a measurable space. An $\mathcal{X}$-valued stochastic process (a process, for short), say X, is an $\mathcal{F} \otimes \mathcal{B}(\mathbb{R}_+)$-measurable mapping from $\Omega \times \mathbb{R}_+$ to $\mathcal{X}$, where $\mathcal{B}(\mathbb{R}_+)$ is the Borel σ-field on $\mathbb{R}_+ := [0, \infty)$. It is customary to denote $X(\omega, t)$ as $X_t(\omega)$ for $(\omega, t) \in \Omega \times \mathbb{R}_+$ and to consider a stochastic process X as a collection of random variables: $X = \{X_t,\, t \geq 0\}$. Frequently, the notation $(X_t)_{t \geq 0}$ is used in place of $\{X_t,\, t \geq 0\}$.

Sometimes, a process is considered as a measurable mapping from $\Omega \times \overline{\mathbb{R}}_+$ to $\mathcal{X}$, where $\overline{\mathbb{R}}_+ := [0, \infty]$.

A process X is $\mathbb{F}$-adapted if for all $t \geq 0$ the random variable X_t is $\mathcal{F}_t$ measurable.

The natural filtration of a process X, denoted by $\mathbb{F}^X$, is the smallest filtration in $(\Omega, \mathcal{F})$ such that it satisfies the usual conditions and X is adapted to it.

In some parts of the book we consider stochastic processes as functions from $\Omega \times [0, T]$ to $\mathcal{X}$, for $0 < T < \infty$. All the concepts and results presented in this appendix carry over to this case in a natural way.

Stochastic processes: path properties Let $\omega \in \Omega$. The path (or trajectory) of a process X corresponding to ω is the map $t \to X_t(\omega)$. We use the notation $X.(\omega)$ to denote the path of X corresponding to ω.

Assume additionally that $(\mathcal{X}, \boldsymbol{\mathcal{X}})$ is a topological space.

A process X is continuous if

$$\mathbb{P}(\{\omega \in \Omega : X.(\omega) \text{ is continuous}\}) = 1.$$

A process X is càdlàg[1] if

$$\mathbb{P}(\{\omega \in \Omega : X.(\omega) \text{ is right-continuous with left limits}\}) = 1.$$

If $\mathcal{X}$ is a topological vector space then for a càdlàg process X we define its jump ΔX_t at time $t > 0$ as

$$\Delta X_t := X_t - X_{t-},$$

where $X_{t-} := \lim_{s \to t, s<t} X_s$ for $t > 0$.

An $\mathbb{R}^d$-valued process X is increasing if X is càdlàg and

$$\mathbb{P}(\{\omega \in \Omega : X.(\omega) \text{ is non-decreasing (coordinate-wise)}\}) = 1.$$

An $\mathbb{R}^d$-valued process X is of finite variation if it is a difference of two increasing processes.

A càdlàg process X admits at most a countable number of jumps on any finite time interval $[s,t]$, for $0 \le s \le t < \infty$. Thus, for an $\mathbb{R}^n$-valued function f on $\mathcal{X}$ the integral $\int_{[s,t]} f(X_u)du$ exists if and only if the integral $\int_{[s,t]} f(X_{u-})du$ exists and

$$\int_{[s,t]} f(X_u)du = \int_{[s,t]} f(X_{u-})du.$$

Stochastic processes: pure jump process A process X is a *pure jump process* if it is càdlàg and has piecewise constant trajectories.

Stochastic processes: optional and predictable σ-fields The σ-field on $\Omega \times \mathbb{R}_+$ generated by all $\mathbb{F}$-adapted càdlàg processes is called the optional σ-field and is denoted by $\mathcal{O}(\mathbb{F})$ (or $\mathcal{O}$, for short, if no confusion regarding the filtration arises).

The σ-field on $\Omega \times \mathbb{R}_+$ that is generated by all $\mathbb{F}$-adapted left-continuous processes (as well as by continuous processes) is called the $\mathbb{F}$-predictable (or predictable) σ-field and is denoted by $\mathcal{P}(\mathbb{F})$ (or $\mathcal{P}$, for short, if no confusion regarding the filtration arises).

[1] From the French: "*continue à droit, limite à gauche*".

Stochastic processes: $\mathbb{F}$-optionality and $\mathbb{F}$-predictability A process X is $\mathbb{F}$-optional if it is $\mathcal{O}(\mathbb{F})$-measurable. A process X is $\mathbb{F}$-predictable if it is $\mathcal{P}(\mathbb{F})$-measurable.

Stopping times A $[0,\infty]$-valued random variable τ is called an $\mathbb{F}$-stopping time if $\{\tau \leq t\} \in \mathcal{F}_t$ for all $t \geq 0$.

For an $\mathbb{F}$-stopping time τ we define

$$\mathcal{F}_\tau := \{A \in \mathcal{F}_\infty : A \cap \{\tau \leq t\} \in \mathcal{F}_t,\ t \geq 0\}$$

and

$$\mathcal{F}_{\tau-} := \mathcal{F}_0 \vee \sigma(\{A \cap \{t < \tau\} : A \in \mathcal{F}_t,\ t > 0\}),$$

where for any family $\mathcal{A}$ of subsets of Ω we denote by $\sigma(\mathcal{A})$ the sigma field generated by $\mathcal{A}$. Conventionally $\mathcal{F}_{0-} = \mathcal{F}_0$.

Stochastic intervals Let θ and η be two functions on Ω taking values in $[0,\infty]$ and such that $\theta \leq \eta$. The corresponding stochastic intervals are defined as follows:

$$\begin{aligned}
[\![\theta,\eta]\!] &:= \{(\omega,t) \in \Omega \times [0,\infty) : \theta(\omega) \leq t \leq \eta(\omega)\},\\
[\![\theta,\eta[\![&:= \{(\omega,t) \in \Omega \times [0,\infty) : \theta(\omega) \leq t < \eta(\omega)\},\\
]\!]\theta,\eta]\!] &:= \{(\omega,t) \in \Omega \times [0,\infty) : \theta(\omega) < t \leq \eta(\omega)\},\\
]\!]\theta,\eta[\![&:= \{(\omega,t) \in \Omega \times [0,\infty) : \theta(\omega) < t < \eta(\omega)\}.
\end{aligned}$$

We define $[\![\theta]\!] := [\![\theta,\theta]\!]$ and call it the graph of θ.

Predictable stopping times An $\mathbb{F}$-stopping time is $\mathbb{F}$-predictable, or simply predictable, if there exists a sequence τ_n, $n = 0,1,\ldots$, of $\mathbb{F}$-stopping times such that the following two properties hold:

(i) $\lim_{n\to\infty} \tau_n = \tau \quad \mathbb{P}$-a.s.,
(ii) $\tau_n < \tau$ for $n = 0,1,\ldots$, on the set $\{\tau > 0\}$.

Equivalently, the stopping time τ is predictable if $[\![\tau,\infty[\![\in \mathcal{P}(\mathbb{F})$ (or, equivalently, if $[\![\tau]\!] \in \mathcal{P}(\mathbb{F})$).

Filtration: quasi-left continuity The filtration $\mathbb{F}$ is said to be quasi-left-continuous if, for any predictable stopping time τ, we have $\mathcal{F}_\tau = \mathcal{F}_{\tau-}$.

Martingales, submartingales, supermartingales An $\mathbb{R}^d$-valued stochastic process X is an $\mathbb{F}$-martingale (or $(\mathbb{F},\mathbb{P})$-martingale) if

1. $\mathbb{E}(|X_t|) < \infty$ for all $t \geq 0$.
2. $\mathbb{E}(X_{t+s}|\mathcal{F}_t) = X_t$ for all $t,s \geq 0$.

Let $Y \in \mathbb{R}^d$ be an integrable random variable. Then the process X defined as

$$X_t := \mathbb{E}(Y|\mathcal{F}_t), \quad t \geq 0,$$

is called a Doob martingale (generated by Y).

An $\mathbb{R}^d$-valued stochastic process X is an $\mathbb{F}$-supermartingale (resp. $\mathbb{F}$-submartingale) if

(i) X is $\mathbb{F}$-adapted.
(ii) $\mathbb{E}(|X_t|) < \infty$ for all $t \geq 0$.
(iii) $\mathbb{E}(X_{t+s}|\mathcal{F}_t) \leq X_t$ (resp. $\mathbb{E}(X_{t+s}|\mathcal{F}_t) \geq X_t$) for all $t, s \geq 0$.

Frequently, we write martingale, supermartingale, and submartingale, in place of $\mathbb{F}$-martingale, $\mathbb{F}$-supermartingale and $\mathbb{F}$-submartingale, if no confusion may arise regarding the filtration $\mathbb{F}$ that is used.

We assume in this book that (super-, sub-) martingales are càdlàg processes.

A martingale X is uniformly integrable (or is a UI-martingale) if the family of random variables $\{X_t,\ t \geq 0\}$ is uniformly integrable.

Localizing sequences and local martingales Let $\mathcal{C}$ be a class of stochastic processes. A collection $\{\tau_n,\ n = 1, 2, \ldots\}$ of $\mathbb{F}$-stopping times is called an $(\mathbb{F}, \mathcal{C})$-localizing sequence (or, just a localizing sequence) for a stochastic process X if

(i) The sequence $\{\tau_n,\ n = 1, 2, \ldots\}$ is increasing and $\lim_{n\to\infty} \tau_n = \infty$, $\quad \mathbb{P}$-a.s.
(ii) For every $n = 1, 2, \ldots$, the stopped (or, localized) process $X^{\tau_n} := (X_{t\wedge\tau_n})_{t\geq 0}$ belongs to the class $\mathcal{C}$.

A process X is a local martingale if there exists a martingale localizing sequence $\{\tau_n,\ n = 1, 2, \ldots\}$ for X, i.e. is for every $n = 1, 2, \ldots$, the localized process X^{τ_n} is a martingale. A process X is a strict local martingale if X is a local martingale but not a martingale.

If a process X is a local martingale then one can choose a localizing sequence $\{\tau_n,\ n = 1, 2, \ldots\}$ such that for every $n = 1, 2, \ldots$, the localized process X^{τ_n} is a UI-martingale. Thus, a localizing sequence for a local martingale will always be chosen in this way.

Continuous local martingale of finite variation Let X be a continuous local martingale with finite variation. Then X is a constant process.

Locally square integrable martingales and oblique brackets Let X be a real-valued martingale. We say that X is square integrable if

$$\sup_{t\geq 0} \mathbb{E}(X_t^2) < \infty.$$

A real-valued process X is called a locally square integrable martingale if there exists a martingale localizing sequence $\{\tau_n,\ n = 1,2,\ldots\}$ for X, that is, for every $n = 1,2,\ldots$, the localized process X^{τ_n} is a square integrable martingale.

A continuous local martingale X such that $\mathbb{E}(X_0^2) < \infty$ is a locally square integrable martingale.

Let X and Y be locally square integrable martingales. Then, there exists a unique (up to an evanescent set) predictable process $\langle X, Y\rangle$ such that $XY - \langle X, Y\rangle$ is a local martingale. The process $\langle X, Y\rangle$ is called the oblique bracket between X and Y (or the predictable quadratic co-variation between X and Y).

We define the predictable quadratic variation process for a locally square integrable martingale X as $\langle X\rangle := \langle X, X\rangle$. Sometimes, in order to emphasize the filtration to relative which is the oblique bracket is computed, one uses the notation $\langle X\rangle^{\mathbb{F}}$.

Let X and Y be two real-valued local martingales. They are called orthogonal if their product XY is a local martingale.

A real-valued local martingale X is said to be purely discontinuous if $X_0 = 0$ and if X is orthogonal to all continuous real-valued local martingales.

Any real-valued local martingale X admits a unique decomposition

$$X = X_0 + X^c + X^d,$$

where X^c and X^d are continuous and purely discontinuous local martingales, respectively, with $X_0^c = X_0^d = 0$. The process X^c is called the continuous part of X.

Immersion of filtrations Let $\mathbb{H}$ and $\mathbb{G}$ be two filtrations on $(\Omega, \mathcal{F}, \mathbb{P})$. We say that a filtration $\mathbb{H}$ is $\mathbb{P}$-*immersed* in the filtration $\mathbb{G}$ if $\mathbb{H} \subset \mathbb{G}$ and if every $(\mathbb{H}, \mathbb{P})$-local martingale is a $(\mathbb{G}, \mathbb{P})$-local martingale.

Equivalent conditions for the immersion of two filtrations can be found in e.g. Jeanblanc et al. (2009). In particular, *local martingale* in the above definition can be replaced with martingale or with square integrable martingale.

Compensator of an increasing process Let X be a locally integrable càdlàg increasing process, $X_0 = 0$. Then there exists a unique predictable locally integrable càdlàg increasing process A, $A_0 = 0$, which is characterized by one of the following three conditions:

(i) $X - A$ is a local martingale.
(ii) $\mathbb{E}X_\tau = \mathbb{E}A_\tau$ for every stopping time τ.
(iii) $\mathbb{E}[(H \cdot X)_\infty] = \mathbb{E}[(H \cdot A)_\infty]$ for all nonnegative predictable processes X.

The process A satisfying the above properties is called the $\mathbb{F}$-compensator (or, just the compensator) of X.

Semimartingales A càdlàg process X-is an $\mathbb{F}$-semimartingale if it is $\mathbb{F}$-adapted, and if

$$X = X_0 + M + B, \tag{A.1}$$

where M is an $\mathbb{F}$-local martingale with $M_0 = 0$, and where B is an $\mathbb{F}$-adapted process of finite variation with $B_0 = 0$. The representation (or, decomposition) (A.1) is not unique in general.

Special semimartingales Let X be an $\mathbb{F}$-semimartingale admitting decomposition (A.1). If the process B is $\mathbb{F}$-predictable then the decomposition (A.1) is unique. In this case the process X is called a special $\mathbb{F}$-semimartingale (or, just a special semimartingale).

The decomposition (A.1) for a special semimartingale X is called the canonical decomposition.

Semimartingales: square brackets An $\mathbb{F}$-adapted subdivision (or just an adapted subdivision) is any sequence $S = \{\tau_n,\ n = 1,2,\ldots\}$ of $\mathbb{F}$-stopping times such that $\tau_0 = 0$, $\tau_n < \tau_{n+1}$ on the set $\{\tau_n < \infty\}$ and $\sup_n \tau_n < \infty$. A sequence $(S_k = \{\tau_n^k,\ n = 1,2,\ldots\})_{k=1}^{\infty}$ of adapted subdivisions S_k is called a Riemann sequence if

$$\sup_n [\tau_{n+1}^k \wedge t - \tau_n^k \wedge t] \underset{k\to\infty}{\longrightarrow} 0 \quad \mathbb{P}\text{-a.s.}$$

for $t \geq 0$.

Let X and Y be two real-valued semimartingales. Then, for any Riemann sequence $(S_k = \{\tau_n^k,\ n = 1,2,\ldots\})_{k=1}^{\infty}$ the processes $\Sigma^k(X,Y)$ defined by

$$\Sigma^k(X,Y)_t = \sum_{n=1}^{\infty} \left(X_{\tau_{n+1}^k \wedge t} - X_{\tau_n^k \wedge t} \right) \left(Y_{\tau_{n+1}^k \wedge t} - X_{\tau_n^k \wedge t} \right)$$

converge, as $k \to \infty$, to a càdlàg process denoted as $[X,Y]$. The convergence is in probability and uniform on each compact interval.

The process $[X,Y]$ is called the square bracket of X and Y (or, the quadratic co-variation between X and Y), and it satisfies the following:

(i) $[X,Y]$ is a process of finite variation.
(ii) $\Delta[X,Y] = \Delta X \Delta Y$.
(iii) If X and Y are local martingales, then the process

$$XY - [X,Y]$$

is a local martingale.
(iv) If X and Y are locally square integrable martingales then the process

$$[X,Y] - \langle X,Y \rangle$$

is a local martingale.

The square bracket (or quadratic variation) of a real-valued semimartingale X is defined as $[X] := [X,X]$. The process $[X]$ is nonnegative.

Orthogonal local martingales Let $U = (U^1, \dots, U^n)$ and $V = (V^1, \dots, V^m)$ be two $\mathbb{F}$-local martingales taking values in $\mathbb{R}^n$ and $\mathbb{R}^m$, respectively. We say that U and V are orthogonal if the processes $U^i V^j$ are $\mathbb{F}$-local martingales or, equivalently, if the square bracket process $[U^i, V^j]$ is null for $i = 1, \dots, n$, $j = 1, \dots, m$.

Random measure A nonnegative function μ on $\Omega \times \mathcal{B}(\mathbb{R}_+) \times \boldsymbol{\mathcal{X}}$ is called a random measure on $\mathbb{R}_+ \times \mathcal{X}$ if, for any $\omega \in \Omega$, the function $\mu(\omega; \cdot, \cdot)$ is a measure on $(\mathbb{R}_+ \times \mathcal{X}, \mathcal{B}(\mathbb{R}_+) \times \boldsymbol{\mathcal{X}})$.

We let

$$\widetilde{\Omega} = \Omega \times \mathbb{R}_+ \times \mathcal{X}, \quad \widetilde{\mathcal{O}} = \mathcal{O} \otimes \boldsymbol{\mathcal{X}}, \quad \widetilde{\mathcal{P}} = \mathcal{P} \otimes \boldsymbol{\mathcal{X}}.$$

Now let W be a real-valued function on $\widetilde{\Omega}$. If W is $\widetilde{\mathcal{O}}$-measurable then it is called $\mathbb{F}$-optional (optional, for short, if no confusion regarding the filtration arises). If W is $\widetilde{\mathcal{P}}$-measurable then it is called $\mathbb{F}$-predictable (predictable, for short, if no confusion regarding the filtration arises).

For an optional function W and for $t \geq 0$, we define

$$W * \mu_t(\omega) := \begin{cases} \int_{[0,t] \times \mathcal{X}} W(\omega, s, x) \mu(\omega; ds, dx) & \text{if } \int_{[0,t] \times \mathcal{X}} |W(\omega, s, x)| \mu(\omega; ds, dx) \\ & \text{is finite,} \\ \infty & \text{if otherwise.} \end{cases}$$

A random measure μ is called $\mathbb{F}$-optional (optional, for short, if no confusion regarding the filtration arises) if the process $W * \mu$ is optional for any $\mathbb{F}$-optional function W. Likewise, a random measure μ is called $\mathbb{F}$-predictable (predictable, for short, if no confusion regarding the filtration arises) if the process $W * \mu$ is $\mathbb{F}$-predictable for any $\mathbb{F}$-predictable function W.

Let a random measure μ be optional. If there exists a strictly positive predictable random function W such that the random variable $W * \mu_\infty$ is integrable then μ is called $\widetilde{\mathcal{P}}$-σ-finite.

Random measures: compensator Let a random measure μ be $\widetilde{\mathcal{P}}$-σ-finite. Then there exists a unique predictable random measure ν which is characterized by either one of the two following properties:

(i) $\mathbb{E}(W * \mu_\infty) = \mathbb{E}(W * \nu_\infty)$ for every nonnegative predictable function W.
(ii) For any predictable function W such that the process $|W| * \mu$ is increasing and locally integrable, the process

$$W * \mu - W * \nu$$

is a local martingale.

A measure ν satisfying the above properties is called the $(\mathbb{F}, \mathbb{P})$-compensator (or, the $\mathbb{F}$-compensator, or just the compensator) or the dual predictable projection onto $\mathbb{F}$ of μ.

Integer-valued random measure A random measure μ that satisfies the conditions

(i) $\mu(\omega;\{t\},\mathcal{X}) \le 1$ for each $\omega \in \Omega$,
(ii) $\mu(\omega;A) \in \{0,1,\ldots,\infty\}$ for each $\omega \in \Omega$ and each $A \in \mathcal{B}(\mathbb{R}_+) \times \boldsymbol{\mathcal{X}}$,
(iii) μ is $\widetilde{\mathcal{P}}$-σ-finite

is called an integer-valued random measure.

Let X be an $\mathbb{F}$-adapted càdlàg process taking values in $\mathbb{R}^d$. Let

$$\mu^X(\omega;dt,dx) = \sum_{s>0} \mathbb{1}_{\{\Delta X_s(\omega)\neq 0\}} \delta_{\{s,\Delta X_s(\omega)\}}(dt,dx).$$

Then μ^X is an integer-valued random measure on $\mathbb{R}_+ \times \mathbb{R}^d$. The measure μ^X is called the jump measure of X and its $(\mathbb{F},\mathbb{P})$-compensator is denoted by ν^X or just by ν if no confusion arises.

Poisson random measure A Poisson random measure is an integer-valued random measure μ such that

(i) the positive measure ν on $\mathbb{R}_+ \times \mathcal{X}$ defined by $\nu(A) = \mathbb{E}(\mu(A))$ is σ-finite,
(ii) for any $t \ge 0$ and any $A \in \mathcal{B}(\mathbb{R}_+) \times \boldsymbol{\mathcal{X}}$ such that $A \subset \mathbb{R}_+ \times \mathcal{X}$ and $\nu(A) < \infty$, the random variable $\mu(\cdot,A)$ is independent of $\mathcal{F}_t$,
(iii) $\nu(\{t\},\mathcal{X}) = 0$ for any $t \ge 0$.

The measure ν is the $(\mathbb{F},\mathbb{P})$-compensator of μ, or just the $\mathbb{F}$-compensator of μ. It is sometimes called the intensity measure of μ.

Suppose that μ is a Poisson measure on $\mathbb{R}_+ \times \mathbb{R}$ with compensator of the form $\nu(dt,dx) = \lambda(t)dt\delta_{\{1\}}(dx)$. Then $N_t = \mu((0,t],\mathbb{R})$, $t \ge 0$, is the one-dimensional time-inhomogeneous Poisson process with rate $\lambda(\cdot)$.

Point processes A point process is a sequence $(T_n)_{n\ge 1}$ of positive random variables which may take the value $+\infty$ and which satisfy

$$T_n < T_{n+1} \quad \text{if} \quad T_n < \infty,$$
$$T_n = T_{n+1} = \infty \quad \text{if} \quad T_n = \infty.$$

The counting process N associated with the sequence (T_n) is the process defined by

$$N_t = \sum_{n\ge 1} \mathbb{1}_{\{T_n \le t\}}, \quad t \ge 0,$$

which is a point process. A nice example of a point process is the one-dimensional Poisson process.

Counting processes A cádlág process N with values in $\mathbb{N} \cup \{\infty\}$, whose paths are nondecreasing functions with all jump sizes equal to 1, is called a counting process. A sequence $(T_n)_{n \geq 1}$ of jump times of N, defined by

$$T_n = \inf\{t > T_{n-1} : \Delta N_t \geq 0\}, \quad T_0 = 0,$$

is called the point process associated with N.

An n-variate process $\widetilde{N} = (N^1, \ldots, N^n)$, where each N^i, $i = 1, \ldots, n$, is a point process, is called an n-variate counting process.

Clearly, one may identify point processes and counting processes. Therefore, frequently, the counting process N associated with a point process $(T_n)_{n \geq 1}$, is also called a point process.

Point processes: intensity Let N be a counting (point) process adapted to a filtration $\mathbb{F}$ and let $(T_n)_{n \geq 1}$ be the associated point process. Let λ be a nonnegative $\mathbb{F}$-progressively-measurable process such that, for all $t \geq 0$,

$$\int_0^t \lambda_s ds < \infty, \quad \mathbb{P}\text{-a.s.}$$

We call λ the $(\mathbb{F}, \mathbb{P})$-intensity of N, or the $\mathbb{F}$-intensity of N, or just the intensity of N, if, for all nonnegative $\mathbb{F}$-predictable processes Y, it holds that

$$\mathbb{E}\Big(\int_0^\infty Y_s dN_s\Big) = \mathbb{E}\Big(\int_0^\infty Y_s \lambda_s ds\Big),$$

where $\int_0^\infty Y_s dN_s := \sum_{n=1}^\infty Y_{T_n} \mathbb{1}_{\{T_n < \infty\}}$.

It turns out that λ is the $\mathbb{F}$-intensity of N if and only if the process M given by

$$M_t = N_t - \int_0^t \lambda_s ds, \quad t \geq 0$$

is an $\mathbb{F}$-local martingale. Frequently, the intensity λ of N is called the jump intensity of N relative to $\mathbb{F}$ (or just the jump intensity of N) or the $\mathbb{F}$-intensity of jumps of N (or just the intensity of jumps of N).

In general, for a point process N, an $\mathbb{F}$-intensity may not exist. However, there always exists a unique $\mathbb{F}$-predictable nondecreasing process Λ such that

$$N_t - \Lambda_t, \quad t \geq 0$$

is an $\mathbb{F}$-local martingale. We call Λ the $\mathbb{F}$-compensator of the point process N. Thus we see that if

$$\Lambda_t = \int_0^t \lambda_u du, \quad t \geq 0,$$

then λ is an $\mathbb{F}$-intensity of a point process N.

Marked point processes: intensity Let $N = (T_n, Z_n)_{n \geq 1}$ be a marked point process (see Section 11.1). The $\mathbb{F}$-intensity λ of a point process $(T_n)_{n \geq 1}$, if exists, is called the $(\mathbb{F}, \mathbb{P})$-intensity, or the $\mathbb{F}$-intensity, or just the intensity, of the marked point process N.

Semimartingale: truncation functions and the triple of (predictable) characteristics A truncation function on $\mathbb{R}^d$ is a bounded function $h : \mathbb{R}^d \to \mathbb{R}^d$, with compact support, satisfying $h(x) = x$ in the neighborhood of 0. The standard truncation function is $h(x) = x\mathbb{1}_{\{|x|\leq 1\}}$.

Let $X = (X^1,\dots,X^d)$ be an $\mathbb{R}^d$-valued semimartingale. Fix the truncation function h and define a process $X(h)$ as

$$X_t(h) = X_t - \sum_{0<s\leq t} (\Delta X_s - h(\Delta X_s)), \quad t \geq 0.$$

The process $X(h)$ is a special semimartingale. Denote its canonical decomposition (see Definition I.4.21 in Jacod and Shiryaev (2003)) as

$$X(h) = X_0 + M(h) + B(h),$$

where $M(h) = (M^1(h),\dots,M^d(h))$ and similarly for $B(h)$.

Let a matrix-valued process $C = (C^{ij})_{1\leq i,j\leq d}$ be defined by

$$C^{ij} = \langle X^{i,c}, X^{j,c}\rangle,$$

where $X^{k,c} = M^{k,c}$, with $M^{k,c}$ the (h-independent) continuous part of the local martingale $M^k(h)$, $k = 1,\dots,d$.

Let ν be the compensator of the jump measure μ^X of X. The triple

$$(B(h), C, \nu)$$

is called the $\mathbb{F}$-characteristic triple (or, the triple of (predictable) characteristics) of X (corresponding to the truncation function h).

The following representation of the semimartingale X,

$$X = X_0 + B(h) + X^c + h * (\mu^X - \nu) + (x - h(x)) * \mu^X,$$

is called the integral (or canonical) representation of X.

Semimartingales: stochastic integrals Let X be a real-valued semimartingale, and let us denote by $\mathfrak{H}$ the class of real-valued simple processes H such that

either $H = Y\mathbb{1}_{\{0\}}$, where Y is bounded and $\mathcal{F}_0$ is measurable,
or $H = Y\mathbb{1}_{]r,s]}$, where $r < s$, Y is bounded, and $\mathcal{F}_r$ is measurable.

For such processes we define the stochastic integral process $H \cdot S = \int_0^\cdot H_t dS_t = \int_{(0,\cdot]} H_t dS_t$:

$$H \cdot S_t = \begin{cases} 0 & \text{if } H = Y\mathbb{1}_{\{0\}}; \\ Y(X_{s\wedge t} - X_{r\wedge t}) & \text{if } H = Y\mathbb{1}_{]r,s]}. \end{cases}$$

The linear map $H \rightsquigarrow H \cdot S$ can be extended from $\mathfrak{H}$ to the class of all processes H that are locally bounded and predictable. The extended map is still denoted $H \rightsquigarrow H \cdot S$ and is called the stochastic integral of H with respect to X.

In particular, $H \rightsquigarrow H \cdot S$ is well defined for left-continuous adapted processes H.

Semimartingales: stochastic integration by parts For a real-valued semimartingale X we denote by X_- its left-continuous version: $X_- = (X_{t-})_{t\geq 0}$, where we set $X_{0-} = X_0$.

Now, let X and Y be two real-valued semimartingales. Then we have

$$XY = X_0Y_0 + X_- \cdot Y + Y_- \cdot X + [X,Y].$$

Change of probability measure and Girsanov's theorem Let $\mathbb{P}'$ be a probability measure on $(\Omega, \mathcal{F})$. Assume that $\mathbb{P}'$ is equivalent to $\mathbb{P}$ and let $Z = d\mathbb{P}'/d\mathbb{P}$ and $Z_t = \mathbb{E}(Z|\mathcal{F}_t)$, for $t \geq 0$.

Let $h(x) = x\mathbb{1}_{\{|x|\leq 1\}}$ be a truncation function, and let X be a real-valued semimartingale on $(\Omega, \mathcal{F}, \mathbb{F}, \mathbb{P})$ with integral decomposition relative to h and $\mathbb{P}$ given as

$$X = X_0 + B(h) + X^c + h * (\mu^X - \nu) + (x - h(x)) * \mu^X.$$

Then X is a real-valued semimartingale on $(\Omega, \mathcal{F}, \mathbb{F}, \mathbb{P}')$ with integral decomposition relative to h and $\mathbb{P}'$ given as

$$X = X_0 + B'(h) + X'^c + h * (\mu^X - \nu') + (x - h(x)) * \mu^X,$$

where

$$X'^c = X^c - \frac{1}{Z_-} \cdot \langle X^c, Z \rangle.$$

The characteristic triplet of X under $\mathbb{P}'$ is $(B'(h), C', \nu')$, where

$$\begin{aligned} B'(h) &= B(h) + \frac{1}{Z_-} \cdot \langle X^c, Z \rangle + ((W-1)h(x)) * \nu, \\ C' &= C, \\ \nu' &= W \cdot \nu, \end{aligned}$$

for some nonnegative predictable function W, with

$$W \cdot \nu(\omega; dt, dx) = \nu(\omega; dt, dx) W(\omega, t, x).$$

For the special case of the Girsanov theorem applied to counting point processes we refer to Theorem VI.T3 in Brémaud (1981).

Measurable kernels and stochastic kernels Let $(\mathcal{Y}, \boldsymbol{\mathcal{Y}})$ and $(\mathcal{Z}, \boldsymbol{\mathcal{Z}})$ be measurable spaces. A measurable kernel (or, just kernel) K from $(\mathcal{Y}, \boldsymbol{\mathcal{Y}})$ to $(\mathcal{Z}, \boldsymbol{\mathcal{Z}})$ is a mapping

$$K : \mathcal{Y} \times \boldsymbol{\mathcal{Z}} \to [0, \infty]$$

such that

1. $K(\cdot, A)$ is measurable for each $A \subset \boldsymbol{\mathcal{Z}}$;
2. $K(y, \cdot)$ is a measure on $(\mathcal{Z}, \boldsymbol{\mathcal{Z}})$ for each $y \in \mathcal{Y}$.

A stochastic (probability) kernel K from $(\mathcal{Y}, \boldsymbol{\mathcal{Y}})$ to $(\mathcal{Z}, \boldsymbol{\mathcal{Z}})$ is a kernel K from $(\mathcal{Y}, \boldsymbol{\mathcal{Y}})$ to $(\mathcal{Z}, \boldsymbol{\mathcal{Z}})$ satisfying

$$K(y, \mathcal{Z}) = 1$$

for each $y \in \mathcal{Y}$.

Lévy kernels A kernel μ from $(\mathbb{R}^d, \mathcal{B}(\mathbb{R}^d))$ to $(\mathbb{R}^d \setminus \{0\}, \mathcal{B}(\mathbb{R}^d \setminus \{0\}))$ is called a Lévy kernel if it satisfies

$$\int_{\mathbb{R}^d \setminus \{0\}} (|y|^2 \wedge 1) \mu(x, dy) < \infty$$

for all $x \in \mathbb{R}^d$.

Copulas A function $C : [0,1]^n \to [0,1]$ is called an n-copula (a copula for short) if and only if the following properties hold:

(i) $C(1, \ldots, 1, u_j, 1, \ldots, 1) = u_j$ for every $j \in \{1, 2, \ldots, n\}$;
(ii) C is isotonic, i.e. $C(u) \leq C(v)$ for all $u, v \in [0,1]^n, u \leq v$;
(iii) C is n-increasing, i.e.

$$\sum_{w \in \{u_1, v_1\} \times \cdots \times \{u_n, v_n\}} (-1)^{\#\{j : w_j = v_j\}} C(w) \geq 0$$

for all $u, v \in [0,1]^n$, $u \leq v$.

Nelsen (2006) provides an excellent overview of the theory of copulas (or, copulae).

Appendix B Markov Processes and Markov Families

B.1 Markov Process

Let $\mathcal{X}$ be a metric space which is endowed with the Borel σ-algebra $\boldsymbol{\mathcal{X}}$. In addition, let $(\Omega, \mathcal{F})$ be a measurable space endowed with a family of filtrations $\{\mathbb{F}_s = \{\mathcal{F}_{s,t}, s \leq t \leq \infty\}, s \geq 0\}$. Thus, for each $s \geq 0$ we have that $\mathcal{F}_{s,u} \subseteq \mathcal{F}_{s,t} \subseteq \mathcal{F}$ for $0 \leq s \leq u \leq t \leq \infty$, where $\mathcal{F}_{s,\infty} = \sigma\left(\bigcup_{t \geq s} \mathcal{F}_{s,t}\right)$. Finally, let X be an $\mathcal{F} \otimes \mathcal{B}(\mathbb{R}_+)$-measurable mapping from $\Omega \times \mathbb{R}_+$ to $\mathcal{X}$. As mentioned in Appendix A, we think of X as of a collection $X = (X_t)_{t \geq 0}$ of measurable mappings defined on $(\Omega, \mathcal{F})$ with values in $\mathcal{X}$.

Definition B.1 Let $s \geq 0$. The collection of objects $(\Omega, \mathcal{F}, \mathbb{F}_s, (X_t)_{t \geq s}, \mathbb{Q}_s)$ is a *Markov process starting at time s and taking values in* $\mathcal{X}$ if

1. $\mathbb{Q}_s$ is a probability measure on $(\Omega, \mathcal{F}_{s,\infty})$.
2. The process $(X_t)_{t \geq s}$ is adapted to $\mathbb{F}_s$.
3. $\mathbb{Q}_s(X_t \in B | \mathcal{F}_{s,u}) = \mathbb{Q}_s(X_t \in B | X_u)$ $\mathbb{Q}_s$-a.s. for any $s \leq u \leq t$ and $B \in \boldsymbol{\mathcal{X}}$.

The measure μ_s defined by

$$\mu_s(B) := \mathbb{Q}_s(X_s \in B), \quad B \in \boldsymbol{\mathcal{X}}$$

is called the *initial law* of $(\Omega, \mathcal{F}, \mathbb{F}_s, (X_t)_{t \geq s}, \mathbb{Q}_s)$.

If no confusion arises, we will typically skip the qualifier "taking values in $\mathcal{X}$" when referring to Markov processes.

Remark B.2 Occasionally we consider Markov processes as a collection

$$\{(\Omega, \mathcal{F}, \mathbb{F}, (X_t)_{t \geq s}, \mathbb{P}) : s \geq 0\},$$

where $\mathbb{F} = \mathbb{F}_0$ and $\mathbb{P}$ is a probability measure on $(\Omega, \mathcal{F}_{0,\infty})$.

Standing assumption (B1) We consider only conservative Markov processes, so that $\mathbb{Q}_s(X_t \in \mathcal{X}) = 1$ for any $s \leq t$.

A key concept in the theory of Markov processes is that of transition function.

Definition B.3 A function $P(u,x,t,B)$, defined for every $0 \leq u \leq t$, $x \in \mathcal{X}$, $B \in \boldsymbol{\mathcal{X}}$, is called a *transition function* if

1. For fixed u,t,x, the function $P(u,x,t,\cdot)$ is a probability measure.
2. For fixed u,t,B, the function $P(u,\cdot,t,B)$ is measurable (with respect to the σ-algebra $\boldsymbol{\mathcal{X}}$).
3. For fixed u,x, we have $P(u,x,u,\cdot) = \delta_x(\cdot)$, the unit mass at x.
4. The following Chapman–Kolmogorov property is satisfied: for arbitrary $u \leq t \leq v$, $x \in \mathcal{X}$, $B \in \boldsymbol{\mathcal{X}}$ it holds that
$$P(u,x,v,B) = \int_{\mathcal{X}} P(t,y,v,B)P(u,x,t,dy).$$

Accordingly, we extend Definition B.1 as follows:

Definition B.4 Let $s \geq 0$. The collection of objects $(\Omega, \mathcal{F}, \mathbb{F}_s, (X_t)_{t \geq s}, \mathbb{Q}_s, P)$ is a *Markov process starting at time s and admitting transition function P* if

1. $(\Omega, \mathcal{F}, \mathbb{F}_s, (X_t)_{t \geq s}, \mathbb{Q}_s)$ is a Markov process starting at time s.
2. $\mathbb{Q}_s(X_t \in B | \mathcal{F}_{s,u}) = P(u, X_u, t, B)$ $\mathbb{Q}_s$-a.s. for any $s \leq u \leq t$ and $B \in \boldsymbol{\mathcal{X}}$.

Remark B.5 It needs to be stressed that there exist Markov processes that do not admit transition functions. We refer to Section 8.1 in Wentzell (1981) for a more detailed discussion of this issue.

Standing assumption (B2) We consider only Markov processes that admit transition functions. Additionally, we assume that the transition function P satisfies property 2 in Definition B.4 for any $s \geq 0$, that is for Markov processes starting at any time $s \geq 0$.

In many applications the property of time-homogeneity of the underlying Markovian model is postulated.

Definition B.6 We say that a transition function P is *time-homogeneous* if
$$P(u,x,t,B) = P(0,x,t-u,B), \quad 0 \leq u \leq t, x \in \mathcal{X}, B \in \boldsymbol{\mathcal{X}}. \tag{B.1}$$

In such a case we shall write $P(x,t,B)$ instead of $P(0,x,t,B)$. If the transition function P of a Markov process $(\Omega, \mathcal{F}, \mathbb{F}_s, (X_t)_{t \geq s}, \mathbb{Q}_s, P)$ is time-homogeneous, we say that the Markov process itself is time-homogeneous.

Remark B.7 We observe that for a time-homogeneous Markov process
$$(\Omega, \mathcal{F}, \mathbb{F}_s, (X_t)_{t \geq s}, \mathbb{Q}_s, P)$$
and for any $0 \leq s \leq u$ and $r \geq 0$ we have
$$\mathbb{Q}_s(X_{u+r} \in B | X_u = x) = P(x,r,B) = \mathbb{Q}_s(X_{s+r} \in B | X_s = x).$$

Accordingly, by convention and without loss of generality, in the time-homogeneous case we take $s = 0$ and consider a time-homogeneous Markov process as a collection of objects $(\Omega, \mathcal{F}, \mathbb{F}, (X_t)_{t \geq 0}, \mathbb{Q}, P)$, where $\mathbb{F} = \mathbb{F}_0$ and $\mathbb{Q} = \mathbb{Q}_0$.

Canonical Space, Canonical Probability Space, and Canonical Markov Process According to Theorem 4.3, Chapter I, in Blumenthal and Getoor (1968), every Markov process is equivalent to what they call a Markov process of function-space type, what is now commonly known as a canonical process. A canonical process is essentially a process defined in the canonical way on the canonical space. The details follow.

We take Ω as

$$\Omega = \Omega(\mathcal{X}) := \mathcal{X}^{\mathbb{R}_+} \quad \text{and} \quad \mathcal{F} = \sigma\{X_u^{-1}(B) : B \in \boldsymbol{\mathcal{X}}, u \geq 0\}, \tag{B.2}$$

where $\mathbb{R}_+ = [0,\infty)$ and where X is defined by the *canonical mapping*

$$X_\cdot(\omega) = \omega(\cdot), \tag{B.3}$$

so that $X_u(\omega) = \omega(u)$.

We endow the space $(\Omega, \mathcal{F})$ with the family of canonical filtrations

$$\{\mathbb{F}_s = \{\mathcal{F}_{s,t},\ s \leq t \leq \infty\},\ s \geq 0\}, \tag{B.4}$$

where, for $t \in [s,\infty)$,

$$\mathcal{F}_{s,t} := \sigma\{X_u^{-1}(B) : B \in \boldsymbol{\mathcal{X}}, s \leq u \leq t\}$$

and

$$\mathcal{F}_{s,\infty} = \sigma\{X_u^{-1}(B) : B \in \boldsymbol{\mathcal{X}}, s \leq u\}.$$

Remark B.8 We follow the typically adopted convention, and, without changing the notation, we appropriately modify filtrations $\mathbb{F}_s$ so that they satisfy the usual conditions, which are obvious analogues of conditions 1. and 2. stated at the beginning of Appendix A. Accordingly, the canonical filtration $\mathbb{F}_s$ is the natural filtration of the canonical process X started at time s. As such, it might be conventionally denoted as $\mathbb{F}_s^X$. We will not use the latter notation in this appendix though.

Definition B.9 The pair $(\Omega, \mathcal{F})$ is called the *canonical space over* $(\mathcal{X}, \boldsymbol{\mathcal{X}})$. For any $s \geq 0$ and a probability measure $\mathbb{Q}_s$ on $(\Omega, \mathcal{F}_{s,\infty})$ we call the collection $(\Omega, \mathcal{F}, \mathbb{F}_s, \mathbb{Q}_s)$ a *canonical probability space over* $(\mathcal{X}, \boldsymbol{\mathcal{X}})$. If $(X_t)_{t\geq s}$ is the canonical mapping on $(\Omega, \mathcal{F})$ defined as in (B.3) then we call the collection $(\Omega, \mathcal{F}, \mathbb{F}_s, (X_t)_{t\geq s}, \mathbb{Q}_s)$ a *canonical process over* $(\mathcal{X}, \boldsymbol{\mathcal{X}})$ starting at time s. If no confusion could arise, we will skip the qualifier "over $(\mathcal{X}, \boldsymbol{\mathcal{X}})$."

Remark B.10 It is important to observe that it is the measure $\mathbb{Q}_s$ which characterizes a canonical process. That is, two canonical processes

$$(\Omega, \mathcal{F}, \mathbb{F}_s, (X_t)_{t\geq s}, \mathbb{Q}_s) \quad \text{and} \quad (\Omega, \mathcal{F}, \mathbb{F}_s, (X_t)_{t\geq s}, \mathbb{P}_s)$$

are identical if and only if the measures $\mathbb{Q}_s$ and $\mathbb{P}_s$ are equal.

Remark B.11 Let $(\Omega, \mathcal{F}, \mathbb{F}_s, (X_t)_{t\geq s}, \mathbb{Q}_s)$ be a canonical process over $(\mathcal{X}, \boldsymbol{\mathcal{X}})$ starting at time s, and let $(\bar{\Omega}, \bar{\mathcal{F}}, \bar{\mathbb{F}}_s, \bar{\mathbb{Q}}_s)$ be a canonical space over $(\bar{\mathcal{X}}, \bar{\boldsymbol{\mathcal{X}}})$. Let h: $(\Omega, \mathcal{F}) \mapsto (\bar{\Omega}, \bar{\mathcal{F}})$ be a measurable mapping and define a process Z on $(\Omega, \mathcal{F}, \mathbb{F}_s, \mathbb{Q}_s)$ by

$$Z_t(\omega) = (h(\omega))_t, \quad t \geq s.$$

Then, in general, $(\Omega, \mathcal{F}, \mathbb{F}_s, (Z_t)_{t\geq s}, \mathbb{Q}_s)$ is not a canonical process over $(\mathcal{X}, \bar{\boldsymbol{\mathcal{X}}})$, and $\mathbb{F}^Z \subsetneq \mathbb{F}_s$. As we will see in Remark B.17 it may happen that $Z_t(\omega) = \bar{Z}_t(\bar{\omega})$ for $t \geq s$, where $(\bar{\Omega}, \bar{\mathcal{F}}, \bar{\mathbb{F}}_s, (\bar{Z}_t)_{t\geq s}, \bar{\mathbb{Q}}_s)$ is a canonical process.

In the rest of this appendix the measurable space $(\Omega, \mathcal{F})$ will be understood in the canonical sense. It needs to be stressed, though, that the canonical definition of $(\Omega, \mathcal{F})$ is not needed for the consistency theory of Markov families and Markov processes developed in this book.

We are now ready to define a canonical Markov process.

Definition B.12 Let $s \geq 0$. We call a collection of objects

$$\mathcal{MP}_s := (\Omega, \mathcal{F}, \mathbb{F}_s, (X_t)_{t\geq s}, \mathbb{Q}_s, P)$$

a canonical Markov process over $(\mathcal{X}, \boldsymbol{\mathcal{X}})$ *starting at time s* if

1. The collection $(\Omega, \mathcal{F}, \mathbb{F}_s, (X_t)_{t\geq s}, \mathbb{Q}_s)$ is a canonical process over $(\mathcal{X}, \boldsymbol{\mathcal{X}})$.
2. The collection $(\Omega, \mathcal{F}, \mathbb{F}_s, (X_t)_{t\geq s}, \mathbb{Q}_s, P)$ is a Markov process as defined in Definition B.4.

If no confusion could arise, we will typically skip the qualifier "over $(\mathcal{X}, \boldsymbol{\mathcal{X}})$" when referring to canonical Markov processes.

The next result asserts that there exists a canonical Markov process over $(\mathcal{X}, \boldsymbol{\mathcal{X}})$ with given transition function P and with given initial law μ_s.

Theorem B.13 *Let P be a transition function and let* μ_s *be a probability measure on* $(\mathcal{X}, \boldsymbol{\mathcal{X}})$. *Then there exists an* $\mathcal{MP}_s$ *with transition function P and with initial law* μ_s.

Proof This result follows immediately from Theorem 1.3.5 in Gikhman and Skorokhod (2004). □

Remark B.14 In this book, in the study of structured dependence we consider canonical Markov processes unless stated otherwise and we refer to them as to Markov processes. For $s = 0$ we will simply write $\mathcal{MP}$. A canonical time-homogenous Markov process is denoted as $\mathcal{MPH}$.

We end this section with the definition of multivariate versions of the objects introduced earlier.

Definition B.15 Let $\mathcal{X} = \times_{i=1}^n \mathcal{X}_i$ and $\boldsymbol{\mathcal{X}} = \otimes_{i=1}^n \boldsymbol{\mathcal{X}}_i$, where $(\mathcal{X}_i, \boldsymbol{\mathcal{X}}_i)$, $i = 1, \ldots, n$, are metric spaces endowed with Borel σ-algebras. Then, $\mathcal{MP}_s$ is called a *canonical multivariate Markov process over* $(\mathcal{X}, \boldsymbol{\mathcal{X}})$ *starting at time s*, and it is denoted as $\mathcal{MMP}_s$. A canonical multivariate time-homogeneous Markov process over $(\mathcal{X}, \boldsymbol{\mathcal{X}})$ is denoted as $\mathcal{MMH}$.

Note that for $\mathcal{MMP}_s = (\Omega, \mathcal{F}, \mathbb{F}_s, (X_t)_{t\geq s}, \mathbb{Q}_s, P)$ we have

$$X_t = (X_t^1, \ldots, X_t^n),$$

where, for $i = 1, \ldots, n$, for $t \geq s$, and for $\omega = (\omega^1, \ldots, \omega^n) \in \Omega$, we have

$$X_t^i(\omega) = \omega^i(t) \in \mathcal{X}_i. \tag{B.5}$$

Accordingly, for $i = 1, \ldots, n$, we define the family of marginal canonical filtrations

$$\{\mathbb{F}_s^i = \{\mathcal{F}_{s,t}^i, s \leq t \leq \infty\} : s \geq 0\}, \tag{B.6}$$

where, for $t \in [s, \infty)$,

$$\mathcal{F}_{s,t}^i = \sigma\{(X_u^i)^{-1}(B^i) : B^i \in \boldsymbol{\mathcal{X}}_i, s \leq u \leq t\}$$

and

$$\mathcal{F}_{s,\infty}^i = \sigma\{(X_u^i)^{-1}(B^i) : B^i \in \boldsymbol{\mathcal{X}}_i, s \leq u\}.$$

Remark B.16 Note that the marginal canonical filtration $\mathbb{F}_s^i$ is the natural filtration of the coordinate X^i of the canonical process X starting at time s. As such, it could be conventionally denoted as $\mathbb{F}_s^{X^i}$. We will not use the latter notation in this appendix, though.

Remark B.17 It is important to note that the ith coordinate process, i.e. $(\Omega, \mathcal{F}, \mathbb{F}_s^i, (X_t^i)_{t\geq s}, \mathbb{Q}_s)$, is not a canonical process for $n \geq 2$. Now, define

$$\widehat{\Omega}^i = \mathcal{X}_i^{\mathbb{R}_+} \quad \text{and} \quad \widehat{\mathcal{F}}^i = \sigma\{(\widehat{X}_u^i)^{-1}(B^i) : B^i \in \boldsymbol{\mathcal{X}}_i, u \geq 0\},$$

where $\widehat{X}_u^i : \widehat{\Omega}^i \to \mathcal{X}^i$ is defined by

$$\widehat{X}_u^i(\widehat{\omega}^i) = \widehat{\omega}^i(u), \tag{B.7}$$

and, for $t \in [s, \infty)$,

$$\widehat{\mathcal{F}}_{s,t}^i = \sigma\{(\widehat{X}_u^i)^{-1}(B^i) : B^i \in \boldsymbol{\mathcal{X}}_i, s \leq u \leq t\}$$

and

$$\widehat{\mathcal{F}}_{s,\infty}^i = \sigma\{(\widehat{X}_u^i)^{-1}(B^i) : B^i \in \boldsymbol{\mathcal{X}}_i, s \leq u\}.$$

Also, define $h^i : \Omega \to \widehat{\Omega}^i$ by $h^i(\omega) = \omega^i$. Finally, define a probability measure $\widehat{\mathbb{Q}}_s^i$ on $(\widehat{\Omega}^i, \widehat{\mathcal{F}}_{s,\infty}^i)$ by $\widehat{\mathbb{Q}}_s^i := \mathbb{Q}_s \circ (h^i)^{-1}$. Then, the process $(\widehat{\Omega}^i, \widehat{\mathcal{F}}^i, \widehat{\mathbb{F}}_s^i, (\widehat{X}_t^i)_{t\geq s}, \widehat{\mathbb{Q}}_s^i)$ is a canonical process, and we have $X_t^i(\omega) = h^i(\omega)(t) = \omega^i(t) = \widehat{X}_t^i(\omega^i)$.

B.2 Markov Family

Let $(\Omega, \mathcal{F})$ be a canonical space over $(\mathcal{X}, \boldsymbol{\mathcal{X}})$ endowed with the family $\{\mathbb{F}_s, s \geq 0\}$ of canonical filtrations.

Definition B.18 The collection of objects

$$\mathcal{MF} := \{(\Omega, \mathcal{F}, \mathbb{F}_s, (X_t)_{t\geq s}, \mathbb{P}_{s,x}) : s \geq 0, x \in \mathcal{X}\}$$

is a *(canonical) Markov family* if

1. $X_t(\omega) = \omega(t)$ for $t \geq 0$ and $\omega \in \Omega$.

2. $\{\mathbb{P}_{s,x}, x \in \mathcal{X}\}$ is family of probability measures on $(\Omega, \mathcal{F}_{s,\infty})$ for $s \geq 0$.
3. The function $P : \mathbb{R}_+ \times \mathcal{X} \times \mathbb{R}_+ \times \boldsymbol{\mathcal{X}} \to [0,1]$ defined for $0 \leq s \leq t$ by

$$P(s,x,t,B) := \mathbb{P}_{s,x}(X_t \in B) \tag{B.8}$$

   is measurable with respect to x for any fixed s,t, $0 \leq s \leq t$, and $B \in \boldsymbol{\mathcal{X}}$.
4. $\mathbb{P}_{s,x}(X_s = x) = 1$ for any $s \geq 0$ and for any $x \in \mathcal{X}$.
5. For any $0 \leq s \leq t \leq u < \infty$, $x \in \mathcal{X}$, and $B \in \boldsymbol{\mathcal{X}}$, we have

$$\mathbb{P}_{s,x}(X_u \in B | \mathcal{F}_{s,t}) = \mathbb{P}_{t,X_t}(X_u \in B), \; \mathbb{P}_{s,x}\text{-}a.s.$$

Remark B.19 We consider only conservative Markov families, that is, Markov families such that $P(\cdot,\cdot,\cdot,\mathcal{X}) \equiv 1$. As usual, by $\mathbb{E}_{s,x}$ we denote the expectation operator with respect to the probability measure $\mathbb{P}_{s,x}$.

The next lemma is a direct consequence of Definition B.18, and therefore we skip its proof.

Lemma B.20 *The function P defined in* (B.8) *is a transition function.*

Definition B.21 The function P defined in (B.8) is called the *transition function associated with the family* $\{(\Omega, \mathcal{F}, \mathbb{F}_s, (X_t)_{t \geq s}, \mathbb{P}_{s,x}) : s \geq 0, x \in \mathcal{X}\}$. Accordingly, we will write

$$\mathcal{MF} = \{(\Omega, \mathcal{F}, \mathbb{F}_s, (X_t)_{t \geq 0}, \mathbb{P}_{s,x}, P) : s \geq 0, x \in \mathcal{X}\}$$

whenever the presence of a transition function is needed.

Remark B.22 It is common in the literature on Markov processes to complete the canonical filtrations. The discussion of the key concepts and ideas presented in this book does not require any such completions, for the most part. Nevertheless, relevant completions of the canonical filtrations are needed for some results presented in this book (in particular in Section 2.2.1). Having this in mind, we shall now briefly summarize the relevant completion procedure.
First, one universally completes the state space from $(\mathcal{X}, \boldsymbol{\mathcal{X}})$ to $(\mathcal{X}, \boldsymbol{\mathcal{X}}^*)$, where $\boldsymbol{\mathcal{X}}^*$ is the universal completion of $\boldsymbol{\mathcal{X}}$. Then, one takes filtrations $\{\mathbb{F}_s^* = \{\mathcal{F}_{s,t}^*, s \leq t \leq \infty\}, s \geq 0\}$, where $\mathcal{F}_{s,\infty}^*$ denotes the completion of $\mathcal{F}_{s,\infty}$ with respect to all measures $\mathbb{P}_{s,\eta}$ and where

$$\mathbb{P}_{s,\eta}(A) := \int_{\mathcal{X}} \mathbb{P}_{s,x}(A) \eta(dx).$$

Here η is a finite measure on $(\mathcal{X}, \boldsymbol{\mathcal{X}}^*)$ and $\mathcal{F}_{s,t}^*$ denotes the completion of $\mathcal{F}_{s,t}$ in $\mathcal{F}_{s,\infty}^*$ with respect to all measures $\mathbb{P}_{s,\eta}$. We refer to Gikhman and Skorokhod (2004) or to Blumenthal and Getoor (1968) for details.
In view of the corollary that immediately follows Theorem I.3.1 in Gikhman and Skorokhod (2004), we see that if

$$\{(\Omega, \mathcal{F}, \mathbb{F}_s, (X_t)_{t \geq s}, \mathbb{P}_{s,x}) : s \geq 0, x \in \mathcal{X}\}$$

is a (canonical) Markov family over $(\mathcal{X}, \boldsymbol{\mathcal{X}})$ then

$$\{(\Omega, \mathcal{F}, \mathbb{F}_s^*, (X_t)_{t \geq s}, \mathbb{P}_{s,x}) : s \geq 0, x \in \mathcal{X}\}$$

is a (canonical) Markov family over $(\mathcal{X}, \boldsymbol{\mathcal{X}}^*)$.

The next definition parallels the concept of time-homogeneous Markov processes.

Definition B.23 Let $\{(\Omega, \mathcal{F}, \mathbb{F}_s, (X_t)_{t\geq s}, \mathbb{P}_{s,x}, P) : s \geq 0, x \in \mathcal{X}\}$ be a Markov family. We say that the family is *time-homogeneous* if its associated transition function is time-homogeneous.

Note that if the family $\{(\Omega, \mathcal{F}, \mathbb{F}_s, (X_t)_{t\geq s}, \mathbb{P}_{s,x}, P) : s \geq 0, x \in \mathcal{X}\}$ is time-homogeneous then we have

$$\mathbb{P}_{s,x}(X_t \in B) = \mathbb{P}_{0,x}(X_{t-s} \in B), \quad 0 \leq s \leq t, x \in \mathcal{X}, B \in \boldsymbol{\mathcal{X}}. \tag{B.9}$$

It is standard and sufficient in this case to define the Markov family as the collection of objects[1]

$$\mathcal{MFH} := \{(\Omega, \mathcal{F}, \mathbb{F}, (X_t)_{t\geq 0}, \mathbb{P}_x, P) : x \in \mathcal{X}\},$$

with the family of measures $\{\mathbb{P}_x, x \in \mathcal{X}\}$ given as

$$\mathbb{P}_x(A) = \mathbb{P}_{0,x}(A), \quad x \in \mathcal{X}, A \in \mathcal{F}, \tag{B.10}$$

and with conditions 4 and 5 in Definition B.18 replaced with

4′. $\mathbb{P}_x(X_0 = x) = 1$ for any $x \in \mathcal{X}$.
5′. $\mathbb{P}_x(X_u \in B | \mathcal{F}_{0,t}) = \mathbb{P}_{X_t}(X_{u-t} \in B)$, $\mathbb{P}_x$-a.s., for any $0 \leq t \leq u < \infty$, $x \in \mathcal{X}$, and $B \in \boldsymbol{\mathcal{X}}$.

B.3 Associations between Markov Processes and Markov Families

Next, we will relate Markov processes to Markov families. First, we observe that if

$$\mathcal{MF} = \{(\Omega, \mathcal{F}, \mathbb{F}_s, (X_t)_{t\geq s}, \mathbb{P}_{s,x}, P) : s \geq 0, x \in \mathcal{X}\}$$

is a Markov family then, for each $(s,x) \in [0,\infty) \times \mathcal{X}$, the collection of objects $(\Omega, \mathcal{F}, \mathbb{F}_s, (X_t)_{t\geq s}, \mathbb{P}_{s,x}, P)$ is an $\mathcal{MP}_s$ and it admits the initial law $\mu_s = \delta_x$. Thus, a Markov family $\mathcal{MF}$ may be considered as a collection of Markov processes starting at various times $s \geq 0$ from various initial positions $x \in \mathcal{X}$.

Moreover, let $s \geq 0$ and let γ_s be a probability measure on $(\mathcal{X}, \boldsymbol{\mathcal{X}})$; we define a probability measure $\mathbb{Q}_s^{\gamma_s}$ on $(\Omega, \mathcal{F}_{s,\infty})$ by

$$\mathbb{Q}_s^{\gamma_s}(A) := \int_{\mathcal{X}} \mathbb{P}_{s,x}(A)\gamma_s(dx). \tag{B.11}$$

Then the collection of objects $(\Omega, \mathcal{F}, \mathbb{F}_s, (X_t)_{t\geq s}, \mathbb{Q}_s^{\gamma_s}, P)$ is an $\mathcal{MP}_s$ in the sense of Definition B.4 and it admits the initial law $\mu_s = \gamma_s$. Accordingly, we introduce the concept of a Markov process associated with $\mathcal{MF}$.

[1] Recall that $\mathbb{F} = \mathbb{F}_0$.

Definition B.24 Fix $s \geq 0$. Let γ_s be a probability measure on $(\mathcal{X}, \boldsymbol{\mathcal{X}})$ and $\mathcal{MF} = \{(\Omega, \mathcal{F}, \mathbb{F}_s, (X_t)_{t \geq s}, \mathbb{P}_{s,x}, P) : s \geq 0, x \in \mathcal{X}\}$ be a Markov family. Let $\mathbb{Q}_s^{\gamma_s}$ be as in (B.11). Then, the collection of objects $(\Omega, \mathcal{F}, \mathbb{F}_s, (X_t)_{t \geq s}, \mathbb{Q}_s^{\gamma_s}, P)$ is called the *Markov process associated with* $\mathcal{MF}$ *starting at time* s *and with initial law* $\mu_s = \gamma_s$. In the case $s = 0$ we simply call such a collection the *Markov process associated with this* $\mathcal{MF}$ *and with initial law* $\mu_0 = \gamma_0$.

Clearly, as already observed above, for each $s \geq 0$ and $x \in \mathcal{X}$, the collection of objects $(\Omega, \mathcal{F}, \mathbb{F}_s, (X_t)_{t \geq s}, \mathbb{P}_{s,x}, P)$ is the Markov process associated with the $\mathcal{MF}$ as above, starting at time s from position x, i.e. with initial law $\mu_s = \delta_x$.

A natural question is whether a given Markov process is associated with some Markov family. The next result provides an affirmative answer to this question.

Theorem B.25 *Let* $s \geq 0$ *and let a Markov process*

$$\mathcal{MP}_s = (\Omega, \mathcal{F}, \mathbb{F}, (X_t)_{t \geq s}, \mathbb{Q}_s, P)$$

be given. Then, there exists a unique Markov family $\mathcal{MF}$ *such that* $\mathcal{MP}_s$ *is associated with it in the sense of Definition B.24.*

Proof Given a transition function P one can construct a unique canonical Markov family $\mathcal{MF} = \{(\Omega, \mathcal{F}, \mathbb{F}_r, (X_t)_{t \geq r}, \mathbb{P}_{r,x}, P) : r \geq 0, x \in \mathcal{X}\}$; see Theorem 1.3.5 in Gikhman and Skorokhod (2004).

Now we will prove that $\mathcal{MP}_s = (\Omega, \mathcal{F}, \mathbb{F}, (X_t)_{t \geq s}, \mathbb{Q}_s, P)$ is associated with this $\mathcal{MF}$. Towards this end it is enough to show that, taking $\mathbb{Q}_s^{\gamma_s}$ as in (B.11), with γ_s defined by $\gamma_s(B) := \mathbb{Q}_s(X_s \in B)$ for $B \in \boldsymbol{\mathcal{X}}$, we obtain

$$\mathbb{Q}_s = \mathbb{Q}_s^{\gamma_s}, \tag{B.12}$$

which is sufficient to conclude that the collection $(\Omega, \mathcal{F}, \mathbb{F}, (X_t)_{t \geq s}, \mathbb{Q}_s, P)$ is a Markov process associated with the $\mathcal{MF}$ in the sense of Definition B.24.

It remains to prove (B.12). For this, it suffices to show that (B.12) holds for cylinder sets of the form

$$C_m = \{(X_{t_1}, \ldots, X_{t_m}) \in B_{1:m}\},$$

where $t_m > t_{m-1} > \cdots > t_1 \geq s$, $B_{1:m} = B_1 \times \cdots \times B_m$, $B_i \in \boldsymbol{\mathcal{X}}$, $i = 1, \ldots, m$, and $m = 1, 2, \ldots$ For such sets the following equality holds $\mathbb{Q}_s$-a.s.:

$$\mathbb{Q}_s(C_m | X_s) = \int_{B_{1:m}} P(t_{m-1}, x_{m-1}, t_m, dx_m) \cdots P(t_1, x_1, t_2, dx_2) P(s, X_s, t_1, dx_1).$$

Moreover,

$$\mathbb{P}_{s,x}(C_m) = \int_{B_{1:m}} P(t_{m-1}, x_{m-1}, t_m, dx_m) \cdots P(t_1, x_1, t_2, dx_2) P(s, x, t_1, dx_1)$$

for every $x \in \mathcal{X}$. Thus, $\mathbb{Q}_s(C_m|X_s) = \mathbb{P}_{s,X_s}(C_m)$ $\mathbb{Q}_s$-a.s. and, consequently, using the definitions of γ_s and $\mathbb{Q}_s^{\gamma_s}$ we have

$$\begin{aligned}\mathbb{Q}_s(C_m) &= \mathbb{E}_{\mathbb{Q}_s}(\mathbb{E}_{\mathbb{Q}_s}(\mathbb{1}_{C_m}|X_s)) = \mathbb{E}_{\mathbb{Q}_s}(\mathbb{P}_{s,X_s}(C_m)) = \int_{\mathcal{X}} \mathbb{P}_{s,x}(C_m)\mathbb{Q}_s(X_s \in dx) \\ &= \int_{\mathcal{X}} \mathbb{P}_{s,x}(C_m)\gamma_s(dx) = \mathbb{Q}_s^{\gamma_s}(C_m),\end{aligned}$$

which proves (B.12) for cylinder sets. This completes the proof of the theorem. □

Definition B.26 The unique family $\mathcal{MF}$ furnished by Theorem B.25 is called the *Markov family associated with the process* $\mathcal{MP}_s$.

Remark B.27 It is clear that if a given Markov family is time-homogeneous then all Markov processes associated with it are time-homogeneous. Likewise, if a given Markov process is time-homogeneous then the unique Markov family associated with it is time-homogeneous. In particular, if

$$\mathcal{MFH} = \{(\Omega, \mathcal{F}, \mathbb{F}, (X_t)_{t\geq 0}, \mathbb{P}_x, P) : x \in \mathcal{X}\}$$

is a time-homogeneous Markov family then, for each $x \in \mathcal{X}$, the collection of objects $(\Omega, \mathcal{F}, \mathbb{F}, (X_t)_{t\geq 0}, \mathbb{P}_x, P)$ is a time-homogenous $\mathcal{MP}$ starting (at time 0) from position x. Thus, a time-homogeneous Markov family $\mathcal{MFH}$ may be considered as a collection of time-homogenous Markov processes starting (at time 0) from various initial positions $x \in \mathcal{X}$.

Appendix C Finite Markov Chains: Auxiliary Technical Framework

The results presented here are the key underlying technical results used in Chapter 3.

C.1 Martingale Characterization of a Finite Markov Chain

Let us consider a càdlàg process V defined on a probability space $(\Omega, \mathcal{F}, \mathbb{P})$ and taking values in a finite set $\mathcal{V}$. Let $\mathbb{F}$ be a filtration on $(\Omega, \mathcal{F}, \mathbb{P})$ satisfying the usual conditions, and assume that V is adapted to $\mathbb{F}$.

For any two distinct states $v, w \in \mathcal{V}$, we define an $\mathbb{F}$-optional integer-valued measure $N_{vw}(dt)$ on $(\mathbb{R}_+, \mathcal{B}(\mathbb{R}_+))$ by

$$N_{vw}((0,t]) = \sum_{0<s\leq t} \mathbb{1}_{\{V_{s-}=v, V_s=w\}}, \quad t \geq 0. \tag{C.1}$$

Manifestly, $N_{vw}((0,t])$ represents the number of jumps from state v to state w that the process V executes over the time interval $(0,t]$.

The associated process N_{vw}, defined as

$$N_{vw}(t) := N_{vw}((0,t]), \quad t \geq 0, \tag{C.2}$$

is a counting (point) process.

Typically, the measure $N_{vw}(dt)$ and the process $N_{vw}(t), t \geq 0$, are identified and the measure $N_{vw}(dt)$ is called a counting measure. Accordingly, the $(\mathbb{F}, \mathbb{P})$-compensator $\nu_{vw}(t)$ of the process $N_{vw}(t)$ is identified with the $(\mathbb{F}, \mathbb{P})$-compensator $\nu_{vw}(dt)$ of the measure $N_{vw}(dt)$ by

$$\nu_{vw}(t) = \nu_{vw}((0,t]), \quad t \geq 0. \tag{C.3}$$

Definition C.1 Let $v, w \in \mathcal{V}$. Suppose that the process $N_{vw}(t)$, $t \geq 0$, admits an $(\mathbb{F}, \mathbb{P})$-intensity, say λ_{vw}. Then, we say that the measure $N_{vw}(dt)$ admits the $(\mathbb{F}, \mathbb{P})$-intensity λ_{vw}.

Next, let us define a deterministic matrix-valued function Γ on $[0, \infty)$ by

$$\Gamma(t) = [\gamma_{vw}(t)]_{v,w\in\mathcal{V}}, \tag{C.4}$$

where the γ_{vw} satisfy the following properties:

P1 For $v, w \in \mathcal{V}$, γ_{vw} is a real-valued, locally integrable function on $[0, \infty)$.

P2 For almost every $t \in [0, \infty)$ and $v, w \in \mathcal{V}$, $v \neq w$, we have

$$\gamma_{vw}(t) \geq 0$$

and

$$\gamma_{vv}(t) = -\sum_{w \neq v} \gamma_{vw}(t).$$

Below, the matrix-valued function Γ will play the role of an infinitesimal generator function (see Appendix E).

The following result gives necessary and sufficient conditions for a càdlàg process V with values in $\mathcal{V}$ to be a Markov chain.

Proposition C.2 *Let $\Gamma(t) = [\gamma_{vw}(t)]_{v,w \in \mathcal{V}}$, $t \geq 0$, where the functions γ_{vw} satisfy properties **P1** and **P2**. Then, a process V taking values in a finite set $\mathcal{V}$ is a Markov chain with respect to a filtration $\mathbb{F}$ and with infinitesimal generator function $\Gamma(\cdot)$ if and only if, for any $v, w \in \mathcal{V}$, $v \neq w$, the compensator with respect to $(\mathbb{F}, \mathbb{P})$ of the counting measure $N_{vw}(dt)$ is of the form*

$$\nu_{vw}((0,t]) = \int_0^t \mathbb{1}_{\{V_s = v\}} \gamma_{vw}(s) ds. \tag{C.5}$$

Proof It was shown in Lemma 5.1 in Bielecki et al. (2008b) that a process V is a Markov chain (with respect to $\mathbb{F}$) with infinitesimal generator function Γ if and only if the compensators with respect to $\mathbb{F}$ of the counting measure $N_{vw}(dt)$, $v, w \in \mathcal{V}$, are of the form

$$\nu_{vw}((0,t]) = \int_0^t \mathbb{1}_{\{V_{s-} = v\}} \gamma_{vw}(s) ds. \tag{C.6}$$

Now, analysis of the proof of Lemma 5.1 in Bielecki et al. (2008b) indicates that the left-hand limits V_{t-} used in that lemma can, in fact, be replaced with V_t, which proves the present result. □

The above result is readily seen to be equivalent to the following. A process V taking values in a finite set $\mathcal{V}$ is a Markov chain with respect to a filtration $\mathbb{F}$ and with infinitesimal generator function Γ if and only if, for any real-valued function f on $\mathcal{V}$, the process M^f defined as

$$M_t^f = f(V_t) - f(V_0) - \int_0^t (\Gamma(u) f)(V_u) du, \quad t \geq 0,$$

is an $\mathbb{F}$-martingale, where f is understood as a vector of size card($\mathcal{V}$) in the expression $\Gamma(u)f$ (see the proof of Lemma 5.1 in Bielecki et al. (2008b)).

Invoking Definition C.1 we arrive at the following:

Corollary C.3 *Let $\Gamma(t) = [\gamma_{vw}(t)]_{v,w\in\mathcal{V}}$, $t \geq 0$, where the γ_{vw} satisfy properties* ***P1*** *and* ***P2****. Then, a process V taking values in a finite set $\mathcal{V}$ is a Markov chain with respect to a filtration $\mathbb{F}$ and with infinitesimal generator function Γ if and only if, for any $v,w \in \mathcal{V}$, $v \neq w$, the measure N_{vw} admits the $(\mathbb{F},\mathbb{P})$-intensity λ_{vw} and this intensity is of the form*

$$\lambda_{vw}(t) = \mathbb{1}_{\{V_t=v\}}\gamma_{vw}(t), \quad dt \otimes d\mathbb{P}\text{-}a.e. \tag{C.7}$$

C.2 Marginal Characteristics

According to the bivariate convention signaled at the beginning of Chapter 3, we take $X = (X^1,X^2) \in \mathcal{X} = \mathcal{X}_1 \times \mathcal{X}_2$ to be a Markov chain with generator function $\Lambda = [\lambda_{xy}]_{x,y\in\mathcal{X}}$. We take $\mathbb{F} = \mathbb{F}^X$.

In analogy to (C.1), for any two states $x = (x_1,x_2), y = (y_1,y_2) \in \mathcal{X}$ such that $x \neq y$, we define an $\mathbb{F}$-optional random measure on $[0,\infty)$ by

$$N_{xy}((0,t]) = \sum_{0<s\leq t} \mathbb{1}_{\{(X^1_{s-}=x_1,X^2_{s-}=x_2),(X^1_s=y_1,X^2_s=y_2)\}}, \tag{C.8}$$

and we shall denote by ν_{xy} the compensator of N_{xy} with respect to $\mathbb{F}$ $(\mathbb{F}^X,\mathbb{P})$.

Next, for any two states $x_1,y_1 \in \mathcal{X}_1$ such that $x_1 \neq y_1$, we define the following $\mathbb{F}^X$-optional random measure on $[0,\infty)$:

$$N^1_{x_1y_1}((0,t]) = \sum_{0<s\leq t} \mathbb{1}_{\{X^1_{s-}=x_1,X^1_s=y_1\}}, \tag{C.9}$$

and we denote by $\nu^1_{x_1y_1}$ the compensator of $N^1_{x_1y_1}$ with respect to $(\mathbb{F}^X,\mathbb{P})$. It is easy to see that

$$N^1_{x_1y_1}((0,t]) = \sum_{x_2,y_2\in\mathcal{X}_2} N_{(x_1,x_2),(y_1,y_2)}((0,t]) \tag{C.10}$$

and, consequently (owing to the uniqueness of compensators),

$$\nu^1_{x_1y_1}((0,t]) = \sum_{x_2,y_2\in\mathcal{X}_2} \nu_{(x_1,x_2),(y_1,y_2)}((0,t]). \tag{C.11}$$

In view of Proposition C.2, we see that, for any two distinct states $x = (x_1,x_2)$, $y = (y_1,y_2) \in \mathcal{X}$,

$$\nu_{(x_1,x_2),(y_1,y_2)}(dt) = \mathbb{1}_{\{(X^1_t,X^2_t)=(x_1,x_2)\}}\lambda_{(x_1x_2)(y_1y_2)}(t)dt. \tag{C.12}$$

Thus the $(\mathbb{F}^X,\mathbb{P})$-compensator of $N^1_{x_1y_1}$ is given by

$$\nu^1_{x_1y_1}((0,t]) = \int_0^t \sum_{x_2,y_2\in\mathcal{X}_2} \mathbb{1}_{\{(X^1_u,X^2_u)=(x_1,x_2)\}}\lambda_{(x_1x_2)(y_1y_2)}(u)du,$$

and hence

$$\Big(\sum_{x_2,y_2\in\mathcal{X}_2} \mathbb{1}_{\{(X^1_t,X^2_t)=(x_1,x_2)\}}\lambda_{(x_1x_2)(y_1y_2)}(t)\Big)_{t\geq 0} \tag{C.13}$$

is the $(\mathbb{F}^X, \mathbb{P})$-intensity of the measure $N^1_{x_1 y_1}$, which counts jumps of X^1 from state x_1 to state y_1; informally, this is the $(\mathbb{F}^X, \mathbb{P})$-intensity of jumps of X^1 from x_1 to y_1.

Analogously one can show that the process

$$\Big(\sum_{x_1, y_1 \in \mathcal{X}_1} \mathbb{1}_{\{(X_t^1, X_t^2) = (x_1, x_2)\}} \lambda_{(x_1 x_2)(y_1 y_2)}(t) \Big)_{t \geq 0} \tag{C.14}$$

is the $(\mathbb{F}^X, \mathbb{P})$-intensity of the measure $N^2_{x_2 y_2}$ defined as

$$N^2_{x_2, y_2}((0,t]) = \sum_{0 < s \leq t} \mathbb{1}_{\{X^2_{s-} = x_2, X^2_s = y_2\}}, \quad t \geq 0,$$

which counts jumps of X^2 from state x_2 to state y_2; informally, this is the $(\mathbb{F}^X, \mathbb{P})$-intensity of jumps of X^2 from state x_2 to state y_2.

Let us denote by $\widehat{\nu}^i_{x_i y_i}$ the compensator of the measure $N^i_{x_i y_i}$ with respect to $\mathbb{F}^{X^i}$, for $i = 1, 2$.

Lemma C.4 (i) *The $(\mathbb{F}^{X^1}, \mathbb{P})$-compensator of $N^1_{x_1, y_1}$ has the form*

$$\widehat{\nu}^1_{x_1 y_1}(dt) = \mathbb{1}_{\{X_t^1 = x_1\}} \sum_{x_2, y_2 \in \mathcal{X}_2} \lambda_{(x_1 x_2)(y_1 y_2)}(t)\, \mathbb{E}_{\mathbb{P}}(\mathbb{1}_{\{X_t^2 = x_2\}} | \mathcal{F}_t^{X^1})\, dt. \tag{C.15}$$

(ii) *The $(\mathbb{F}^{X^2}, \mathbb{P})$-compensator of $N^2_{x_2, y_2}$ has the form*

$$\widehat{\nu}^2_{x_2 y_2}(dt) = \mathbb{1}_{\{X_t^2 = x_2\}} \sum_{x_1, y_1 \in \mathcal{X}_1} \lambda_{(x_1 x_2)(y_1 y_2)}(t)\, \mathbb{E}_{\mathbb{P}}(\mathbb{1}_{\{X_t^1 = x_1\}} | \mathcal{F}_t^{X^2})\, dt. \tag{C.16}$$

Proof We give the proof of (i). The proof of (ii) is analogous. It follows from Lemma 4.3 in Bielecki et al. (2008b) that

$$\begin{aligned} \widehat{\nu}^1_{x_1 y_1}(dt) &= \sum_{x_2, y_2 \in \mathcal{X}_2} \mathbb{E}_{\mathbb{P}}(\mathbb{1}_{\{(X_t^1, X_t^2) = (x_1, x_2)\}} \lambda_{(x_1 x_2)(y_1 y_2)}(t) | \mathcal{F}_{t-}^{X^1}) dt \\ &= \sum_{x_2, y_2 \in \mathcal{X}_2} \lambda_{(x_1 x_2)(y_1 y_2)}(t) \mathbb{E}_{\mathbb{P}}(\mathbb{1}_{\{X_t^1 = x_1\}} \mathbb{1}_{\{X_t^2 = x_2\}} | \mathcal{F}_{t-}^{X^1}) dt. \end{aligned} \tag{C.17}$$

The process X is quasi-left-continuous, since it is a Markov chain. Hence, X^1 is also quasi-left-continuous, so its natural filtration $\mathbb{F}^1$ is quasi-left-continuous; hence $\mathcal{F}_t^1 = \mathcal{F}_{t-}^1$ (see Rogers and Williams (2000), III.11). Thus, by (C.17) we have (C.15). □

Appendix D Crash Course on Conditional Markov Chains and on Doubly Stochastic Markov Chains

We present here several technical results that are needed in Chapters 4 and 8. In particular, we will construct an $(\mathbb{F},\mathbb{G})$-conditional Markov chain for a specially designed filtration $\mathbb{G}$ that is sufficient for our needs. The presentation below is based on Jakubowski and Niewęgłowski (2010a) and Bielecki et al. (2017a).

Let $T > 0$ be a fixed finite time horizon, and let $(\Omega, \mathcal{F}, \mathbb{P})$ be the underlying complete probability space, which is endowed with two filtrations, $\mathbb{F} = (\mathcal{F}_t)_{t\in[0,T]}$ and $\mathbb{G} = (\mathcal{G}_t)_{t\in[0,T]}$, that are assumed to satisfy the usual conditions.

D.1 $(\mathbb{F},\mathbb{G})$-CMC: Definition and First Properties

Typically, the processes considered here are defined on $(\Omega, \mathcal{F}, \mathbb{P})$ and are restricted to the time interval $[0,T]$. In addition, we fix a finite set $\mathcal{X}$. For the purpose of construction of a conditional Markov chain (CMC, for short), in Section D.4, it will be convenient to assume that $\mathcal{X}$ is a subset of a vector space. This is done without any loss of generality, and the reason is that in the process of construction of a CMC we will use differences $y-x$ for $x,y \in \mathcal{X}$.

Definition D.1 An $\mathcal{X}$-valued, $\mathbb{G}$-adapted, càdlàg process X is called an $(\mathbb{F},\mathbb{G})$-*conditional Markov chain* if, for every $x_1,\dots,x_k \in \mathcal{X}$ and for every $0 \le t \le t_1 \le \cdots \le t_k \le T$, it satisfies the following property:

$$\begin{aligned}&\mathbb{P}(X_{t_1} = x_1,\dots,X_{t_k} = x_k | \mathcal{F}_t \vee \mathcal{G}_t)\\ &\quad = \mathbb{P}(X_{t_1} = x_1,\dots,X_{t_k} = x_k | \mathcal{F}_t \vee \sigma(X_t)).\end{aligned} \tag{D.1}$$

Remark D.2 (i) We call filtration $\mathbb{G}$ the *base* filtration and filtration $\mathbb{F}$ the *reference* filtration.
(ii) Clearly, any $(\mathbb{F},\mathbb{G})$-conditional Markov chain is also an $(\mathbb{F},\mathbb{F}^X)$-conditional Markov chain.

(iii) It needs to be stressed that an $(\mathbb{F},\mathbb{G})$-conditional Markov chain might not be a classical Markov chain (in any filtration). However, if $\mathbb{G}$ is independent of $\mathbb{F}$ then the above definition reduces to the case of a classical Markov chain with respect to the filtration $\mathbb{G}$, i.e. a $\mathbb{G}$-Markov chain. In other words, a classical $\mathbb{G}$-Markov chain is an $(\mathbb{F},\mathbb{G})$-conditional Markov chain for the reference filtration, independently of the base filtration.

In what follows we shall write $(\mathbb{F},\mathbb{G})$-CMC to abbreviate an $(\mathbb{F},\mathbb{G})$-conditional Markov chain.

D.1.1 Intensity of an $(\mathbb{F},\mathbb{G})$-CMC

Let X be an $(\mathbb{F},\mathbb{G})$-CMC. For each $x \in \mathcal{X}$ we define the corresponding state indicator process of X,

$$H_t^x = \mathbb{1}_{\{X_t = x\}}, \quad t \in [0,T]. \tag{D.2}$$

Accordingly, we define a column vector $H_t = (H_t^x, x \in \mathcal{X})^\top$, where $\top$ denotes transposition. Similarly, for $x,y \in \mathcal{X}, x \neq y$, we define a process H^{xy} that counts the number of transitions from x to y,

$$H_t^{xy} = \#\{u \leq t : X_{u-} = x \text{ and } X_u = y\} = \int_{]0,t]} H_{u-}^x dH_u^y, \quad t \in [0,T]. \tag{D.3}$$

The following definition generalizes the concept of the generator matrix (or intensity matrix) of a Markov chain.

Definition D.3 We say that an $\mathbb{F}$-adapted (matrix-valued) process $\Lambda_t = [\lambda_t^{xy}]_{x,y \in \mathcal{X}}$ such that

$$\lambda_t^{xy} \geq 0 \quad \text{for all } x,y \in \mathcal{X}, x \neq y \tag{D.4}$$

and

$$\sum_{y \in \mathcal{X}} \lambda_t^{xy} = 0 \quad \text{for all } x \in \mathcal{X} \tag{D.5}$$

is an *$\mathbb{F}$-intensity matrix process ($\mathbb{F}$-intensity for short) of X* if the $\mathbb{R}^d$-valued process $M = (M^x, x \in \mathcal{X})^\top$ defined as

$$M_t = H_t - \int_0^t \Lambda_u^\top H_u du, \quad t \in [0,T], \tag{D.6}$$

is an $\mathbb{F} \vee \mathbb{G}$-local martingale.[1]

Remark D.4 We remark that even though the above definition is stated for an $(\mathbb{F},\mathbb{G})$-CMC process X, it applies to an $\mathcal{X}$-valued pure jump process as well. That is to say, the $\mathcal{X}$-valued $\mathbb{F} \vee \mathbb{G}$-CMC process X in Definition D.3 can be replaced with an $\mathcal{X}$-valued $\mathbb{F} \vee \mathbb{G}$-pure jump process X and then any process Λ for which the process M given in (D.6) is an $\mathbb{F} \vee \mathbb{G}$-local martingale is called the *$\mathbb{F}$-intensity matrix process of X*.

[1] More accurately, we should use the terms $(\mathbb{F},\mathbb{P})$-intensity, and $(\mathbb{F} \vee \mathbb{G},\mathbb{P})$-local martingale. For simplicity, and where no confusion arises, we omit reference to probability measure.

It is important to observe that an $(\mathbb{F},\mathbb{G})$-CMC may or may not admit an $\mathbb{F}$-intensity. The next example illustrates the case of an $(\mathbb{F},\mathbb{G})$-CMC admitting an $\mathbb{F}$-intensity.

Example D.5 Time-changed discrete Markov chain Consider a process $\bar{C}$ which is a discrete-time Markov chain with values in $\mathcal{X} = \{1,\dots,K\}$ and with transition probability matrix Q. Let N be a Cox process (see Asmussen (2003)), i.e. a counting process N such that for $0 \le s \le u \le t \le T$, we have

$$\mathbb{P}(N_t - N_u = k | \mathcal{F}_s^N \vee \mathcal{F}_T) = \exp\left(-\int_u^t \tilde{\lambda}_v dv\right) \frac{(\int_u^t \tilde{\lambda}_v dv)^k}{k!}, \quad k = 0,1,\dots, \tag{D.7}$$

where $\tilde{\lambda}$ is an $\mathbb{F}$-adapted càdlàg process. Suppose that $\tilde{\lambda}$ is independent of $\bar{C}$. Moreover, assume that $\bar{C}$ and N are conditionally independent given $\mathcal{F}_T$. Then the process

$$C_t := \bar{C}_{N_t}, \quad t \in [0,T]$$

is an $(\mathbb{F},\mathbb{F}^C)$-CMC. Moreover the $\mathbb{F}$-intensity matrix process Λ of C is given by

$$\lambda_t^{xy} = (Q-I)_{x,y}\tilde{\lambda}_t, \quad t \in [0,T].$$

For some additional aspects of this example we refer to Example D.43.

In the case of classical Markov chains with a finite state space, the intensity matrix may not exist if the matrix of transition probabilities is not differentiable (e.g. when X is not quasi-left-continuous). In the case of $(\mathbb{F},\mathbb{G})$-CMCs the situation is similar. That is, there exist $(\mathbb{F},\mathbb{G})$-CMCs that do not admit $\mathbb{F}$-intensities. We illustrate this possibility by means of the following example.

Example D.6 CMC admitting no intensity Suppose that $(\Omega,\mathcal{F},\mathbb{P})$ supports a standard real-valued Brownian motion W and a random variable E with unit exponential distribution and independent of W. Define a nonnegative process γ by

$$\gamma_t := \sup_{u \in [0,t]} W_u, \quad t \in [0,T].$$

Thus, γ is an increasing and continuous process. It is well known (see Section 1.7 in Itô and McKean (1974)) that the trajectories of γ are not absolutely continuous with respect to the Lebesgue measure on the real line.
Next, define a process X by

$$X_t := \mathbb{1}_{\{\tau \le t\}}, \quad t \ge 0,$$

where

$$\tau := \inf\{t > 0 : \gamma_t > E\}.$$

Note that, for any $t \in [0,T]$, we have

$$\mathbb{P}(\tau > t | \mathcal{F}_t^W) = \mathbb{P}(E \ge \gamma_t | \mathcal{F}_t^W) = \mathbb{E}(\mathbb{P}(E \ge \gamma_t | \mathcal{F}_\infty^W) | \mathcal{F}_t^W) = e^{-\gamma_t}.$$

This means that the process γ is the $\mathbb{F}^W$-hazard process of τ (see e.g. Bielecki and Rutkowski (2004)). It is well known (see e.g. Proposition 5.1.3 in Bielecki and Rutkowski (2004)) that the process $\widehat{M}$ given as

$$\widehat{M}_t = \mathbb{1}_{\{\tau \le t\}} - \int_{]0,t\wedge\tau]} d\gamma_u = X_t - \int_{]0,t]} (1-X_t) d\gamma_u \tag{D.8}$$

is an $\mathbb{F}^W \vee \mathbb{F}^X$-martingale.

Clearly, the process X takes values in $\mathcal{X} = \{0,1\}$. To show that X is an $(\mathbb{F}^W,\mathbb{F}^X)$-CMC we first note that state 1 is an absorbing state of X. Thus, for every $x_1,\ldots,x_k \in \mathcal{X}$ and for every $0 \le t \le t_1 \le \cdots \le t_k \le T$, taking $A = \{X_{t_k} = x_k, \ldots, X_{t_1} = x_1\}$ we have

$$\begin{aligned} &\mathbb{P}(A|\mathcal{F}_t^W \vee \mathcal{F}_t^X) \\ &= \begin{cases} 0 & \text{if } \exists k < n,\ x_k = 1,\ x_{k+1} = 0, \\ \mathbb{P}(X_{t_1} = 1|\mathcal{F}_t^W \vee \mathcal{F}_t^X) & \text{if } \forall k = 1,\ldots n,\ x_k = 1, \\ \mathbb{P}(X_{t_n} = 0|\mathcal{F}_t^W \vee \mathcal{F}_t^X) & \text{if } \forall k = 1,\ldots n,\ x_k = 0, \\ \mathbb{P}(X_{t_{k+1}} = 1, X_{t_k} = 0|\mathcal{F}_t^W \vee \mathcal{F}_t^X) & \text{if } \exists k < n,\ x_k = 0,\ x_{k+1} = 1. \end{cases} \end{aligned} \tag{D.9}$$

In addition, for $t \le v \le u \le T$,

$$\begin{aligned} \mathbb{P}(X_u = 1, X_v = 0|\mathcal{F}_t^W \vee \mathcal{F}_t^X) &= \mathbb{1}_{\{X_t=0\}}\mathbb{P}(X_u = 1, X_v = 0|\mathcal{F}_t^W \vee \mathcal{F}_t^X) \\ &= \mathbb{1}_{\{X_t=0\}}\mathbb{P}(X_u = 1, X_v = 0|\mathcal{F}_t^W \vee \sigma(X_t)), \\ \mathbb{P}(X_u = 1|\mathcal{F}_t^W \vee \mathcal{F}_t^X) &= \mathbb{1}_{\{X_t=1\}} + \mathbb{1}_{\{X_t=0\}}\mathbb{P}(X_u = 1|\mathcal{F}_t^W \vee \mathcal{F}_t^X) \\ &= \mathbb{1}_{\{X_t=1\}} + \mathbb{1}_{\{X_t=0\}}\mathbb{P}(X_u = 1|\mathcal{F}_t^W \vee \sigma(X_t)). \end{aligned}$$

Hence,

$$\begin{aligned} \mathbb{P}(X_u = 1, X_v = 0|\mathcal{F}_t^W \vee \mathcal{F}_t^X) &= \mathbb{P}(X_u = 1, X_v = 0|\mathcal{F}_t^W \vee \sigma(X_t)), \\ \mathbb{P}(X_u = 1|\mathcal{F}_t^W \vee \mathcal{F}_t^X) &= \mathbb{P}(X_u = 1|\mathcal{F}_t^W \vee \sigma(X_t)), \\ \mathbb{P}(X_u = 0|\mathcal{F}_t^W \vee \mathcal{F}_t^X) &= \mathbb{P}(X_u = 0|\mathcal{F}_t^W \vee \sigma(X_t)). \end{aligned}$$

Those relations and (D.9) imply the property (D.1), which proves that X is an $(\mathbb{F}^W,\mathbb{F}^X)$-CMC.

In order to see that X does not admit an $\mathbb{F}^W$-intensity matrix, let us consider the following (vector) indicator process $H = (H^0, H^1)$ associated with X:

$$H_t^0 = (1 - X_t), \quad H_t^1 = X_t, \quad t \in [0,T].$$

The martingale property of $\widehat{M}$ yields that processes M^0 and M^1 defined, for $t \in [0,T]$, by

$$M_t^0 = H_t^0 - \int_{]0,t]} H_t^0 d(-\gamma_u) = 1 - \widehat{M}_t, \tag{D.10}$$

$$M_t^1 = H_t^1 - \int_{]0,t]} H_t^0 d\gamma_u = \widehat{M}_t, \tag{D.11}$$

are $(\mathbb{F}^W, \mathbb{F}^X)$-martingales. Upon denoting $\widetilde{M} = (M^0, M^1)^\top$, we see that equalities (D.10) and (D.11) can be written as

$$\widetilde{M}_t = H_t - \int_0^t (d\Gamma_t)^\top H_t, \tag{D.12}$$

where

$$\Gamma_t := \begin{bmatrix} -\gamma_t & \gamma_t \\ 0 & 0 \end{bmatrix}.$$

Now, suppose that there exists an $\mathbb{F}$-intensity matrix Λ of X such that the process M given in (D.6) is an $\mathbb{F}^W \vee \mathbb{F}^X$-local martingale. This implies that the process N given by

$$N_t = \widetilde{M}_t - M_t = \int_0^t (d\Gamma_u)^\top H_u - \int_0^t \Lambda_u^\top H_u du, \quad t \in [0,T]$$

is also an $\mathbb{F}^W \vee \mathbb{F}^X$-local martingale. Since N is a continuous process of finite variation it must be constant, so that $N_t = N_0 = 0$. This is a contradiction because it would imply that, for $t \in [0,T]$,

$$\int_0^t (d\Gamma_u)^\top H_u = \int_0^t \Lambda_u^\top H_u du,$$

which is impossible as Γ is not absolutely continuous with respect to Lebesgue measure on the real line. We conclude that the process X does not admit an $\mathbb{F}$-intensity matrix.

We will now discuss the question of uniqueness of the $\mathbb{F}$-intensity.

Definition D.7 We say that two processes Λ and $\widehat{\Lambda}$ are *equivalent with respect to* X if

$$\int_0^t (\Lambda_u - \widehat{\Lambda}_u)^\top H_u du = 0, \quad t \in [0,T]. \tag{D.13}$$

Proposition D.8 *Let X be an $(\mathbb{F}, \mathbb{G})$-CMC.*

(i) *If Λ and $\widehat{\Lambda}$ are $\mathbb{F}$-intensities of X then they are equivalent with respect to X. In particular, the $\mathbb{F}$-intensity of X is unique up to equivalence with respect to X.*
(ii) *Let Λ be an $\mathbb{F}$-intensity of X. If $\widehat{\Lambda}$ is an $\mathbb{F}$-adapted process equivalent to Λ with respect to X then $\widehat{\Lambda}$ is an $\mathbb{F}$-intensity of X.*

Proof (i) By assumption, the process M given by (D.6) and the quantity $\widehat{M}$ defined as

$$\widehat{M}_t = H_t - \int_0^t \widehat{\Lambda}_u^\top H_u du, \quad t \in [0,T]$$

are $\mathbb{F} \vee \mathbb{G}$-local martingales. We have

$$\widehat{M}_t - M_t = \int_0^t (\Lambda_u - \widehat{\Lambda}_u)^\top H_u du.$$

Thus $\widehat{M} - M$ is a continuous finite-variation $\mathbb{F} \vee \mathbb{G}$-local martingale starting from 0, and hence it is a constant null process. Thus (D.13) holds.

(ii) Note that (D.13) implies that, for the $\mathbb{F}\vee\mathbb{G}$-local martingale M given by (D.6) it holds that

$$M_t = H_t - \int_0^t \Lambda_u^\top H_u du + \int_0^t (\Lambda_u - \widehat{\Lambda}_u)^\top H_u du = H_t - \int_0^t \widehat{\Lambda}_u^\top H_u du, \quad t \in [0,T].$$

Thus $\widehat{\Lambda}$ is an $\mathbb{F}$-intensity of X. □

The above proposition shows that an $\mathbb{F}$-intensity of X is unique up to equivalence with respect to X. In Example 4.11 we exhibit an $(\mathbb{F},\mathbb{G})$-CMC X admitting two intensities Λ and Γ which are different.

We have already seen in Example D.6 that an $\mathbb{F}$-intensity may not exist for an $(\mathbb{F},\mathbb{G})$-CMC. Theorem D.9 below provides more insight into the issue of existence of an $\mathbb{F}$-intensity for an $(\mathbb{F},\mathbb{G})$-CMC. In particular, this theorem demonstrates that an $\mathbb{F}$-intensity matrix of an $(\mathbb{F},\mathbb{G})$-CMC X is related to the $\mathbb{F}\vee\mathbb{G}$-compensators of the processes H^{xy}, $x,y \in \mathcal{X}$, $x \neq y$. We omit the proof of the theorem since this is a special case of Theorem D.21.

Theorem D.9 *Let X be an $(\mathbb{F},\mathbb{G})$-CMC.*

(i) *Suppose that X admits an $\mathbb{F}$-intensity matrix process Λ. Then, for every $x,y \in \mathcal{X}$, $x \neq y$, the process K^{xy} defined by*

$$K_t^{xy} = H_t^{xy} - \int_0^t H_u^x \lambda_u^{xy} du, \quad t \in [0,T] \tag{D.14}$$

is an $\mathbb{F}\vee\mathbb{G}$-local martingale; in other words, the process $\int_0^\cdot H_u^x \lambda_u^{xy} du$ is the $\mathbb{F}\vee\mathbb{G}$-compensator of H^{xy}.

(ii) *Suppose that we are given a family of nonnegative $\mathbb{F}$-adapted processes $(\lambda^{xy}, x,y \in \mathcal{X}, x \neq y)$ such that, for every $x,y \in \mathcal{X}$, $x \neq y$, the process K^{xy} given in* (D.14) *is an $\mathbb{F}\vee\mathbb{G}$-local martingale. Then, the matrix-valued process $\Lambda_t = [\lambda_t^{xy}]_{x,y\in\mathcal{X}}$, with diagonal elements defined by*

$$\lambda^{xx} = - \sum_{y\in\mathcal{X}, y\neq x} \lambda^{xy}, \quad x \in \mathcal{X}$$

is an $\mathbb{F}$-intensity matrix of X.

We see that the $\mathbb{F}$-intensity may not exist since the $\mathbb{F}\vee\mathbb{G}$-compensators of H^{xy} may not be absolutely continuous with respect to Lebesgue measure on $[0,T]$. However, the absolute continuity of $\mathbb{F}\vee\mathbb{G}$-compensators of all the processes H^{xy}, for $x,y \in \mathcal{X}$, $x \neq y$, is not sufficient for the existence of an $\mathbb{F}$-intensity. This is due to the fact that the density with respect to Lebesgue measure on $[0,T]$ of $\mathbb{F}\vee\mathbb{G}$-compensator is, in general, $\mathbb{F}\vee\mathbb{G}$-adapted, whereas the $\mathbb{F}$-intensity is only $\mathbb{F}$-adapted.

We end this section by introducing two concepts that will be exploited in the remaining sections of this appendix.

Definition D.10 A $\mathbb{G}$-adapted càdlàg process $X = (X_t)_{t\in[0,T]}$ is called an $(\mathbb{F},\mathbb{G})$-doubly stochastic Markov chain ($(\mathbb{F},\mathbb{G})$-DMC) with state space $\mathcal{X}$ if, for any

$0 \le s \le t \le T$ and for every $y \in \mathcal{X}$, it holds that

$$\mathbb{P}(X_t = y \mid \mathcal{F}_T \vee \mathcal{G}_s) = \mathbb{P}(X_t = y \mid \mathcal{F}_t \vee \sigma(X_s)).$$

Definition D.11 We say that a process X is an $(\mathbb{F},\mathbb{G})$-conditional doubly stochastic Markov chain ($(\mathbb{F},\mathbb{G})$-CDMC) with an $\mathbb{F}$-intensity if it is both an $(\mathbb{F},\mathbb{G})$-CMC with an $\mathbb{F}$-intensity and an $(\mathbb{F},\mathbb{G})$-DMC admitting an intensity (see Definition D.19 below).

In Appendix D.4 we provide a construction of an $(\mathbb{F},\mathbb{G})$-CMC for a filtration $\mathbb{G}$ as in (D.48) and with a given intensity matrix process Λ. In Section D.2 we prove that there exists an $(\mathbb{F},\mathbb{G})$-DMC process admitting an $\mathbb{F}$-intensity and an $(\mathbb{F},\mathbb{G})$-CDMC processs admitting an $\mathbb{F}$-intensity.

D.2 The Processes $(\mathbb{F},\mathbb{G})$-DMC, $(\mathbb{F},\mathbb{G})$-CMC, and $(\mathbb{F},\mathbb{G})$-CDMC

In this section we first revisit the concept of the doubly stochastic Markov chain, first introduced in Jakubowski and Niewęgłowski (2010a). Then, we study the relationship between conditional Markov chains and doubly stochastic Markov chains. This relationship leads to the concept of $(\mathbb{F},\mathbb{G})$-CDMCs, which is crucial for the theory of the consistency of CMCs and for the theory of CMC structures.

Throughout this section we assume the existence of an $(\mathbb{F},\mathbb{G})$-CMC on $(\Omega,\mathcal{F},\mathbb{P})$. In Section D.4 we will construct a probability measure $\mathbb{P}$ on $(\Omega,\mathcal{F})$ and a process X on $(\Omega,\mathcal{F})$ such that under $\mathbb{P}$ the process X is an $(\mathbb{F},\mathbb{G})$-CMC with filtration $\mathbb{G}$ as in (D.48).

D.2.1 The Process $(\mathbb{F},\mathbb{G})$-DMC

The concept of an $(\mathbb{F},\mathbb{G})$-doubly-stochastic Markov chain ($(\mathbb{F},\mathbb{G})$-DMC) generalizes the notion of an $\mathbb{F}$-doubly-stochastic Markov chain, as well as the notion of a continuous-time $\mathbb{G}$-Markov chain.

Definition D.12 A $\mathbb{G}$-adapted, càdlàg, and $\mathcal{X}$-valued process $X = (X_t)_{t\in[0,T]}$ is called an $(\mathbb{F},\mathbb{G})$-doubly-stochastic Markov chain with state space $\mathcal{X}$ if, for any $0 \le s \le t \le T$ and for every $y \in \mathcal{X}$, it holds that

$$\mathbb{P}(X_t = y \mid \mathcal{F}_T \vee \mathcal{G}_s) = \mathbb{P}(X_t = y \mid \mathcal{F}_t \vee \sigma(X_s)). \tag{D.15}$$

Given that an $(\mathbb{F},\mathbb{G})$-CMC exists, the existence of an $(\mathbb{F},\mathbb{G})$-DMC is provided by Proposition D.42.

Remark D.13 We refer to Jakubowski and Niewęgłowski (2010a) for examples of processes which are $(\mathbb{F},\mathbb{F}^X)$-DMCs. We note that in Jakubowski and Niewęgłowski

(2010a) it was assumed that the chain X starts from some point $x \in \mathcal{X}$ with probability one, whereas here we allow for the initial state X_0 to be a non-constant random variable.

The following result provides a characterization of $(\mathbb{F},\mathbb{G})$-DMCs.

Proposition D.14 *An $\mathcal{X}$-valued process X is an $(\mathbb{F},\mathbb{G})$-DMC if and only if there exists a stochastic matrix-valued random field $\widetilde{P}(s,t) = (\widetilde{p}_{xy}(s,t))_{x,y\in\mathcal{X}}$, $0 \le s \le t \le T$, such that:*

(i) *For every $s \in [0,T]$, the process $\widetilde{P}(s,\cdot)$ is $\mathbb{F}$-adapted on $[s,T]$.*
(ii) *For any $0 \le s \le t \le T$ and for every $x,y \in \mathcal{X}$, we have*

$$\mathbb{1}_{\{X_s=x\}}\mathbb{P}(X_t = y \mid \mathcal{F}_T \vee \mathcal{G}_s) = \mathbb{1}_{\{X_s=x\}}\widetilde{p}_{xy}(s,t). \tag{D.16}$$

Proof We first prove the sufficiency. Using (D.16) and summing over $x \in \mathcal{X}$, we obtain

$$\mathbb{P}(X_t = y \mid \mathcal{F}_T \vee \mathcal{G}_s) = \sum_{x\in\mathcal{X}} \mathbb{1}_{\{X_s=x\}}\widetilde{p}_{xy}(s,t). \tag{D.17}$$

So, taking conditional expectations with respect to $\mathcal{F}_t \vee \sigma(X_s)$ on both sides of (D.17), observing that $\mathcal{F}_t \vee \sigma(X_s) \subset \mathcal{F}_T \vee \mathcal{G}_s$, and using the tower property of conditional expectations, we obtain

$$\begin{aligned}\mathbb{P}(X_t = y \mid \mathcal{F}_t \vee \sigma(X_s)) &= \mathbb{E}\Bigg(\sum_{x\in\mathcal{X}} \mathbb{1}_{\{X_s=x\}}\widetilde{p}_{xy}(s,t) \mid \mathcal{F}_t \vee \sigma(X_s)\Bigg)\\ &= \sum_{x\in\mathcal{X}} \mathbb{1}_{\{X_s=x\}}\widetilde{p}_{xy}(s,t),\end{aligned}$$

where the second equality follows from the measurability of $\sum_{x\in\mathcal{X}} \mathbb{1}_{\{X_s=x\}}\widetilde{p}_{xy}(s,t)$ with respect to $\mathcal{F}_t \vee \sigma(X_s)$. This and (D.17) imply

$$\mathbb{P}(X_t = y \mid \mathcal{F}_T \vee \mathcal{G}_s) = \mathbb{P}(X_t = y \mid \mathcal{F}_t \vee \sigma(X_s)),$$

which is (D.15).

We now prove the necessity. First we observe that, using similar arguments to those in Lemma 3 of Jakubowski and Niewęgłowski (2008b) (see also Lemma 2.1 in Bielecki et al. (2008c)), we have, for $t \ge s$,

$$\begin{aligned}&\mathbb{P}(X_t = y \mid \mathcal{F}_t \vee \sigma(X_s))\\ &\quad= \sum_{x\in\mathcal{X}} \mathbb{1}_{\{X_s=x\}}\left(q_{xy}(s,t)\mathbb{1}_{\{\mathbb{P}(X_s=x|\mathcal{F}_t)>0\}} + \mathbb{1}_{\{y=x\}}\mathbb{1}_{\{\mathbb{P}(X_s=x|\mathcal{F}_t)=0\}}\right) \quad \mathbb{P}\text{-a.s.},\end{aligned} \tag{D.18}$$

where $q_{xy}(s,t)$ is given by

$$q_{xy}(s,t) = \frac{\mathbb{P}(X_t = y, X_s = x \mid \mathcal{F}_t)}{\mathbb{P}(X_s = x|\mathcal{F}_t)}.$$

Consequently, for

$$p_{xy}(s,t) := q_{xy}(s,t)\mathbb{1}_{\{\mathbb{P}(X_s=x|\mathcal{F}_t)>0\}} + \mathbb{1}_{\{x=y\}}\mathbb{1}_{\{\mathbb{P}(X_s=x|\mathcal{F}_t)=0\}} \tag{D.19}$$

we have

$$\mathbb{P}(X_t = y \mid \mathcal{F}_t \vee \sigma(X_s)) = \sum_{x \in \mathcal{X}} \mathbb{1}_{\{X_s = x\}} p_{xy}(s,t).$$

It is enough now to let $\widetilde{p}_{xy}(s,t) = p_{xy}(s,t)$, for $x, y \in \mathcal{X}, 0 \le s \le t \le T$, and to use (D.15) to conclude that (i) and (ii) in the statement of the proposition are satisfied.

The proof is complete. □

As we saw in the proof of Proposition D.14, we have

$$\mathbb{1}_{\{X_s = x\}} \mathbb{P}(X_t = y \mid \mathcal{F}_T \vee \mathcal{G}_s) = \mathbb{1}_{\{X_s = x\}} p_{xy}(s,t). \tag{D.20}$$

This fact justifies the following definition:

Definition D.15 The matrix-valued random field $P = (P(s,t),\ 0 \le s \le t \le T)$, where

$$P(s,t) = [p_{xy}(s,t)]_{x,y \in \mathcal{X}}, \quad 0 \le s \le t \le T, \tag{D.21}$$

with $p_{xy}(s,t)$ given by

$$p_{xy}(s,t) = q_{xy}(s,t) \mathbb{1}_{\{\mathbb{P}(X_s = x | \mathcal{F}_t) > 0\}} + \mathbb{1}_{\{x = y\}} \mathbb{1}_{\{\mathbb{P}(X_s = x | \mathcal{F}_t) = 0\}} \tag{D.22}$$

and with

$$q_{xy}(s,t) = \frac{\mathbb{P}(X_t = y, X_s = x \mid \mathcal{F}_t)}{\mathbb{P}(X_s = x | \mathcal{F}_t)} \tag{D.23}$$

is called the *conditional transition probability matrix field* (*c-transition field*) of the $(\mathbb{F}, \mathbb{G})$-DMC X.

We note that, in view of results in Rao (1972), the c-transition field P of X has desired measurability properties. Namely, for every $s \in [0,T]$ and for almost every $\omega \in \Omega$ the function $P(s,\cdot)$ is measurable on $[s,T]$ and, for every $t \in [0,T]$ and almost every $\omega \in \Omega$, the function $P(\cdot,t)$ is measurable on $[0,t]$.

Remark D.16 For future reference, we note that the formula (D.16) can be written in terms of the c-transition field P as follows:

$$\mathbb{E}(H_t \mid \mathcal{F}_T \vee \mathcal{G}_s) = P(s,t)^\top H_s. \tag{D.24}$$

It is known that in the case of classical Markov chains the transition semigroup and the initial distribution of the chain characterize the finite-dimensional distributions of the chain and thus the law of the chain. The next proposition shows that, in the case of an $(\mathbb{F}, \mathbb{G})$-DMC X, the c-transition field P of X and the conditional law of X_0 given $\mathcal{F}_T$ characterize the conditional law of X given $\mathcal{F}_T$.

Proposition D.17 *If X is an $(\mathbb{F}, \mathbb{G})$-DMC with c-transition field P then, for arbitrary $0 = t_0 \le t_1 \le \cdots \le t_n \le t \le T$ and $(x_1, \ldots, x_n) \in \mathcal{X}^n$, we have*

$$\begin{aligned} &\mathbb{P}(X_{t_1} = x_1, \ldots, X_{t_n} = x_n \mid \mathcal{F}_T) \\ &\quad = \sum_{x_0 \in \mathcal{X}} \mathbb{P}(X_0 = x_0 | \mathcal{F}_T) \prod_{k=0}^{n-1} p_{x_k x_{k+1}}(t_k, t_{k+1}). \end{aligned} \tag{D.25}$$

Moreover, if

$$\mathbb{P}(X_0 = x_0|\mathcal{F}_T) = \mathbb{P}(X_0 = x_0|\mathcal{F}_0) \quad \textit{for every } x_0 \in \mathcal{X} \tag{D.26}$$

then, for arbitrary $0 \le t_1 \le \cdots \le t_n \le t \le T$ *and* $(x_1,\ldots,x_n) \in \mathcal{X}^n$, *it holds that*

$$\mathbb{P}(X_{t_1} = x_1,\ldots,X_{t_n} = x_n \mid \mathcal{F}_T) = \mathbb{P}(X_{t_1} = x_1,\ldots,X_{t_n} = x_n \mid \mathcal{F}_t). \tag{D.27}$$

Proof Let us fix arbitrary $x_1,\ldots,x_n \in \mathcal{X}$ and $0 \le t_1 \le \cdots \le t_n \le t \le T$ and define the set $A \in \mathcal{F}$ by

$$A = \{X_{t_1} = x_1,\ldots,X_{t_n} = x_n\}.$$

Note that by using analogous arguments to those in Lemma 3.1 in Jakubowski and Niewęgłowski (2010a) we have

$$\mathbb{P}(A|\mathcal{F}_T \vee \mathcal{G}_0)\mathbb{1}_{\{X_0=x_0\}} = \mathbb{1}_{\{X_0=x_0\}} \prod_{k=0}^{n-1} p_{x_k x_{k+1}}(t_k,t_{k+1}).$$

Consequently,

$$\mathbb{P}(A \mid \mathcal{F}_T) = \sum_{x_0\in\mathcal{X}} \mathbb{P}(X_0 = x_0|\mathcal{F}_T) \prod_{k=0}^{n-1} p_{x_k x_{k+1}}(t_k,t_{k+1}),$$

which proves (D.25). Thus, in view of (D.26), the following equality is satisfied:

$$\mathbb{P}(A|\mathcal{F}_T) = \sum_{x_0\in\mathcal{X}} \mathbb{P}(X_0 = x_0|\mathcal{F}_0) \prod_{k=0}^{n-1} p_{x_k x_{k+1}}(t_k,t_{k+1}).$$

Since P is a c-transition field we obtain that $\mathbb{P}(A|\mathcal{F}_T)$ is $\mathcal{F}_t$-measurable as a product of $\mathcal{F}_t$-measurable random variables. Thus, the tower property of conditional expectations yields (D.27). □

Corollary D.18 *Let X be an $(\mathbb{F},\mathbb{G})$-DMC with X_0 satisfying* (D.26). *Then $\mathbb{F}$ is $\mathbb{P}$-immersed in $\mathbb{F} \vee \mathbb{F}^X$.*

Proof In view of Proposition D.17, process X satisfies (D.27). This, by Lemma 2 in Jakubowski and Niewęgłowski (2008b), is equivalent to the $\mathbb{P}$-immersion of $\mathbb{F}$ in $\mathbb{F} \vee \mathbb{F}^X$. □

In analogy to the concept of an $\mathbb{F}$-intensity for $(\mathbb{F},\mathbb{G})$-CMCs, we will consider the concept of intensity with regard to $(\mathbb{F},\mathbb{G})$-DMCs. Definition D.19 below introduces the concept of such an intensity. This definition is stated in a form which is consistent with the way in which the original definition of intensity for DMCs was introduced in Jakubowski and Niewęgłowski (2010a). Later on, we will show that this definition can be equivalently stated in a form similar to Definition D.3.

Definition D.19 We say that an $\mathbb{F}$-adapted matrix-valued process $\Gamma = (\Gamma_s)_{s\ge0} = ([\gamma_s^{xy}]_{x,y\in\mathcal{X}})_{s\ge0}$ is an intensity of an $(\mathbb{F},\mathbb{G})$-DMC X with c-transition field P if the following conditions are satisfied:

1. We have that

$$\int_0^T \sum_{x\in\mathcal{X}} \|\gamma_s^{xx}\| ds < \infty. \tag{D.28}$$

2. The following holds:

$$\gamma_s^{xy} \geq 0 \quad x,y\in\mathcal{X}, x\neq y, \quad \gamma_s^{xx} = -\sum_{y\in\mathcal{X}:y\neq x} \gamma_s^{xy}, \quad x\in S. \tag{D.29}$$

3. The Kolmogorov backward equation holds: for all $v \leq t$,

$$P(v,t) - \mathrm{I} = \int_v^t \Gamma_u P(u,t) du. \tag{D.30}$$

4. The Kolmogorov forward equation holds: for all $v \leq t$,

$$P(v,t) - \mathrm{I} = \int_v^t P(v,u)\Gamma_u du. \tag{D.31}$$

Remark D.20 The above Kolmogorov equations admit a unique solution provided that Γ satisfies (D.28) (those formulae are given by the Peano–Baker series or by the Magnus expansion; see, e.g. Blanes et al. (2009)).

Martingale Characterization of an $(\mathbb{F},\mathbb{G})$-DMC

Let us define a filtration $\widehat{\mathbb{G}} = (\widehat{\mathcal{G}}_t)_{t\in[0,T]}$ as follows:

$$\widehat{\mathcal{G}}_t := \mathcal{F}_T \vee \mathcal{G}_t, \quad t\in[0,T]. \tag{D.32}$$

It turns out that the $(\mathbb{F},\mathbb{G})$-DMC property of the process $X = (X_t)_{t\in[0,T]}$ is fully characterized by the $\widehat{\mathbb{G}}$-local martingale property of some processes related to X.

We start with a general result regarding the equivalence of local martingale properties.

Theorem D.21 *Given our underlying probability space* $(\Omega,\mathcal{F},\mathbb{P})$*, let* $\mathbb{A}$ *be a filtration and* X *be an* $\mathcal{X}$*-valued* $\mathbb{A}$*-adapted stochastic process. Assume that* $\Gamma_t = [\gamma_t^{xy}]_{x,y\in\mathcal{X}}$, $t\in[0,T]$, *is an* $\mathbb{A}$*-adapted matrix-valued process satisfying* (D.28) *and* (D.29)*. Then the following conditions are equivalent:*

(i) *The processes* $\widehat{M}^x$, $x\in\mathcal{X}$, *defined by*

$$\widehat{M}_t^x := H_t^x - \int_{]0,t]} \gamma_u^{X_u,x} du \tag{D.33}$$

where $H_t^x = \mathbb{1}_{\{X_t=x\}}$, *are* $\mathbb{A}$*-local martingales.*

(ii) *Let* $H_t^{xy} = \int_{]0,t]} H_{u-}^x dH_u^y$. *The processes* K^{xy}, $x,y\in\mathcal{X}$, $x\neq y$, *defined by*

$$K_t^{xy} := H_t^{xy} - \int_{]0,t]} H_s^x \gamma_s^{xy} ds, \tag{D.34}$$

are $\mathbb{A}$*-local martingales.*

(iii) *The process L defined by*

$$L_t := Z_t^\top H_t, \tag{D.35}$$

where Z is a unique solution to the random integral equation

$$dZ_t = -\Gamma_t Z_t dt, \quad Z_0 = \mathrm{I}, \tag{D.36}$$

is an $\mathbb{A}$-local martingale.

For a proof see Lemmas 4.3 and 4.4 in Jakubowski and Niewęgłowski (2010a).

The martingale characterizations of the $(\mathbb{F},\mathbb{G})$-DMC property are given in the next theorem.

Theorem D.22 *Let X be an $\mathcal{X}$-valued $\mathbb{G}$-adapted stochastic process and let $(\Gamma_t)_{t\in[0,T]}$ be an $\mathbb{F}$-adapted matrix-valued process satisfying* (D.28) *and* (D.29). *Then, the following conditions are equivalent:*

(i) *The process X is an $(\mathbb{F},\mathbb{G})$-DMC with intensity process Γ.*
(ii) *The processes $\widehat{M}^x$, $x \in \mathcal{X}$, defined by* (D.33) *are $\widehat{\mathbb{G}}$-local martingales.*
(iii) *The processes K^{xy}, $x,y \in \mathcal{X}$, $x \neq y$, defined by* (D.34) *are $\widehat{\mathbb{G}}$-local martingales.*
(iv) *The process L defined by* (D.35) *is a $\widehat{\mathbb{G}}$-local martingale.*
(v) *For any $t \in [0,T]$, the process N^t defined by*

$$N_s^t = P(s,t)^\top H_s \quad \text{for } 0 \le s \le t \tag{D.37}$$

is a $\widehat{\mathbb{G}}$-martingale, where $P(s,t) := Z_s Y_t$ with

$$dY_t = Y_t \Gamma_t dt, \quad Y_0 = \mathrm{I}, \quad t \in [0,T].$$

Proof The proof of the equivalence of (i)–(iv) goes along the lines of the proof of Theorem 4.1 in Jakubowski and Niewęgłowski (2010a). Clearly the equivalence between conditions (ii), (iii), and (iv) is a simple consequence of Theorem D.21. The equivalence of (iv) and (v) follows from the formula $N_s^t = Y_t^\top L_s$ for $0 \le s \le t$ and Lemma A4 in Bielecki et al. (2017a), which implies that Y_t is a uniformly bounded $\widehat{\mathcal{G}}_0$-measurable invertible matrix. □

The following result is a direct counterpart of Proposition D.8 and therefore we omit its proof.

Proposition D.23 *Let X be an $(\mathbb{F},\mathbb{G})$-DMC.*

(i) *If Γ and $\widehat{\Gamma}$ are intensities of X then they are equivalent relative to X. In particular, an intensity of X is unique up to equivalence relative to X.*
(ii) *Let Γ be an intensity of X. If $\widehat{\Gamma}$ is an $\mathbb{F}$-adapted process equivalent to Γ relative to X then $\widehat{\Gamma}$ is an intensity of X.*

Since an $(\mathbb{F},\mathbb{G})$-DMC X is an $\mathcal{X}$-valued càdlàg process, it is a pure jump process. This observation sheds new light on the intensity of X, as the following corollary shows.

Corollary D.24 *An intensity of an $(\mathbb{F},\mathbb{G})$-DMC X is an $\mathbb{F}$-intensity of X in the sense of Definition D.3.*

Proof The process $\widehat{M}$ is a $\widehat{\mathbb{G}}$-local martingale by Theorem D.22(ii). In fact it is also an $\mathbb{F} \vee \mathbb{G}$-local martingale. To see this, we take a localizing sequence of $\widehat{\mathbb{G}}$-stopping times $\tau_n := \inf\{t \geq 0 : \int_0^t \sum_{y \in \mathcal{X}} |\gamma_s^{yy}| ds \geq n\}$. Since $\widehat{M}$ is also $\mathbb{F} \vee \mathbb{G}$-adapted and the $(\tau_n)_{n \geq 1}$ are also $\mathbb{F} \vee \mathbb{G}$ stopping times, we see that $\widehat{M}$ is an $\mathbb{F} \vee \mathbb{G}$-local martingale (see e.g. Theorem 3.7 in Föllmer and Protter (2011)). This implies that the $\mathbb{F}$-adapted process Γ is an $\mathbb{F}$-intensity of X. □

The next theorem concerns the invariance of the DMC property under an appropriate shrinkage of the filtration.

Theorem D.25 *Assume that X is an $(\mathbb{F},\mathbb{G})$-DMC with an $\mathbb{F}$-intensity Λ. Let $\mathbb{H} \subset \mathbb{F}$ and Λ be $\mathbb{H}$-adapted. Then X is both an $(\mathbb{H},\mathbb{G})$-DMC and an $(\mathbb{H},\mathbb{F}^X)$-DMC with $\mathbb{H}$-intensity Λ.*

Proof Theorem D.22 implies that $M = (M^x)_{x \in \mathcal{X}}$ given by (D.33) is an $(\mathcal{F}_T \vee \mathcal{G}_t)_{t \in [0,T]}$-local martingale. Since Λ is $\mathbb{H}$-adapted and $\mathbb{H} \subset \mathbb{F}$ we conclude that $M = (M^x)_{x \in \mathcal{X}}$ is $(\mathcal{H}_T \vee \mathcal{G}_t)_{t \in [0,T]}$-adapted. Hence $M = (M^x)_{x \in \mathcal{X}}$ is an $(\mathcal{H}_T \vee \mathcal{G}_t)_{t \in [0,T]}$-local martingale. Thus, in view of Theorem D.22, process X is an $(\mathbb{H},\mathbb{G})$-DMC with $\mathbb{H}$-intensity matrix Λ. Since $M = (M^x)_{x \in \mathcal{X}}$ is $(\mathcal{H}_T \vee \mathcal{F}_t^X)_{t \in [0,T]}$-adapted, the proof that X is an $(\mathbb{H},\mathbb{F}^X)$-DMC goes analogously. □

D.2.2 The Process $(\mathbb{F},\mathbb{G})$-CDMC

In this section we introduce and investigate the concept of an $(\mathbb{F},\mathbb{G})$-CDMC, which is the key concept in our theory of consistency for a multivariate $(\mathbb{F},\mathbb{G})$-CMC (see Definition D.1 and Section 4.1).

Proposition D.26 *Let X be an $(\mathbb{F},\mathbb{G})$-DMC. Assume that*

$$\mathbb{F} \text{ is } \mathbb{P}\text{-immersed in } \mathbb{F} \vee \mathbb{G}. \tag{D.38}$$

Then X is an $(\mathbb{F},\mathbb{G})$-CMC. In addition, if X considered as an $(\mathbb{F},\mathbb{G})$-DMC admits an intensity Γ then X considered as an $(\mathbb{F},\mathbb{G})$-CMC admits an $\mathbb{F}$-intensity $\Lambda = \Gamma$.

For a proof see Proposition 4.13 in Bielecki et al. (2017a).

Theorem D.27 *Suppose that X is an $(\mathbb{F},\mathbb{G})$-CMC admitting an $\mathbb{F}$-intensity Λ. In addition, suppose that X is also an $(\mathbb{F},\mathbb{G})$-DMC with an intensity Γ. Then Γ is an $\mathbb{F}$-intensity of X and Λ is an intensity of X.*

Proof It follows from Corollary D.24 that Γ is an $\mathbb{F}$-intensity. Thus, by Proposition D.8, Λ and Γ are equivalent relative to X. Consequently, by Proposition D.23 the process Λ is an intensity of X. □

Let us note that in view of Theorem D.27 the intensity of X considered as an $(\mathbb{F},\mathbb{G})$-DMC coincides, in the sense of Definition D.7, with the $\mathbb{F}$-intensity Λ of X considered as an $(\mathbb{F},\mathbb{G})$-CMC. Consequently, we introduce the following definition.

Definition D.28 We say that a process X is an $(\mathbb{F},\mathbb{G})$-CDMC with an $\mathbb{F}$-intensity if it is both an $(\mathbb{F},\mathbb{G})$-CMC with an $\mathbb{F}$-intensity and an $(\mathbb{F},\mathbb{G})$-DMC admitting an intensity.

Remark D.29 It is worth noting that Theorem D.27 and Proposition D.17 imply that if X is an $(\mathbb{F},\mathbb{G})$-CDMC with an $\mathbb{F}$-intensity then this $\mathbb{F}$-intensity and the $\mathcal{F}_T$-conditional distribution of X_0 determine the $\mathcal{F}_T$-conditional distribution of X.

Clearly, if X is an $(\mathbb{F},\mathbb{G})$-CDMC with an $\mathbb{F}$-intensity then it is an $(\mathbb{F},\mathbb{F}^X)$-CDMC with $\mathbb{F}$-intensity. It turns out that if X is an $(\mathbb{F},\mathbb{G})$-CDMC satisfying Assumption 4.4 and if its $\mathbb{F}$-intensity is adapted to a smaller filtration, say $\mathbb{H}$, then X is an $(\mathbb{H},\mathbb{G})$-CDMC satisfying Assumption 4.4.

Theorem D.30 *Let X be an $(\mathbb{F},\mathbb{G})$-CDMC with an $\mathbb{F}$-intensity Λ and with initial condition satisfying*

$$\mathbb{P}(X_0 = x|\mathcal{F}_T) = \mathbb{P}(X_0 = x|\mathcal{F}_0), \quad x \in \mathcal{X}. \tag{D.39}$$

Let $\mathbb{H} = \{\mathcal{H}_t\}_{t\in[0,T]}$ be a filtration on $(\Omega,\mathcal{F})$ satisfying $\mathbb{H} \subset \mathbb{F}$ and $\mathcal{H}_0 = \mathcal{F}_0$. Finally, assume that Λ is $\mathbb{H}$-adapted. Then X is an $(\mathbb{H},\mathbb{F}^X)$-CDMC with an $\mathbb{H}$-intensity Λ and with X_0 satisfying

$$\mathbb{P}(X_0 = x|\mathcal{H}_T) = \mathbb{P}(X_0 = x|\mathcal{H}_0), \quad x \in \mathcal{X}. \tag{D.40}$$

Proof From Theorem D.25 it follows that X is an $(\mathbb{H},\mathbb{F}^X)$-DMC with an $\mathbb{H}$-intensity Λ. Taking conditional expectations with respect to $\mathcal{H}_T$ in (D.39) and using the assumption that $\mathcal{H}_0 = \mathcal{F}_0$, we obtain (D.40). Using Corollary D.18 we conclude that $\mathbb{H}$ is $\mathbb{P}$-immersed in $\mathbb{H} \vee \mathbb{F}^X$. Thus, in view of Proposition D.26 the process X is an $(\mathbb{H},\mathbb{F}^X)$-CMC and hence an $(\mathbb{H},\mathbb{F}^X)$-CDMC. □

We end this section with an important technical result.

Lemma D.31 *Let Z be an $(\mathbb{F},\mathbb{F}^Z)$-CDMC and U be an $(\mathbb{F},\mathbb{F}^U)$-CDMC, each with values in some finite state space $\mathcal{Y}$, and with intensities Γ^Z and Γ^U, respectively. Then, the conditional law of Z given $\mathcal{F}_T$ coincides with the conditional law of U given $\mathcal{F}_T$ if and only if*

$$\Gamma^Z = \Gamma^U \quad du \otimes d\mathbb{P}\text{-a.e.}, \tag{D.41}$$

$$\mathbb{P}(Z_0 = y|\mathcal{F}_T) = \mathbb{P}(U_0 = y|\mathcal{F}_T), \quad y \in \mathcal{Y}. \tag{D.42}$$

The proof can be found in Bielecki et al. (2017a) (see Lemma A.6, therein).

D.3 Sufficient Conditions for a Pure Jump Process to be an $(\mathbb{F},\mathbb{G})$-CMC

Throughout this section we continue to assume the existence of an $(\mathbb{F},\mathbb{G})$-CMC on $(\Omega,\mathcal{F},\mathbb{P})$.

So, let X be an $(\mathbb{F},\mathbb{G})$-CMC. Clearly, X is a pure jump process. In this section we will provide sufficient conditions under which pure jump processes with values in $\mathcal{X}$ are also $(\mathbb{F},\mathbb{G})$-CMCs. This will prove useful for the construction of an $(\mathbb{F},\mathbb{G})$-CMC in Section D.4.

Towards this end, we first consider the following important result.

Proposition D.32 *Let X be an $(\mathbb{F},\mathbb{G})$-CMC. A matrix-valued process $\Lambda_t = [\lambda_t^{xy}]_{x,y\in\mathcal{X}}$, $t\in[0,T]$, is an $\mathbb{F}$-intensity matrix process of X if and only if the $\mathbb{F}\vee\mathbb{G}$-dual predictable projection under $\mathbb{P}$ of the jump measure μ^X of X, i.e. the $(\mathbb{F}\vee\mathbb{G},\mathbb{P})$-compensator of μ^X, is given follows:*

$$\begin{aligned}\nu(\omega,dt,dz) &= \sum_{x\in\mathcal{X}} H_t^x\Big(\sum_{y\in\mathcal{X}\setminus\{x\}}\delta_{y-x}(dz)\lambda_t^{xy}\Big)dt\\ &= \sum_{x\in\mathcal{X}} \mathbb{1}_{\{X_t=x\}}\Big(\sum_{y\in\mathcal{X}\setminus\{x\}}\delta_{y-x}(dz)\lambda_t^{xy}\Big)dt. \qquad \text{(D.43)}\end{aligned}$$

Proof Suppose that the $(\mathbb{F}\vee\mathbb{G},\mathbb{P})$-compensator of μ^X is given by (D.43). Fix $i,j\in\mathcal{X}$, $i\neq j$. Note that the process H^{ij} that counts jumps of X from i to j (see (D.3)) can be written in terms of μ^X as

$$H_t^{ij} = \int_0^t\int_{\mathcal{X}-\mathcal{X}} \mathbb{1}_{\{X_{s-}=i\}}\mathbb{1}_{\{j-i\}}(z)\mu^X(ds,dz), \quad t\in[0,T],$$

where

$$\mathcal{X}-\mathcal{X} := \{y-x : y\in\mathcal{X}, x\in\mathcal{X}\}.$$

The process H^{ij} is an increasing process, and its $(\mathbb{F}\vee\mathbb{G},\mathbb{P})$-compensator, say $\widehat{H}^{ij}$, is given by

$$\begin{aligned}\widehat{H}_t^{ij} &= \int_0^t\int_{\mathcal{X}-\mathcal{X}} \mathbb{1}_{\{X_{s-}=i\}}\mathbb{1}_{\{j-i\}}(z)\nu(\omega,ds,dz)\\ &= \int_0^t\int_{\mathcal{X}-\mathcal{X}} \mathbb{1}_{\{X_{s-}=i\}}\mathbb{1}_{\{j-i\}}(z)\Big(\sum_{y\in\mathcal{X}\setminus\{i\}}\delta_{y-i}(dz)\lambda_s^{iy}\Big)ds\\ &= \int_0^t \mathbb{1}_{\{X_{s-}=i\}}\Big(\sum_{y\in\mathcal{X}\setminus\{i\}}\mathbb{1}_{\{j-i\}}(y-i)\lambda_s^{iy}\Big)ds\\ &= \int_0^t \mathbb{1}_{\{X_{s-}=i\}}\lambda_s^{ij}ds = \int_0^t \mathbb{1}_{\{X_s=i\}}\lambda_s^{ij}ds\\ &= \int_0^t H_s^i\lambda_s^{ij}ds, \quad t\in[0,T],\end{aligned}$$

where $H_s^i = \mathbb{1}_{\{X_s=i\}}$ (see (D.2)).

Thus the process K^{ij} defined as

$$K_t^{ij} := H_t^{ij} - \widehat{H}_t^{ij} = H_t^{ij} - \int_0^t H_s^i \lambda_s^{ij} ds, \quad t \in [0,T] \tag{D.44}$$

is an $(\mathbb{F} \vee \mathbb{G}, \mathbb{P})$-local martingale. Consequently, from this and from the special case of Theorem D.21, stated in Theorem D.9, we conclude that X has an $\mathbb{F}$-intensity matrix Λ.

Now suppose that X admits an $\mathbb{F}$-intensity matrix process Λ. The special case of Theorem D.21, stated in Theorem D.9, implies that the process K^{ij} given in (D.44) is an $(\mathbb{F} \vee \mathbb{G}, \mathbb{P})$-local martingale. Fix $k \in \mathcal{X} - \mathcal{X}$. We have

$$\mu^X((0,t],\{k\}) = \sum_{i,j \in \mathcal{X}, j \neq i} \mathbb{1}_{\{k\}}(j-i) H_t^{ij} = \sum_{i \in \mathcal{X}} \sum_{j \in \mathcal{X} \setminus \{i\}} \mathbb{1}_{\{k\}}(j-i) H_t^{ij}.$$

Hence, using (D.44) and summing up, we have that the $(\mathbb{F} \vee \mathbb{G}, \mathbb{P})$-compensator of the process $(\mu^X((0,t],\{k\}))_{t \in [0,T]}$ is given by

$$\begin{aligned} &\sum_{i \in \mathcal{X}} \sum_{j \in \mathcal{X} \setminus \{i\}} \mathbb{1}_{\{k\}}(j-i) \int_0^t H_s^i \lambda_s^{ij} ds \\ &\quad = \int_0^t \sum_{i \in \mathcal{X}} \mathbb{1}_{\{X_s = i\}} \Big(\sum_{j \in \mathcal{X} \setminus \{i\}} \mathbb{1}_{\{k\}}(j-i) \lambda_s^{ij} \Big) ds. \end{aligned}$$

However, from (D.43) we see that

$$\nu((0,t],\{k\}) = \int_0^t \sum_{i \in \mathcal{X}} \mathbb{1}_{\{X_s = i\}} \Big(\sum_{j \in \mathcal{X} \setminus \{i\}} \mathbb{1}_{\{k\}}(j-i) \lambda_s^{ij} \Big) ds.$$

Thus

$$\mu^X((0,t],\{k\}) - \nu((0,t],\{k\}), \quad t \in [0,T]$$

is an $(\mathbb{F} \vee \mathbb{G}, \mathbb{P})$-local martingale. Taking a localizing sequence T_n we obtain that, for every positive measurable function $g : \mathcal{X} \to \mathbb{R}_+$ and arbitrary $\mathbb{F} \vee \mathbb{G}$-stopping time σ,

$$\mathbb{E} \int_0^{\sigma \wedge T_n} g(x) \mu^X(dt,dx) = \mathbb{E} \int_0^{\sigma \wedge T_n} g(x) \nu(dt,dx).$$

Letting $n \to \infty$ we obtain

$$\mathbb{E} \int_0^\infty g(x) \mathbb{1}_{[0,\sigma]}(t) \mu^X(dt,dx) = \mathbb{E} \int_0^\infty g(x) \mathbb{1}_{[0,\sigma]}(t) \nu(dt,dx).$$

Since functions W of the form $W(\omega,t,x) = g(x)\mathbb{1}_{[0,\sigma(\omega)]}(t)$ generate the σ-field of predictable functions on $\Omega \times [0,T] \times \mathcal{X}$, by a monotone-class argument we obtain

$$\mathbb{E}(W * \mu_\infty) = \mathbb{E}(W * \nu_\infty),$$

for all nonnegative predictable functions W on $\Omega \times [0,T] \times \mathcal{X}$. Therefore the random measure ν given by (D.43) is an $(\mathbb{F} \vee \mathbb{G}, \mathbb{P})$-compensator of μ^X, which completes the proof. □

Remark D.33 In light of Remark D.4, with a slight abuse of terminology, we shall refer to process X, considered as a $\mathbb{G}$-adapted, $\mathcal{X}$-valued, pure jump process with $\mathbb{F} \vee \mathbb{G}$ compensator ν given in (D.43), as a $\mathbb{G}$-adapted, $\mathcal{X}$-valued, pure jump process admitting the $\mathbb{F}$-intensity process Λ.

In view of this remark and of Proposition D.32 we see that the problem of the construction of an $(\mathbb{F}, \mathbb{G})$-CMC with an $\mathbb{F}$-intensity matrix process Λ is equivalent to the problem of the construction of a $\mathbb{G}$-adapted, $\mathcal{X}$-valued, pure jump process satisfying properly (D.1) and with $(\mathbb{F} \vee \mathbb{G}, \mathbb{P})$-compensator ν of the form (D.43).

Theorem D.34 below, borrowed from Bielecki et al. (2017a) (see Theorem 2.11 and its proof therein), shows that a $\mathbb{G}$-adapted, $\mathcal{X}$-valued, pure jump process admitting $\mathbb{F}$-intensity process Λ is, under some additional conditions, an $(\mathbb{F}, \mathbb{G})$-CMC with the same $\mathbb{F}$-intensity process Λ. For the notion of immersion between two filtrations and the notion of the orthogonality of local martingales used in the statement of the theorem we refer to Appendix A.

Theorem D.34 *Assume that $\mathbb{F}$ and $\mathbb{G}$ satisfy the immersion property* (D.38). *Let X be a $\mathbb{G}$-adapted, $\mathcal{X}$-valued, pure jump process admitting the $\mathbb{F}$-intensity process Λ. Moreover, suppose that*

the process M given by a formula analogous to (D.6), *but relative to the process X considered here, is orthogonal to all real valued $\mathbb{F}$-local martingales.* (D.45)

Then X is an $(\mathbb{F}, \mathbb{G})$-CMC with $\mathbb{F}$-intensity process Λ.

A sufficient condition for (D.45) to hold is given in the next proposition.

Proposition D.35 *Suppose that the coordinates of M (equivalently, the coordinates of X) and the real-valued $\mathbb{F}$-local martingales do not have common jump times. Then M is orthogonal to any $\mathbb{R}^d$-valued $\mathbb{F}$-local martingale.*

Proof Let Z be a real-valued $\mathbb{F}$-local martingale. Since M^x is a real-valued $\mathbb{F} \vee \mathbb{G}$-local martingale of finite variation we have

$$[Z, M^x]_t = \sum_{0<u\le t} \Delta Z_u \Delta M^x_u = \sum_{0<u\le t} \Delta Z_u \Delta H^x_u, \quad t \in [0, T].$$

Now, note that X jumps if and only if one of the processes $H^x, x \in \mathcal{X}$ jumps. Thus if X and Z do not have common jumps then $[Z, M^x]$ is the null process. Consequently Z and M^x are orthogonal local martingales, and thus M is orthogonal to any $\mathbb{R}^d$-valued $\mathbb{F}$-local martingale. □

We complete this section with an interesting example of filtrations $\mathbb{F}$ and $\mathbb{G}$ that satisfy (D.38) and condition (D.45) of Theorem D.34.

Example D.36 Let X be an $\mathcal{X}$-valued pure jump process in its own filtration. Let W be a Brownian motion in the filtration $\mathbb{F}^W \vee \mathbb{F}^X$. Then one can see that (D.38) and (D.45) of Theorem D.34 are satisfied by $\mathbb{G} = \mathbb{F}^X$ and $\mathbb{F} = \mathbb{F}^W$.

Indeed, note that W is also $\mathbb{F}^W$-Brownian-motion, thus any square integrable $\mathbb{F}^W$-martingale V can be represented as

$$V_t = V_0 + \int_0^t \phi_u dW_u, \quad t \in [0,T]$$

for some $\mathbb{F}^W$-predictable process ϕ. The assumption that W is a Brownian motion in $\mathbb{F}^W \vee \mathbb{F}^X$ implies that V is also an $\mathbb{F}^W \vee \mathbb{F}^X$-martingale. This proves that $\mathbb{F}^W$ is immersed in $\mathbb{F}^W \vee \mathbb{F}^X$. So (D.38) holds. Condition (D.45) is satisfied, since all $\mathbb{F}^W$-martingales are continuous.

D.4 Construction of an $(\mathbb{F},\mathbb{G})$-CMC with a Given Intensity Matrix Process

In this section we provide a construction of a probability measure $\mathbb{P}$ on $(\Omega,\mathcal{F})$ and a process $(X_t)_{t\in[0,T]}$ on $(\Omega,\mathcal{F})$ such that, under the measure $\mathbb{P}$, X is an $(\mathbb{F},\mathbb{G})$-CMC with $\mathbb{F}$-intensity matrix process Λ, and with $\xi = X_0$ satisfying

$$\mathbb{P}(\xi = x|\mathcal{F}_T) = \mathbb{P}(\xi = x|\mathcal{F}_0), \quad x \in \mathcal{X}. \tag{D.46}$$

To motivate condition (D.46) we refer to the discussion after Assumption 4.4.

In Bielecki et al. (2015) the authors constructed a CMC that starts from a given state with probability unity. Even though the in case of ordinary Markov chains the construction of a chain starting from a given state with probability unity directly leads to the construction of a chain with arbitrary initial distribution, this in general is not the case when one deals with CMCs. Consequently, the construction of an $(\mathbb{F},\mathbb{G})$-CMC with an arbitrary initial distribution requires a new methodology.

In our present construction we start from a probability space, say $(\Omega,\mathcal{F},\mathbb{Q})$, on which we are given:

I1 A (reference) filtration $\mathbb{F}$.

I2 An $\mathcal{X}$-valued random variable ξ such that for any $x \in \mathcal{X}$, we have that

$$\mathbb{Q}(\xi = x|\mathcal{F}_T) = \mu_x \tag{D.47}$$

for some $\mathcal{F}_0$-measurable random variable μ_x taking values in $[0,1]$.

I3 A family $\mathcal{N} = (N^{xy})_{x,y\in\mathcal{X},y\neq x}$ of mutually independent Poisson processes that are independent of $\mathcal{F}_T \vee \sigma(\xi)$ and with strictly positive (time-independent deterministic) intensities $(a^{xy})_{x,y\in\mathcal{X},y\neq x}$.

In what follows we take

$$\mathcal{G}_t = \Big(\bigvee_{\substack{x,y\in\mathcal{X}\\ y\neq x}} \mathcal{F}_t^{N^{xy}} \Big) \vee \sigma(\xi), \quad t \in [0,T], \tag{D.48}$$

where $\mathbb{G} = (\mathcal{G}_t)_{t\in[0,T]}$. It is known that $\mathcal{G}_{t+} = \mathcal{G}_t$ for $t \in [0,T)$ (see Proposition 3.3 in Amendinger (2000)).

In order to proceed, we take the filtration $\widehat{\mathbb{G}} = (\widehat{\mathcal{G}}_t)_{t\in[0,T]}$ to be as in (D.32), with $\mathbb{G}$ as in (D.48). In what follows we will not require that $\widehat{\mathbb{G}}$ is right-continuous.

We will now construct a $\widehat{\mathbb{G}}$-Markov chain with a given infinitesimal generator. This chain will be constructed as a solution of an appropriate stochastic differential equation. This construction is an intermediate step in our process of constructing an $(\mathbb{F},\mathbb{G})$-CMC X with a given $\mathbb{F}$-intensity matrix process Λ, with X_0 satisfying (D.46) for a measure $\mathbb{P}$ to be constructed later.

We start with an auxiliary lemma.

Lemma D.37 *Let Y be a real-valued càdlàg process, and let Z be a Poisson process, both defined on $(\Omega,\mathcal{F},\mathbb{Q})$. Suppose that Y and Z are independent under $\mathbb{Q}$. Then*

$$\mathbb{Q}(\{\omega\in\Omega : \exists t\in[0,T] \text{ s.t. } \Delta Y_t(\omega)\Delta Z_t(\omega)\neq 0\}) = 0.$$

Proof First note that both Y and Z have countable numbers of jumps on $[0,T]$, and let us denote their jump times as $(T_n)_{n\geq 1}$ and $(S_n)_{n\geq 1}$, respectively. The independence of Y and Z implies that $(T_n)_{n\geq 1}$ and $(S_n)_{n\geq 1}$ are independent. Since each random variable S_n is gamma distributed and thus has a density, for any $n,k\geq 1$ it holds that

$$\mathbb{Q}(T_n = S_k) = 0.$$

Since

$$A := \{\omega : \exists t\in[0,T] \text{ such that } \Delta Y_t(\omega)\Delta Z_t(\omega)\neq 0\} = \bigcup_{n,k\geq 1}\{\omega : T_n(\omega) = S_k(\omega)\},$$

we have

$$\mathbb{Q}(A) \leq \sum_{n,k\geq 1}\mathbb{Q}(T_n = S_k) = 0. \qquad \square$$

As promised, we will now construct a $\widehat{\mathbb{G}}$-Markov chain with a given infinitesimal generator.

Proposition D.38 *Let $\mathcal{N} = (N^{xy})_{x,y\in\mathcal{X},\, y\neq x}$ be as in **I3**, and let $A = [a^{xy}]_{x,y\in\mathcal{X}}$, with diagonal elements defined as $a^{xx} := -\sum_{y\in\mathcal{X},y\neq x} a^{xy}$. Let ξ be an $\mathcal{X}$-valued random variable that is independent of $\mathcal{N}$. Then there exists a unique strong solution X of the following SDE:*

$$dX_t = \sum_{\substack{x,y\in\mathcal{X}\\ x\neq y}}(y-x)\mathbb{1}_{\{x\}}(X_{t-})dN_t^{xy}, \quad t\in[0,T], \quad X_0 = \xi. \tag{D.49}$$

Moreover, X is a $(\widehat{\mathbb{G}},\mathbb{Q})$-Markov chain with infinitesimal generator A. In addition, A is an $\mathbb{F}$-intensity of X under $\mathbb{Q}$ in the sense of Definition D.3.

Proof In view of **I3** and Lemma D.37, the processes N^{xy} and $N^{xy'}$, $y \neq y'$, do not jump together $\mathbb{Q}$-a.s. From standard results on SDEs we know that there exists a unique strong solution X of (D.49). Clearly, X is an $\mathcal{X}$-valued càdlàg process. Moreover, the process H^{xy} defined for $x, y \in \mathcal{X}$, $x \neq y$ by

$$H_t^{xy} = \int_0^t H_{u-}^x dN_u^{xy}, \quad t \in [0,T],$$

where

$$H_t^x = \mathbb{1}_{\{X_t = x\}}$$

counts the number of transitions of X from state x to state y. Recall that Watanabe's characterization theorem (see Theorem 2.3 in Watanabe (1964) and the remark following it) states that N^{xy} is a Poisson process with respect to some filtration $\mathbb{H}$ if and only if

(a) The paths of N^{xy} are increasing step functions with jumps of size 1.
(b) The process $(N_t^{xy} - a^{xy}t)_{t \geq 0}$ is an $\mathbb{H}$-martingale for some positive constants a^{xy}.

Hence, and by the independence of N^{xy} from $\mathcal{F}_T \vee \sigma(\xi)$, we obtain that N^{xy} (which is an $\mathbb{F}^{N^{xy}}$-Poisson process) is also a $\widehat{\mathbb{G}}$-Poisson process with intensity a^{xy}. Thus, by the boundedness and $\widehat{\mathbb{G}}$-predictability of $(H_{t-}^x)_{t \in [0,T]}$, the process L^{xy} given as

$$\begin{aligned} L_t^{xy} &= \int_0^t H_{u-}^x (dN_u^{xy} - a^{xy} du) = H_t^{xy} - \int_0^t H_{u-}^x a^{xy} du \\ &= H_t^{xy} - \int_0^t H_u^x a^{xy} du, \quad t \in [0,T] \end{aligned} \tag{D.50}$$

is a $\widehat{\mathbb{G}}$-martingale. Consequently, application of the relevant characterization theorem (Jakubowski and Niewęgłowski (2010a), Theorem 4.1) yields that X is a $\widehat{\mathbb{G}}$-Markov chain with infinitesimal generator A.

To finish the proof we observe that X given by (D.49) is a pure jump process with finite variation. The $(\widehat{\mathbb{G}}, \mathbb{Q})$-compensator of the jump measure of X is given in terms of the matrix A (see (D.50)). Moreover, since X is adapted to the filtration $\mathbb{F} \vee \mathbb{G} \subseteq \widehat{\mathbb{G}}$, we see that X is a pure jump process with the $(\mathbb{F} \vee \mathbb{G}, \mathbb{Q})$-compensator of its jump measure given in terms of the matrix A. Now, A is $\mathbb{F}$-adapted (since it is deterministic), so, in view of the terminology introduced earlier (see Definition D.3 and Remark D.4), A is an $\mathbb{F}$-intensity of X under $\mathbb{Q}$. □

Now we introduce the canonical conditions for a matrix-valued process Λ.

Definition D.39 We say that a matrix-valued process $\Lambda = [\lambda^{xy}]_{x,y\in\mathcal{X}}$ satisfies canonical conditions relative to the pair $(\mathcal{X},\mathbb{F})$ if:

D1 Λ is $\mathbb{F}$-progressively measurable and it fulfills

$$\lambda_t^{xy} \geq 0 \quad \text{for all } x,y\in\mathcal{X}, x\neq y. \tag{D.51}$$

D2 The processes λ^{xy}, $x,y\in\mathcal{X}$, $x\neq y$, have countably many jumps $\mathbb{Q}$-a.s, and their trajectories admit left limits.

Any $\mathbb{F}$-adapted càdlàg process $\Lambda_t = [\lambda_t^{xy}]_{x,y\in\mathcal{X}}$ for which (D.51) holds satisfies canonical conditions. We are now ready to proceed with the construction of a CMC via a change of measure. For this, we provide a construction of a probability measure $\mathbb{P}$, under which the process X, which follows the dynamics (D.49), is an $(\mathbb{F},\mathbb{G})$-CMC with a given $\mathbb{F}$-intensity matrix Λ and with an $\mathcal{F}_T$-conditional initial distribution satisfying (D.46).

Theorem D.40 *Let Λ be a matrix-valued process that satisfies canonical conditions relative to the pair $(\mathcal{X},\mathbb{F})$. Let X be the unique solution of the SDE* (D.49), *with $X_0 = \xi$ satisfying* (D.47). *For each pair $x,y\in\mathcal{X}, x\neq y$, define the processes κ^{xy} by*

$$\kappa_t^{xy} = \frac{\lambda_{t-}^{xy}}{a^{xy}} - 1, \quad t\in[0,T],$$

and assume that the random variable ϑ given as

$$\vartheta = \prod_{x,y\in\mathcal{X}:x\neq y} \exp\left(-\int_0^T H_{u-}^x a^{xy}\kappa_u^{xy}du\right)\prod_{0<u\leq T}(1+\kappa_u^{xy}\Delta H_u^{xy})$$

satisfies $\mathbb{E}_{\mathbb{Q}}\vartheta = 1$. Finally, define on $(\Omega,\mathcal{F})$ the probability $\mathbb{P}$ by

$$\frac{d\mathbb{P}}{d\mathbb{Q}} = \vartheta. \tag{D.52}$$

Then

(i) $\mathbb{P}_{|\mathcal{F}_T} = \mathbb{Q}_{|\mathcal{F}_T}$.

(ii) *X is an $(\mathbb{F},\mathbb{G})$-CMC under $\mathbb{P}$ with $\mathbb{F}$-intensity matrix process Λ, and with initial distribution satisfying*

$$\mathbb{P}(X_0 = x|\mathcal{F}_T) = \mathbb{P}(X_0 = x|\mathcal{F}_0) = \mathbb{Q}(X_0 = x|\mathcal{F}_T) = \mathbb{Q}(\xi = x|\mathcal{F}_T) = \mu_x, \quad x\in\mathcal{X}. \tag{D.53}$$

Proof In view of Theorem D.34, in order to prove (ii) it suffices to prove that:

(a) Under the measure $\mathbb{P}$ the pure jump process X has an $\mathbb{F}$-intensity Λ.
(b) The filtration $\mathbb{F}$ is $\mathbb{P}$-immersed in $\mathbb{F}\vee\mathbb{G}$.

(c) All real-valued $(\mathbb{F},\mathbb{P})$-martingales are orthogonal (under $\mathbb{P}$) to martingales M^x, $x\in\mathcal{X}$.
(d) Equation (D.53) holds.

We will prove these claims in separate steps. In the process of doing so, we will also demonstrate (i) (see Step 2 below).

Step 1: Here we will show that Λ is an $\mathbb{F}$-intensity of X under $\mathbb{P}$. Towards this end, we consider a $\widehat{\mathbb{G}}$-adapted process η given as[2]

$$\eta_t=\prod_{x,y\in\mathcal{X}:x\neq y}\exp\left(-\int_0^t H^x_{u-}a^{xy}\kappa^{xy}_u du\right)\prod_{0<u\le t}(1+\kappa^{xy}_u\Delta H^{xy}_u),\quad t\in[0,T],$$

so that

$$d\eta_t=\eta_{t-}\sum_{x,y\in\mathcal{X}:x\neq y}\kappa^{xy}_t dL^{xy}_t,\quad \eta_0=1,$$

where L^{xy} is a $(\widehat{\mathbb{G}},\mathbb{Q})$-martingale given by (D.50). Consequently, since κ^{xy} is a left-continuous and $\mathbb{F}$-adapted process, the process η is a $(\widehat{\mathbb{G}},\mathbb{Q})$-local martingale. Now, note that $\eta_T=\vartheta$ and thus $\mathbb{E}_{\mathbb{Q}}\eta_T=1=\eta_0$. Consequently, η is an $(\widehat{\mathbb{G}},\mathbb{Q})$-martingale (on $[0,T]$).
Again, since κ^{xy} is a left-continuous and $\mathbb{F}$-adapted process, and since $\mathbb{F}\subset\widehat{\mathbb{G}}$, we conclude that κ^{xy} is $\widehat{\mathbb{G}}$-predictable. Thus, by the Girsanov theorem, we conclude that the $(\widehat{\mathbb{G}},\mathbb{P})$-compensator of H^{xy} has density with respect to the Lebesgue measure on $[0,T]$ given as follows[3]

$$\mathbb{1}_{\{x\}}(X_{t-})a^{xy}(1+\kappa^{xy}_t)=\mathbb{1}_{\{x\}}(X_{t-})a^{xy}\left(1+\frac{\lambda^{xy}_{t-}}{a^{xy}}-1\right)=\mathbb{1}_{\{x\}}(X_{t-})\lambda^{xy}_{t-}$$

for $t\in[0,T]$. So, for any $x\neq y$, the process $\widehat{K}^{xy}$ defined by

$$\widehat{K}^{xy}_t:=H^{xy}_t-\int_0^t\mathbb{1}_{\{x\}}(X_{u-})\lambda^{xy}_{u-}du,\quad t\in[0,T]$$

is a $\widehat{\mathbb{G}}$-local martingale under $\mathbb{P}$. Since X is a càdlàg process and since the function λ^{xy} satisfies condition **D2** we see that

$$\widehat{K}^{xy}_t=H^{xy}_t-\int_0^t H^x_u\lambda^{xy}_u du,\quad t\in[0,T] \tag{D.54}$$

is a $\widehat{\mathbb{G}}$-local martingale under $\mathbb{P}$. Note that $\mathbb{F}\vee\mathbb{G}\subset\widehat{\mathbb{G}}$ and that the process $\widehat{K}^{xy}$ is $\mathbb{F}\vee\mathbb{G}$-adapted. Taking

$$\tau_n:=\inf\left\{t\ge 0: H^{xy}_t\ge n \text{ or } \int_0^t\lambda^{xy}_u du\ge n\right\}$$

gives us a localizing sequence of $\widehat{\mathbb{G}}$-stopping-times for $\widehat{K}^{xy}$ which are also $\mathbb{F}\vee\mathbb{G}$-stopping-times. So, in view of Theorem 3.7 in Föllmer and Protter (2011), we have

[2] As usual, we use the convention that a product over an empty set is equal to 1.
[3] The usual convention is that $U_{0-}:=0$ for any real-valued process U.

that $\widehat{K}^{xy}$ is also a $\mathbb{F} \vee \mathbb{G}$-local martingale. Thus according to Remark D.33 we can use the special case of Theorem D.21 stated in Theorem D.9 to conclude that Λ is an $\mathbb{F}$-intensity of X under $\mathbb{P}$.

Step 2: Here we will prove (i). In Step 1 we proved that η is a $(\widehat{\mathbb{G}}, \mathbb{Q})$-martingale. By the definition of $\mathbb{P}$ and by the tower property of conditional expectations we conclude that, for an arbitrary $\psi \in L^\infty(\Omega, \mathcal{F}, \mathbb{P})$,

$$\mathbb{E}_{\mathbb{P}}(\psi) = \mathbb{E}_{\mathbb{Q}}(\psi\eta_T) = \mathbb{E}_{\mathbb{Q}}(\mathbb{E}_{\mathbb{Q}}(\psi\eta_T|\widehat{\mathcal{G}}_0)) = \mathbb{E}_{\mathbb{Q}}(\psi\mathbb{E}_{\mathbb{Q}}(\eta_T|\widehat{\mathcal{G}}_0)) = \mathbb{E}_{\mathbb{Q}}(\psi).$$

Step 3: We will show that $\mathbb{F}$ is $\mathbb{P}$-immersed in $\mathbb{F} \vee \mathbb{G}$. In view of Proposition 5.9.1.1 in Jeanblanc et al. (2009) it suffices to show that, for any $\psi \in L^\infty(\Omega, \mathcal{F}, \mathbb{P})$ and any $t \in [0, T]$, it holds that

$$\mathbb{E}_{\mathbb{P}}(\psi|\mathcal{F}_t \vee \mathcal{G}_t) = \mathbb{E}_{\mathbb{P}}(\psi|\mathcal{F}_t), \quad \mathbb{P}\text{-a.s.} \tag{D.55}$$

Now, we observe that

$$\mathbb{P}(\eta_t > 0) = \mathbb{E}_{\mathbb{Q}}(\mathbb{1}_{\{\eta_t>0\}}\eta_T) \geq \mathbb{E}_{\mathbb{Q}}(\mathbb{1}_{\{\eta_T>0\}}\eta_T) = \mathbb{E}_{\mathbb{Q}}(\eta_T) = 1,$$

so that $\mathbb{P}(\eta_t > 0) = 1$. Moreover, η_t is $\mathcal{F}_t \vee \mathcal{G}_t$-measurable by **I3**, (D.50) and condition **D1**. Thus we have

$$\mathbb{E}_{\mathbb{P}}(\psi|\mathcal{F}_t \vee \mathcal{G}_t) = \frac{\mathbb{E}_{\mathbb{Q}}(\psi\eta_T|\mathcal{F}_t \vee \mathcal{G}_t)}{\mathbb{E}_{\mathbb{Q}}(\eta_T|\mathcal{F}_t \vee \mathcal{G}_t)} = \frac{\mathbb{E}_{\mathbb{Q}}(\psi\mathbb{E}_{\mathbb{Q}}(\eta_T|\widehat{\mathcal{G}}_t)|\mathcal{F}_t \vee \mathcal{G}_t)}{\eta_t}$$

$$= \frac{\mathbb{E}_{\mathbb{Q}}(\psi\eta_t|\mathcal{F}_t \vee \mathcal{G}_t)}{\eta_t} = \mathbb{E}_{\mathbb{Q}}(\psi|\mathcal{F}_t \vee \mathcal{G}_t) = \mathbb{E}_{\mathbb{Q}}(\psi|\mathcal{F}_t), \quad \mathbb{P}\text{-a.s.},$$

where the third equality holds in view of the fact that η is a $(\widehat{\mathbb{G}}, \mathbb{Q})$-martingale, and where the last equality holds since $\mathbb{F}$ is $\mathbb{Q}$-immersed in $\mathbb{F} \vee \mathbb{G}$ (see Corollary A.2 in Bielecki et al. (2017a)). Hence, using (i) we conclude that

$$\mathbb{E}_{\mathbb{P}}(\psi|\mathcal{F}_t \vee \mathcal{G}_t) = \mathbb{E}_{\mathbb{P}}(\psi|\mathcal{F}_t), \quad \mathbb{P}\text{-a.s.}$$

Consequently, (D.55) holds.

Step 4: Now we will show the required orthogonality, that is we prove claim (c). To this end it suffices to prove that no real-valued $(\mathbb{F}, \mathbb{P})$-martingale has common jumps with X under $\mathbb{P}$ (see Proposition D.35). Let us take Z to be an arbitrary real-valued $(\mathbb{F}, \mathbb{P})$-martingale. Then, in view of (i), Z is an $(\mathbb{F}, \mathbb{Q})$-martingale. By (I3), we have that $(\mathbb{F}, \mathbb{Q})$-martingales and Poisson processes in the family $\mathcal{N}$ are independent under $\mathbb{Q}$. Thus, by Lemma D.37, the $\mathbb{Q}$-probability that process Z has common jumps with any process from family $\mathcal{N}$ is zero. Consequently, in view of (D.49), the $(\mathbb{F}, \mathbb{Q})$-martingale Z does not jump together with X, $\mathbb{Q}$-a.s. Therefore, by the absolute continuity of $\mathbb{P}$ with respect to $\mathbb{Q}$, the $\mathbb{P}$-probability that Z jumps at the same time as X is zero.

Step 5: Finally, we will show that (D.53) holds. To this end, let us take an arbitrary real-valued function h on $\mathcal{X}$. The abstract Bayes rule yields

$$\mathbb{E}_{\mathbb{P}}(h(X_0)|\mathcal{F}_T) = \frac{\mathbb{E}_{\mathbb{Q}}(h(X_0)\eta_T|\mathcal{F}_T)}{\mathbb{E}_{\mathbb{Q}}(\eta_T|\mathcal{F}_T)} = \frac{\mathbb{E}_{\mathbb{Q}}(\mathbb{E}_{\mathbb{Q}}(h(X_0)\eta_T|\widehat{\mathcal{G}}_0)|\mathcal{F}_T)}{\mathbb{E}_{\mathbb{Q}}(\mathbb{E}_{\mathbb{Q}}(\eta_T|\widehat{\mathcal{G}}_0)|\mathcal{F}_T)}$$
$$= \mathbb{E}_{\mathbb{Q}}(h(X_0)\mathbb{E}_{\mathbb{Q}}(\eta_T|\widehat{\mathcal{G}}_0)|\mathcal{F}_T) = \mathbb{E}_{\mathbb{Q}}(h(X_0)|\mathcal{F}_T) = \mathbb{E}_{\mathbb{Q}}(h(X_0)|\mathcal{F}_0),$$

where the last equality follows from the fact that, by assumption (D.47), the initial condition of the process X satisfies

$$\mathbb{Q}(X_0 = x|\mathcal{F}_T) = \mathbb{Q}(X_0 = x|\mathcal{F}_0), \quad x \in \mathcal{X}. \tag{D.56}$$

Consequently,

$$\mathbb{E}_{\mathbb{P}}(h(X_0)|\mathcal{F}_0) = \mathbb{E}_{\mathbb{P}}(\mathbb{E}_{\mathbb{P}}(h(X_0)|\mathcal{F}_T)|\mathcal{F}_0) = \mathbb{E}_{\mathbb{P}}(\mathbb{E}_{\mathbb{Q}}(h(X_0)|\mathcal{F}_0)|\mathcal{F}_0)$$
$$= \mathbb{E}_{\mathbb{Q}}(h(X_0)|\mathcal{F}_0) = \mathbb{E}_{\mathbb{P}}(h(X_0)|\mathcal{F}_T).$$

This completes the proof of (D.53), and the proof of the theorem. □

Remark D.41 There exist many different sufficient conditions ensuring that $\mathbb{E}_{\mathbb{Q}}\vartheta = 1$. For example the uniform boundedness of Λ is such a condition.

Thanks to Theorem D.40 we may now obtain the following key result.

Proposition D.42 *Let X be a process constructed using Theorem D.40, so that X is an $(\mathbb{F},\mathbb{G})$-CMC process with an $\mathbb{F}$-intensity process Λ. Then X is an $(\mathbb{F},\mathbb{G})$-DMC with an intensity process $\Gamma = \Lambda$. Thus, X is an $(\mathbb{F},\mathbb{G})$-CDMC with $\mathbb{F}$-intensity.*

For a proof, see the proof of Proposition 4.16 in Bielecki et al. (2017a).

We end this section with an example of an $(\mathbb{F},\mathbb{G})$-CDMC.

Example D.43 Time-changed discrete Markov chain: Example D.5 continued Consider a process $\bar{C}$ which is a discrete-time Markov chain with values in $\mathcal{X} = \{1,\ldots,K\}$ and with transition probability matrix Q. In addition, consider the process N, which is a Cox process with càdlàg $\mathbb{F}$-adapted process $\tilde{\lambda}$ satisfying (D.7). From Jakubowski and Niewęgłowski (2008b), Theorems 7 and 9, we know that under the assumption that the processes $(\bar{C}_k)_{k\geq 0}$ and $(N_t)_{t\in[0,T]}$ are independent and conditionally independent given $\mathcal{F}_T$, the process

$$C_t := \bar{C}_{N_t}$$

is an $(\mathbb{F},\mathbb{F}^C)$-DMC. Moreover C admits the intensity process $\Gamma = [\gamma^{xy}]$ given as $\gamma_t^{xy} = (Q-I)_{xy}\tilde{\lambda}_t$. Thus, by Corollary D.18 and Proposition D.26, the process C is an $(\mathbb{F},\mathbb{F}^C)$-CMC with $\mathbb{F}$-intensity $\Lambda = \Gamma$, and hence it is also an $(\mathbb{F},\mathbb{F}^C)$-CDMC.

Appendix E Evolution Systems and Semigroups of Linear Operators

We gather here selected concepts and definitions regarding the evolution systems of linear operators and semigroups of linear operators that we use throughout the book. For a comprehensive study of these concepts and definitions we refer to Ethier and Kurtz (1986), Jacob (2001, 2002, 2005), and van Casteren (2011a).

Let $m \geq 1$ and consider the Banach space $(\mathbb{B}(\mathbb{R}^m), \|\cdot\|_\infty)$ of real-valued, Borel measurable, and bounded functions on $\mathcal{X} = \mathbb{R}^m$, endowed with the supremum norm. As usual, by $C_0(\mathbb{R}^m)$ we denote the class of real-valued continuous functions vanishing at infinity and endowed with the supremum norm.

Let $\mathcal{T} = (T_{s,t},\ 0 \leq s \leq t)$ denote a two-parameter system of bounded linear operators on $(\mathbb{B}(\mathbb{R}^m), \|\cdot\|_\infty)$. We call the system $\mathcal{T}$ an evolution system if the following conditions are satisfied.

1. For any $0 \leq s$, we have $T_{s,s} = \mathrm{Id}$, where Id is the identity operator on $(\mathbb{B}(\mathbb{R}^m), \|\cdot\|_\infty)$.
2. For any $0 \leq s \leq t$, we have $T_{s,t}\mathbb{1} = \mathbb{1}$, where $\mathbb{1}$ is the constant function taking the value 1.
3. For any $0 \leq s \leq t \leq u$, we have $T_{s,u} = T_{s,t}T_{t,u}$.

Let $\mathcal{T}$ be an evolution system. The *strong infinitesimal generator system* of $\mathcal{T}$ is the system of operators, say $\mathcal{A} = (A_u,\ u \geq 0)$, defined as

$$A_u f = \lim_{h \downarrow 0} \frac{T_{u,u+h} f - f}{h}, \tag{E.1}$$

where the convergence is understood in the strong sense, i.e. in the norm of $(\mathbb{B}(\mathbb{R}^m), \|\cdot\|_\infty)$. The domain of A_u is defined as

$$\mathcal{D}(A_u) := \left\{ f \in \mathbb{B}(\mathbb{R}^m) : \exists g \in \mathbb{B}(\mathbb{R}^m) \text{ such that } g = \lim_{h \downarrow 0} \frac{T_{u,u+h} f - f}{h} \right\}.$$

The *weak generator system* of $\mathcal{T}$ is the system of operators, say $\widetilde{\mathcal{A}} = (\widetilde{A}_u,\ u \geq 0)$, defined as

$$\widetilde{A}_u f = \text{bp-}\lim_{h\downarrow 0} \frac{T_{u,u+h}f - f}{h}, \tag{E.2}$$

where "bp" means that the convergence is understood in the bounded-pointwise sense (see Ethier and Kurtz (1986), Section 3.4, for the definition of bp-convergence). The domain of $\widetilde{A}_u$ is defined as

$$\mathcal{D}(\widetilde{A}_u) := \left\{ f \in \mathbb{B}(\mathbb{R}^m) : \exists g \in \mathbb{B}(\mathbb{R}^m) \text{ such that } g = \text{bp-}\lim_{h\downarrow 0} \frac{T_{u,u+h}f - f}{h} \right\}.$$

Definition E.1 An evolution system $\mathcal{T}$ is called a $C_0(\mathbb{R}^m)$-Feller evolution system, or just $C_0(\mathbb{R}^m)$-Feller evolution, if

1. $T_{s,t}(C_0(\mathbb{R}^m)) \subset C_0(\mathbb{R}^m)$ for any $0 \leq s \leq t$,
2. $\mathcal{T}$ is strongly continuous on $C_0(\mathbb{R}^m)$, that is,

$$\lim_{\substack{(s,t)\to(u,v)\\ 0\leq s\leq t,\ 0\leq u\leq v}} \|T_{s,t}f - T_{u,v}f\|_\infty = 0, \quad \forall f \in C_0(\mathbb{R}^m).$$

If an evolution system $\mathcal{T}$ is strongly continuous on $C_0(\mathbb{R}^m)$ then the strong infinitesimal generator system for $\mathcal{T}$ coincides with the weak infinitesimal generator system for $\mathcal{T}$, that is $\mathcal{A} = \widetilde{\mathcal{A}}$.

If the system $\mathcal{T}$ satisfies the property that $T_{s,t} = T_{0,t-s}$ for all $0 \leq s \leq t$, then we call the system $\mathcal{T}$ a time-homogenous system or semigroup. In this case we define $T_t := T_{0,t}$ for all $t \geq 0$. The strong (resp. weak) infinitesimal generator system reduces to a single operator: $A = A_0$ (resp. $\widetilde{A} = \widetilde{A}_0$), which is called the strong (resp. weak) infinitesimal generator of T_t, $t \geq 0$.

Definition E.2 A semigroup $\mathcal{T}$ is called the $C_0(\mathbb{R}^m)$-Feller semigroup if

1. $T_t(C_0(\mathbb{R}^m)) \subset C_0(\mathbb{R}^m)$ for any $0 \leq t$,
2. $\mathcal{T}$ is strongly continuous on $C_0(\mathbb{R}^m)$, i.e.

$$\lim_{\substack{t\to 0\\ 0\leq t}} \|T_t f - f\|_\infty = 0, \quad \forall f \in C_0(\mathbb{R}^m).$$

If a semigroup $\mathcal{T}$ is strongly continuous on $C_0(\mathbb{R}^m)$ then the strong generator for $\mathcal{T}$ coincides with the weak generator for $\mathcal{T}$, i.e. $A = \tilde{A}$ (see Theorem 2.1.3 in Pazy (1983)).

A semigroup $\mathcal{T} = (T_t,\ t \geq 0)$ on $\mathbb{B}(\mathbb{R}^m)$ is called positive if for all $t \geq 0$ we have $T_t f \geq 0$ for $f \geq 0$. It is called a contraction semigroup if for all $t \geq 0$ we have $\|T_t\| \leq 1$.

It is customary to consider Feller evolution systems and Feller semigroups as operators acting on $C_0(\mathbb{R}^m)$ rather than as operators acting on $\mathbb{B}(\mathbb{R}^m)$. We proceed in this way in this book as well, and we restrict Feller evolution systems and Feller semigroups to $C_0(\mathbb{R}^m)$.

Accordingly, for Feller evolution systems we modify the definition of an evolution system acting on $C_0(\mathbb{R}^m)$ by removing the condition that for any $0 \le s \le t$ we have $T_{s,t}\mathbb{1} = \mathbb{1}$. In addition, the domain of the element A_u of the strong generator system of a Feller evolution system $\mathcal{T}$ is defined now as

$$\mathcal{D}(A_u) := \left\{ f \in C_0(\mathbb{R}^m) : \exists g \in C_0(\mathbb{R}^m) \text{ such that } g = \lim_{h \downarrow 0} \frac{T_{u,u+h}f - f}{h} \right\}.$$

According to Theorem 1.33 in Böttcher et al. (2013), the strong generator system coincides with the weak generator system for a Feller evolution system.

The above discussion carries over to Feller semigroups in an obvious way.

Let A be a generator of a semigroup on $C_0(\mathbb{R}^m)$. We say that a set $\mathcal{D} \subset \mathcal{D}(A)$ is a *core* of A if $\mathcal{D}$ is dense in $C_0(\mathbb{R}^m)$ and if for every $f \in \mathcal{D}(A)$ there exists a sequence $(f_n)_{n \ge 1} \subset \mathcal{D}$ such that

$$\lim_{n \to \infty} (\|f - f_n\|_\infty + \|Af - Af_n\|_\infty) = 0.$$

Remark E.3 (i) If the evolution system $\mathcal{T}$ is defined by

$$T_{s,t}f(x) = \int_{\mathbb{R}^m} f(y)P(s,x,t,dy), \quad x \in \mathbb{R}^m,$$

where $P(s,x,t,dy)$ is the transition function for some time-inhomogeneous Markov family (or some time-inhomogeneous Markov process), then we also refer to the generator system of $\mathcal{T}$ as the generator system of this Markov family (or of this Markov process).

(ii) If the semigroup $\mathcal{T}$ is defined by

$$T_t f(x) = \int_{\mathbb{R}^m} f(y)P(x,t,dy), \quad x \in \mathbb{R}^m,$$

where $P(x,t,dy)$ is the transition function for some time-homogeneous Markov family (or some time-homogeneous Markov process), then we also refer to the generator of $\mathcal{T}$ as the generator of this Markov family (or of this Markov process).

(iii) Let

$$\mathcal{MMFH} = \{(\Omega, \mathcal{F}, \mathbb{F}, (X_t)_{t \ge 0}, \mathbb{P}_x, P) : x \in \mathbb{R}^m\}$$

be an $\mathbb{R}^m$-Feller–Markov family. Let $\mathcal{T}$ and A be the corresponding $C_0(\mathbb{R}^m)$-Feller semigroup and its generator, respectively. Then, for $f \in \mathcal{D}(A)$, the following equality holds:

$$\mathbb{E}_x(f(X_t)) = f(x) + \mathbb{E}_x\left(\int_0^t Af(X_s)ds\right), \quad t \ge 0, \, x \in \mathbb{R}^m. \tag{E.3}$$

The above equality is called the Dynkin formula.

Remark E.4 In the case of a time-inhomogeneous Markov chain with a finite state space $\mathcal{X}$, rather than using the notation A_u, $u \ge 0$, and the term "infinitesimal generator system" we use the notation $\Lambda(u)$, $u \ge 0$, and the term "infinitesimal generator function", which is a matrix-valued function. We denote the entries of $\Lambda(u)$ as

$\lambda_{xy}(u)$ for $x, y \in \mathcal{X}$, and we call them the intensities of the transitions of X. Here, we tacitly assume that the corresponding transition function (i.e. transition probability) $P(u,x,t,\{y\})$ is differentiable in t, so that

$$\left.\frac{\partial P(u,x,t,\{y\})}{\partial t}\right|_{t=u} = \lambda_{xy}(u), \quad x \neq y.$$

Moreover, $\lambda_{xx}(u) = -\sum_{y\in\mathcal{X}, y\neq x} \lambda_{xy}(u)$.

Remark E.5 The term "infinitesimal" is frequently skipped when referring to generator systems or to generators.

Appendix F Martingale Problem: Some New Results Needed in this Book

Let $(\Omega, \mathcal{F}, \mathbb{F}, \mathbb{P})$ be the underlying (not necessarily canonical) complete, filtered, probability space. Let $m \geq 1$ be an integer, and recall that by $C_0(\mathbb{R}^m)$ we denote the Banach space of real-valued, continuous functions on $\mathcal{X} = \mathbb{R}^m$ that vanish at infinity, endowed with the supremum norm. By $\mathcal{A} = \{A_u : u \geq 0\}$ we denote a family of (possibly unbounded) linear operators on $C_0(\mathbb{R}^m)$. The domain of A_u is denoted by $\mathcal{D}(A_u)$ and we let $\mathcal{D}(\mathcal{A}) = \bigcap_{u \geq 0} \mathcal{D}(A_u)$. Finally, recall that by $C_c^\infty(\mathbb{R}^m)$ we denote the class of infinitely differentiable real-valued functions on $\mathbb{R}^m$, such that the functions themselves and all their derivatives have compact supports.

Standing assumption We assume that $C_c^\infty(\mathbb{R}^m) \subset \mathcal{D}(\mathcal{A})$.

The above assumption is satisfied in the case of the so-called rich $\mathbb{R}^m$-Feller–Markov evolution families that are considered in Section 2.2.

In what follows we will consider stochastic processes defined on $(\Omega, \mathcal{F}, \mathbb{P})$ with values in $\mathbb{R}^m$ and adapted to $\mathbb{F}$.

Definition F.1 Let μ be a probability measure on $(\mathbb{R}^m, \mathcal{B}(\mathbb{R}^m))$. We say that the martingale problem for $(\mathcal{A}, \mu)$ starting at $s \geq 0$ admits a solution with respect to $\mathbb{P}$ if there exists a stochastic process Z such that

1. $\mathbb{P}(Z_s \in \cdot) = \mu(\cdot)$.
2. For every $f \in C_c^\infty(\mathbb{R}^m)$ the process M^f given by

$$M_t^f := f(Z_t) - \int_s^t A_u f(Z_u) du, \quad t \geq s \tag{F.1}$$

   is well defined[1] and is a $\mathbb{P}$-martingale in the filtration $\mathbb{F}^Z$.

If the above martingale property holds in the filtration $\mathbb{F}$ then we say that the martingale problem for $(\mathcal{A}, \mu)$ starting at $s \geq 0$ admits a solution with respect to $(\mathbb{F}, \mathbb{P})$. The stochastic process Z is called a solution of the martingale problem for $(\mathcal{A}, \mu)$ starting at $s \geq 0$.

Definition F.2 We say that the martingale problem for $(\mathcal{A}, \mu)$ starting at $s \geq 0$ is well posed with respect to $\mathbb{P}$ if it admits a solution with respect to $\mathbb{P}$ and if the solution

[1] Note that, in particular, the above definition requires that the integral $\int_0^t A_u f(Z_u) du$ is well defined for $t \geq 0$.

is unique in the sense of finite-dimensional distributions (i.e. any two solutions have the same finite-dimensional distributions under $\mathbb{P}$). If for every probability law μ on $(\mathbb{R}^m, \mathcal{B}(\mathbb{R}^m))$ and every $s \geq 0$ the martingale problem for $(\mathcal{A}, \mu)$ starting at $s \geq 0$ is well posed with respect to $\mathbb{P}$, then we simply say that the martingale problem for $\mathcal{A}$ is well posed with respect to $\mathbb{P}$. The well-posedness of the martingale problem with respect to $(\mathbb{F}, \mathbb{P})$ is defined analogously.

The next lemma is used in the proof of Theorem F.6.

Lemma F.3 *Suppose that a process Z has càdlàg paths. Then Z solves the martingale problem for $(\mathcal{A}, \mu)$ starting at $s \geq 0$ if the following two conditions hold:*

(i) $\mathbb{P}(Z_s \in B) = \mu(B)$ *for all* $B \in \mathcal{B}(\mathbb{R}^m)$.
(ii) *For all random variables $\eta_s^f(Z)$ of the form*

$$\eta_s^f(Z) = \left(M_{t_{\ell+1}}^f - M_{t_\ell}^f\right) \prod_{k=1}^{\ell} h_k(Z_{t_k}), \tag{F.2}$$

where $0 \leq s < t_1 < \cdots < t_\ell < t_{\ell+1}$, $h_k, k = 1, \ldots, \ell$, $\ell \in \mathbb{N}$ are bounded measurable functions, $f \in C_c^\infty(\mathbb{R}^m)$, and M^f is given by (F.1), *we have*

$$\mathbb{E}\left(\eta_s^f(Z)\right) = 0. \tag{F.3}$$

Proof It is enough to observe that (F.3) implies that

$$\mathbb{E}((M_{t_{\ell+1}}^f - M_{t_\ell}^f)\mathbb{1}_C) = 0 \quad \text{for all } C \in \mathcal{F}_{t_\ell}^Z.$$

This means that M^f is an $\mathbb{F}^Z$-martingale for all $f \in C_c^\infty(\mathbb{R}^m)$. □

In the next definition we take $\widehat{\Omega} = \Omega(\mathbb{R}^m)$ and $\widehat{Z}$ as the canonical mapping on $\widehat{\Omega}$, so that

$$\widehat{Z}(\widehat{\omega}) = \widehat{\omega} \tag{F.4}$$

for $\widehat{\omega} \in \widehat{\Omega}$. Accordingly, we take the canonical space $(\widehat{\Omega}, \widehat{\mathcal{F}})$ and the canonical filtration $\widehat{\mathbb{F}} := \widehat{\mathbb{F}}_0 = \mathbb{F}_0^{\widehat{Z}}$ (see (B.4)).

Definition F.4 We call a probability measure $\widehat{\mathbb{P}}$ on $(\widehat{\Omega}, \widehat{\mathcal{F}})$ a *solution of the D-martingale problem for $(\mathcal{A}, \mu)$ starting at $s \geq 0$* if

$$\widehat{\mathbb{P}}(D_s) = 1,$$

where[2]

$$D_s := \left\{\omega \in \widehat{\Omega} : \omega 1_{[s,\infty)} \in D([0,\infty), \mathbb{R}^m)\right\},$$

and if $\widehat{Z}$ solves the martingale problem for $(\mathcal{A}, \mu)$ starting at $s \geq 0$ with respect to $(\widehat{\mathbb{F}}, \widehat{\mathbb{P}})$ in the sense of Definition F.1. Accordingly, we say that the D-martingale problem for $(\mathcal{A}, \mu)$ starting at $s \geq 0$ is well posed in $D([0,\infty), \mathbb{R}^m)$ if $\widehat{\mathbb{P}}$ is the unique solution of the D-martingale problem for $(\mathcal{A}, \mu)$ starting at $s \geq 0$. We also say that

[2] $D([0,\infty), \mathbb{R}^m)$ is the Skorokhod space over $\mathbb{R}^m$ (see Ethier and Kurtz (1986), Section 3.5).

the D-martingale problem for $\mathcal{A}$ is well posed if, for every probability measure μ on $(\mathbb{R}^m, \mathcal{B}(\mathbb{R}^m))$ and every $s \geq 0$, the D-martingale problem for $(\mathcal{A}, \mu)$ starting at $s \geq 0$ is well posed.

Let Z be a stochastic process on $(\Omega, \mathbb{F}, \mathbb{P})$. Consider the mapping ϕ_Z from Ω to $\widehat{\Omega}$ defined by $\phi_Z(\omega) = \widehat{\omega} := Z_\cdot(\omega)$, and define $\mathbb{P}^Z(\widehat{A}) = \mathbb{P}(\phi_Z^{-1}(\widehat{A})) = \mathbb{P}(\{\omega : Z_\cdot(\omega) \in \widehat{A}\})$ for any $\widehat{A} \in \widehat{\mathcal{F}}$. The measure $\mathbb{P}^Z$ is called the law of Z on $(\widehat{\Omega}, \widehat{\mathcal{F}})$.

Lemma F.5 *Assume that the process Z is càdlàg and that it is a solution of the martingale problem for $(\mathcal{A}, \mu)$ starting at $s \geq 0$, with respect to $(\mathbb{F}, \mathbb{P})$. Then, $\mathbb{P}^Z$ solves the D-martingale problem for $(\mathcal{A}, \mu)$ starting at $s \geq 0$.*

Proof First, note that clearly we have

$$\mathbb{P}^Z(D_s) = 1.$$

It remains to show that the canonical mapping $\widehat{Z}$ solves the martingale problem for (A, μ) starting at s with respect to $(\widehat{\mathbb{F}}, \mathbb{P}^Z)$. To this end, we first observe that

$$\mathbb{P}^Z(\widehat{\omega} \in \widehat{\Omega} : \widehat{\omega}_s \in B) = \mathbb{P}(Z_s \in B) = \mu(B), \quad B \in \mathcal{B}(\mathbb{R}^m),$$

since Z solves the martingale problem for (A, μ) starting at s, with respect to $(\mathbb{F}, \mathbb{P})$. We need to show that, for every $f \in C_c^\infty(\mathbb{R}^m)$, the process

$$\widehat{M}_t^f = f(\widehat{Z}_t) - \int_s^t A_u f(\widehat{Z}_u) du, \quad t \geq s$$

is a $\mathbb{P}^Z$-martingale in the filtration $\widehat{\mathbb{F}}$. In view of Lemma F.3 it remains to show that

$$\mathbb{E}_{\mathbb{P}^Z}\left(\widehat{\eta}_s^f(\widehat{Z})\right) = 0,$$

where

$$\eta_s^f(\widehat{Z}) = \left(\widehat{M}_{t_{\ell+1}}^f - \widehat{M}_{t_\ell}^f\right) \prod_{k=1}^{\ell} h_k(\widehat{Z}_{t_k}).$$

We have

$$\mathbb{E}_{\mathbb{P}^Z}\left(\widehat{\eta}_s^f(\widehat{Z})\right) = \mathbb{E}_{\mathbb{P}}\left(\eta_s^f(Z)\right) = 0,$$

where the second equality is a consequence of the fact that Z solves the martingale problem for (A, μ) starting at s with respect to $(\mathbb{F}, \mathbb{P})$. The proof is complete. □

We now present an important result, which states that the well-posedness of the D-martingale problem implies the Markov property.

Theorem F.6 *Suppose that the D-martingale problem for $\mathcal{A} = \{A_u,\ u \geq 0\}$ is well posed. Fix $s \geq 0$ and a probability measure on μ on $\mathcal{B}(\mathbb{R}^m)$. Assume that a process Z is càdlàg and that it is a solution of the martingale problem for $(\mathcal{A}, \mu)$ starting at s, with respect to $(\mathbb{F}, \mathbb{P})$. Then, the collection of objects $(\Omega, \mathcal{F}, \mathbb{F}, (Z_t)_{t \geq s}, \mathbb{P})$ is a Markov process in the sense of Remark B.2, with $\mathbb{P}(Z_s \in B) = \mu(B)$.*

Proof We need only to prove the Markov property. To this end, let us fix $r \geq s$ and $C \in \mathcal{F}_r$ such that $\mathbb{P}(C) > 0$. Next, define two measures on $(\Omega, \mathcal{F})$:

$$\mathbb{Q}_1(G) := \frac{\mathbb{E}_{\mathbb{P}}(\mathbb{1}_C \mathbb{P}(G|\mathcal{F}_r))}{\mathbb{P}(C)}, \quad \mathbb{Q}_2(G) := \frac{\mathbb{E}_{\mathbb{P}}(\mathbb{1}_C \mathbb{P}(G|Z_r))}{\mathbb{P}(C)}, \quad G \in \mathcal{F}.$$

By the definitions of $\mathbb{Q}_1$ and $\mathbb{Q}_2$ we see that the laws of Z_r under $\mathbb{Q}_1$ and $\mathbb{Q}_2$ coincide. In fact, for any $B \in \mathcal{B}(\mathbb{R}^m)$,

$$\mathbb{Q}_1(Z_r \in B) = \mathbb{P}(Z_r \in B|C). \tag{F.5}$$

Moreover, since Z solves the martingale problem for $(\mathcal{A}, \mu)$ with respect to $(\mathbb{F}, \mathbb{P})$, we have (see (F.2) for the definition of $\eta_r^f(Z)$)

$$\mathbb{E}_{\mathbb{P}}(\eta_r^f(Z)|\mathcal{F}_{t_\ell}) = \mathbb{E}_{\mathbb{P}}\big(M_{t_{\ell+1}}^f - M_{t_\ell}^f|\mathcal{F}_{t_\ell}\big) \prod_{k=1}^{\ell} h_k(Z_{t_k}) = 0,$$

and hence $\mathbb{E}_{\mathbb{P}}(\eta_r^f(Z)|\mathcal{F}_r) = 0$. Thus,

$$\mathbb{E}_{\mathbb{Q}_1}(\eta_r^f(Z)) = \frac{\mathbb{E}_{\mathbb{P}}(\mathbb{1}_C \mathbb{E}_{\mathbb{P}}(\eta_r^f(Z)|\mathcal{F}_r))}{\mathbb{P}(C)} = 0$$

for all $f \in C_c^\infty(\mathbb{R}^m)$.

Analogously, since $\mathbb{E}_{\mathbb{P}}(\eta_r^f(Z)|\mathcal{F}_r) = 0$ implies that $\mathbb{E}_{\mathbb{P}}(\eta_r^f(Z)|Z_r) = 0$, we have $\mathbb{E}_{\mathbb{Q}_2}(\eta_r^f(Z)) = 0$ for all $f \in C_c^\infty(\mathbb{R}^m)$. Let $\nu_C(B) := \mathbb{Q}_2(Z_r \in B)$. Then $\nu_C(B) = \mathbb{P}(Z_r \in B|C)$. Thus, in view of Lemma F.3 and (F.5), Z solves the martingale problem for $(\mathcal{A}, \nu_C)$ starting at r, relative to both $(\Omega, \mathcal{F}, \mathbb{F}, \mathbb{Q}_1)$ and $(\Omega, \mathcal{F}, \mathbb{F}, \mathbb{Q}_2)$. Since Z is càdlàg, the well-posedness of the D-martingale problem for $\mathcal{A}$ (see Definition F.4), combined with Lemma F.5, implies that, for all $B \in \mathcal{B}(\mathbb{R}^m)$ and all $t \geq r$,

$$\mathbb{Q}_1(Z_t \in B) = \mathbb{P}^Z(\{\widehat{\omega} : \widehat{\omega}_t \in B\}) = \mathbb{Q}_2(Z_t \in B).$$

Therefore, by the definitions of $\mathbb{Q}_1$ and $\mathbb{Q}_2$, we have, for $C \in \mathcal{F}_r$,

$$\mathbb{E}_{\mathbb{P}}(\mathbb{1}_C \mathbb{P}(Z_t \in B|\mathcal{F}_r)) = \mathbb{E}_{\mathbb{P}}(\mathbb{1}_C \mathbb{P}(Z_t \in B|Z_r)).$$

Recall that C was taken to be arbitrary, so we obtain

$$\mathbb{P}(Z_t \in B|\mathcal{F}_r) = \mathbb{P}(Z_t \in B|Z_r).$$

Finally, since $r \geq s$ is arbitrary, the proof is complete. □

In this book we also use time-homogeneous versions of the martingale problems introduced above. A martingale problem is time-homogeneous if it starts at time 0 and if $\mathcal{A} = \{A\}$. The definition of the time-homogeneous version of the D-martingale problem reads as follows.

Definition F.7 We call a probability measure $\widehat{\mathbb{P}}$ on $(\widehat{\Omega}, \widehat{\mathcal{F}})$ a *solution of the time-homogeneous D-martingale problem for* (A, μ) if

$$\widehat{\mathbb{P}}(D([0,\infty), \mathbb{R}^m)) = 1,$$

and if $\widehat{Z}$ solves the martingale problem for (A,μ) starting at 0 with respect to $(\widehat{\mathbb{F}},\widehat{\mathbb{P}})$ in the sense of (time-homogeneous versions of) Definition F.1.

Accordingly, we say that the time-homogeneous D-martingale problem for (A,μ) is well posed in $D([0,\infty),\mathbb{R}^m)$ if $\widehat{\mathbb{P}}$ is the unique solution of the time-homogeneous D-martingale problem for (A,μ) starting at 0. We also say that the time-homogeneous D-martingale problem for A is well posed if for every probability measure μ on $(\mathbb{R}^m,\mathcal{B}(\mathbb{R}^m))$ the time-homogeneous D-martingale problem for (A,μ) is well posed.

Appendix G Function Spaces and Pseudo-Differential Operators

We gather here selected concepts and definitions regarding function spaces and pseudo-differential operators (PDOs) that we use throughout the book. For a comprehensive study of these concepts and definitions we refer to Jacob (2001, 2002, 2005) and Rüschendorf et al. (2016).

Some function spaces In most of the book we use real-valued functions. Accordingly, the function spaces listed below are those of real-valued functions. We take $m \geq 1$.

$C_0(\mathbb{R}^m)$ – the space of continuous functions f on $\mathbb{R}^m$ vanishing at infinity, i.e. $\lim_{\|x\|\to\infty} |f(x)| = 0$. In Jacob (2001, 2002, 2005) this space is denoted as $C_\infty(\mathbb{R}^m)$;

$C_0^\infty(\mathbb{R}^m)$ – the space of infinitely differentiable functions f such that f and all its derivatives belong to $C_0(\mathbb{R}^m)$;

$C_c(\mathbb{R}^m)$ – the space of continuous functions f on $\mathbb{R}^m$ with compact support. In Jacob (2001, 2002, 2005) this space is denoted as $C_0(\mathbb{R}^m)$;

$C_c^\infty(\mathbb{R}^m)$ – the space of infinitely differentiable functions f such that f and all its derivatives belong to $C_c(\mathbb{R}^m)$;

$C_b(\mathbb{R}^m)$ – the space of bounded and continuous functions f on $\mathbb{R}^m$;

$C^\infty(\mathbb{R}^m)$ – the space of infinitely differentiable functions f on $\mathbb{R}^m$;

$\mathcal{S}(\mathbb{R}^m)$ the Schwartz space, also called the space of rapidly decreasing functions, is defined as

$$\mathcal{S}(\mathbb{R}^m) = \left\{ f \in C^\infty(\mathbb{R}^m) : \|f\|_{\alpha,\beta} < \infty, \quad \forall \alpha, \beta \in \mathbb{N}^m \right\},$$

where $\|\cdots\|_{\alpha,\beta}$ is a semi-norm defined by

$$\|f\|_{\alpha,\beta} := \sup_{x \in \mathbb{R}^m} \left| x_1^{\alpha_1} \cdots x_m^{\alpha_m} \frac{\partial^{\beta_1 + \cdots + \beta_m}}{\partial_{x_1}^{\beta^1} \cdots \partial_{x_m}^{\beta_m}} f(x) \right|;$$

$L^p(\mathbb{R}^m)$, for $1 \leq p < \infty$, is the space of p-integrable functions f on $\mathbb{R}^m$,

$$\int_{\mathbb{R}^m} |f(x)|^p dx < \infty.$$

The notation $L^p(\mathbb{R}^m;\mathbb{C})$ indicates complex-valued functions. A similar notation is used for all the other classes of functions used in this book, whenever needed.

Positive-definite function A function $\varphi : \mathbb{R}^m \to \mathbb{C}$ is called positive-definite if, for any $k \in \mathbb{N}$ and for any $x_1, \ldots, x_k \in \mathbb{R}^m$, the matrix $(\varphi(x_i - x_j))_{i,j=1,\ldots,k}$ is positive Hermitian, i.e. if

$$\sum_{i,j=1}^{k} \varphi(x_i - x_j)\lambda_i \overline{\lambda_j} \geq 0 \quad \text{for all } \lambda_i, \lambda_j \in \mathbb{C}.$$

Negative-definite functions (in the sense of Schoenberg) A function $\psi : \mathbb{R}^m \to \mathbb{C}$ is called negative-definite if $\psi(0) \geq 0$ and

$$\xi \mapsto \frac{1}{(2\pi)^{n/2}} e^{-t\psi(\xi)}$$

is positive-definite for $t \geq 0$.

Fourier transforms and their inverse Let $f \in L^1(\mathbb{R}^m;\mathbb{C})$. The Fourier transform of f is the function defined as

$$\widehat{f}(\xi) = (2\pi)^{-m/2} \int_{\mathbb{R}^m} e^{-i\langle x,\xi\rangle} f(x)dx, \quad \xi \in \mathbb{R}^m,$$

where $\langle \cdot, \cdot \rangle$ is the scalar product in $\mathbb{R}^m$. If $f \in C_c^\infty(\mathbb{R}^m)$ then its Fourier transform $\widehat{f}$ is in the Schwartz space $\mathcal{S}(\mathbb{R}^m)$.

The function f can be recovered from $\widehat{f}$, provided that $\widehat{f} \in L^1(\mathbb{R}^m;\mathbb{C})$, via the inverse Fourier transform:

$$f(x) = (2\pi)^{-m/2} \int_{\mathbb{R}^m} e^{i\langle x,\xi\rangle} \widehat{f}(\xi)d\xi \quad \text{for almost all } x \in \mathbb{R}^m.$$

The Fourier transform is known to be a bijective continuous operator from $\mathcal{S}(\mathbb{R}^m)$ into itself.

Pseudo-differential operators (PDOs) and their symbol Following Rüschendorf et al. (2016) we present here a generic definition of a time-dependent PDO. A time-dependent pseudo-differential operator $q(t,x,D)$ is an operator of the form

$$q(t,x,D)f(x) := \int_{\mathbb{R}^m} e^{-i\langle x,\xi\rangle} q(t,x,\xi)\widehat{f}(\xi)d\xi, \quad f \in C_c^\infty(\mathbb{R}^m),$$

where

$$q : [0,\infty) \times \mathbb{R}^m \times \mathbb{R}^m \to \mathbb{C}$$

is a function satisfying the following properties:

G1 For every $t \geq 0$, $q(t,\cdot,\cdot)$ is locally bounded.

G2 For every $t \geq 0$ and $\xi \in \mathbb{R}^n$, $q(t,\cdot,\xi)$ is measurable.

G3 For every $t \geq 0$ and $x \in \mathbb{R}^n$, $q(t,x,\cdot)$ is a continuous and negative-definite function.

The function $q(t,x,\xi)$ as above is called the (time-dependent) symbol of the (time-dependent) PDO $q(t,x,D)$.

A pseudo-differential operator $q(x,D)$ with symbol[1] $q(x,\xi)$ is an operator of the form

$$q(x,D)f(x) := \int_{\mathbb{R}^m} e^{-i\langle x,\xi\rangle} q(x,\xi)\widehat{f}(\xi)d\xi, \quad f \in C_c^\infty(\mathbb{R}^m),$$

where

$$q : \mathbb{R}^m \times \mathbb{R}^m \to \mathbb{C}$$

is a function satisfying the appropriate properties, namely

1. The function $q : \mathbb{R}^m \times \mathbb{R}^m \to \mathbb{R}$ is measurable and locally bounded.
2. The function $\xi \to q(x,\xi)$ is continuous and negative-definite for all $x \in \mathbb{R}^m$.

Courrège's theorem We follow Jacob and Schilling (2001). Let $A : C_c^\infty(\mathbb{R}^m) \to C_b(\mathbb{R}^m)$ be a linear operator satisfying the positive maximum principle. Then, there exist measurable functions

(i) $c : \mathbb{R}^m \to [0,\infty)$,
(ii) $b : \mathbb{R}^m \to \mathbb{R}^m$,
(iii) $a : \mathbb{R}^m \to \mathbb{R}^{m\times m}$, which is symmetric positive semi-definite,

and a Lévy kernel $\mu(x,dy)$ from $(\mathbb{R}^m, \mathcal{B}(\mathbb{R}^m))$ to $(\mathbb{R}^m \setminus \{0\}, \mathcal{B}(\mathbb{R}^m \setminus \{0\}))$ satisfying

$$\int_{y\neq 0} (|y|^2 \wedge 1)\mu(x,dy) < \infty$$

for all x, such that the function $q(x,\xi)$ given as

$$q(x,\xi) = c(x) - i\langle b(x),\xi\rangle + \langle \xi, a(x)\xi\rangle + \int_{y\neq 0}\left(1 - e^{-i\langle y,\xi\rangle} - \frac{i\langle y,\xi\rangle}{1+|y|^2}\right)\mu(x,dy) \tag{G.1}$$

is locally bounded, the function $\xi \to q(x,\xi)$ is continuous negative-definite, and, for all $f \in C_c^\infty(\mathbb{R}^m)$ and for all $x \in \mathbb{R}^m$, we have

$$Af(x) = -q(x,D)f(x).$$

[1] Some authors (e.g. Jacob (1998), Hoh (1998)) call such a q a continuous negative symbol.

In addition, A admits the so-called Waldenfels representation:

$$
\begin{aligned}
Af(x) = &-c(x)f(x) + \langle b(x), \nabla f(x)\rangle + \sum_{j,k=1}^{m} a_{jk}(x)\partial_j\partial_k f(x) \\
&+ \int_{y\neq 0} \left(f(x-y) - f(x) + \frac{\langle y, \nabla f(x)\rangle}{1+|y|^2} \right) \mu(x,dy). \qquad \text{(G.2)}
\end{aligned}
$$

Remark G.1 (i) With a slight abuse of terminology, we will also refer to $q(x,\xi)$ as the symbol of A.
(ii) If the operator A generates a sub-Markov process, say X, then c is the so-called killing coefficient of X, b is the drift coefficient of X, a is the diffusion coefficient of X, and $\mu(x,\cdot)$ governs the distribution of the jumps of X. However, in this book we do not consider sub-Markov processes, so instead we consider the case of $c \equiv 0$.

References

Aksamit, Anna, and Jeanblanc, Monique. 2017. *Enlargements of Filtrations with Finance in View*. Springer.

Amendinger, Jürgen. 2000. Martingale representation theorems for initially enlarged filtrations. *Stochastic Process. Appl.*, **89**(1), 101–116.

Asimit, Vali, and Booneny, Tim J. 2017. Insurance with multiple insurers: A game-theoretic approach. *European J. Operational Res.* Available online 20 December 2017.

Asmussen, Søren. 2003. *Applied Probability and Queues*, Second Edition. Academic Press.

Asmussen, Søren, and Albrecher, Hansjörg. 2010. *Ruin Probabilities*, Second Edition. Advanced Series on Statistical Science and Applied Probability, vol. 14. World Scientific Publishing Co. Pte. Ltd.

Assefa, Samson, Bielecki, Tomasz R., Cousin, Areski, Crépey, Stéphane, and Jeanblanc, Monique. 2011. *Credit Risk Frontiers: Subprime Crisis, Pricing and Hedging, CVA, MBS, Ratings and Liquidity*. Wiley, pp. 397–436.

Avram, Florin, Palmowski, Zbigniew, and Pistorius, Martijn. 2008. A two-dimensional ruin problem on the positive quadrant. *Insurance Math. Econom.*, **42**(1), 227–234.

Bacry, E., Delattre, S., Hoffmann, M., and Muzy, J. 2013a. Modelling microstructure noise with mutually exciting point processes. *Quant. Finance*, **13**(1), 65–77.

Bacry, E., Delattre, S., Hoffmann, C., and Muzy, J. 2013b. Some limit theorems for Hawkes processes and application to financial statistics. *Stochastic Proc. Applic.*, **123**(7), 2475–2499.

Bacry, Emmanuel, and Muzy, Jean-François. 2014. Hawkes model for price and trades high-frequency dynamics. *Quant. Finance*, **14**(7), 1147–1166.

Ball, Frank, and Yeo, Geoffrey F. 1993. Lumpability and marginalisability for continuous-time Markov chains. *J. Appl. Probab.*, **30**(3), 518–528.

Ball, Frank, Milne, Robin K., and Yeo, Geoffrey F. 1994. Continuous-time Markov chains in a random environment, with applications to ion channel modelling. *Adv. Appl. Probab.*, **26**(4), 919–946.

Biagini, Francesca, Groll, Andreas, and Widenmann, Jan. 2013. Intensity-based premium evaluation for unemployment insurance products. *Insurance Math. Econom.*, **53**(1), 302–316.

Bielecki, Tomasz R., and Rutkowski, Marek. 2004. *Credit Risk: Modelling, Valuation and Hedging.* Springer.

Bielecki, Tomasz R., Vidozzi, Andrea, and Vidozzi, Luca. 2008a. Markov copulae approach to pricing and hedging of credit index derivatives and ratings triggered step-up bonds. *J. Credit Risk*, **4**(1), 47–76.

Bielecki, Tomasz R., Jakubowski, Jacek, Vidozzi, Andrea, and Vidozzi, Luca. 2008b. Study of dependence for some stochastic processes. *Stoch. Anal. Appl.*, **26**(4), 903–924.

Bielecki, Tomasz R., Crépey, Stephane, Jeanblanc, Monique, and Rutkowski, Marek. 2008c. *Valuation of basket credit derivatives in the credit migrations environment.* In Birge, J. and Linetsky, V. (eds) *Handbooks in Operations Research and Management Science.* Financial Engineering, vol. 15. pp. 471–510.

Bielecki, Tomasz R., Jakubowski, Jacek, and Nieweglowski, Mariusz. 2010. Dynamic modeling of dependence in finance via copulae between stochastic processes. Pages 33–76 of *Copula Theory and Its Applications*. Lecture Notes in Statistical Processes, vol. 198. Springer.

Bielecki, Tomasz R., Cousin, Areski, Crépey, Stéphane, and Herbertsson, Alexander. 2014a. A bottom-up dynamic model of portfolio credit risk with stochastic intensities and random recoveries. *Commun. Statistis – Theory and Methods*, **43**(7), 1362–1389.

Bielecki, Tomasz R., Cousin, Areski, Crépey, Stéphane, and Herbertsson, Alexander. 2014b. Dynamic hedging of portfolio credit risk in a Markov copula model. *J. Optim. Theory Appl.*, **161**(1), 90–102.

Bielecki, Tomasz R., Jakubowski, Jacek, and Nieweglowski, Mariusz. 2015. Conditional Markov chains – construction and properties. Pages 33–42 of *Banach Center Publications 105 (2015), Stochastic Analysis*. Special volume in honour of Jerzy Zabczyk. Polish Academy of Sciences, Institute of Mathematics, Warsaw.

Bielecki, Tomasz R., Jakubowski, Jacek, and Nieweglowski, Mariusz. 2017a. Conditional Markov chains: properties, construction and structured dependence. *Stochastic Process. Appl.*, **127**(4), 1125–1170.

Bielecki, Tomasz R., Jakubowski, Jacek, and Nieweglowski, Mariusz. 2017b. A note on independence copula for conditional Markov chains. Pages 303–321 of *Recent Progress and Modern Challenges in Applied Mathematics, Modeling and Computational Science*. Fields Inst. Commun., vol. 79. Springer.

Bielecki, Tomasz R., Cialenco, Igor, and Feng, Shibi. 2018a. A dynamic model of central counterparty risk. *Int. J. Theore. Appl. Finance*, **21**(8), 1850050 (32 pp.).

Bielecki, Tomasz R., Jakubowski, Jacek, Jeanblanc, Monique, and Nieweglowski, Mariusz. 2018b. Semimartingales and shrinkage of filtration. To appear.

Blanes, S., Casas, F., Oteo, J. A., and Ros, J. 2009. The Magnus expansion and some of its applications. *Phys. Rep.*, **470**(5-6), 151–238.

Blumenthal, R. M., and Getoor, R. K. 1968. *Markov Processes and Potential Theory*. Pure and Applied Mathematics, vol. 29. Academic Press.

Böttcher, Björn, Schilling, René, and Wang, Jian. 2013. *Lévy Matters*, Volume III. Lecture Notes in Mathematics, vol. 2099. Springer.

Brémaud, P. 1975. An extension of Watanabe's theorem of characterization of Poisson processes over the positive real half line. *J. Appl. Probab.*, **12**, 396–399.

Brémaud, Pierre. 1981. *Point Processes and Queues*. Springer-Verlag. Martingale dynamics, Springer Series in Statistics.

Brémaud, Pierre, and Massoulié, Laurent. 1996. Stability of nonlinear Hawkes processes. *Ann. Probab.*, **24**(3), 1563–1588.

Burke, C. J., and Rosenblatt, M. 1958. A Markovian function of a Markov chain. *Ann. Math. Statist.*, **29**, 1112–1122.

Carstensen, Lisbeth. 2010. Hawkes processes and combinatorial transcriptional regulation. Ph.D. thesis, University of Copenhagen.

Çınlar, Erhan. 2011. *Probability and Stochastics*. Graduate Texts in Mathematics, vol. 261. Springer.

Chang, Yu-Sin. 2017. Markov chain structures with applications to systemic risk. Ph.D. thesis, Illinois Iinstitute of Technology.

Clark, N.J., and Dixon, P.M. 2017. Modeling and estimation for self-exciting spatio-temporal models of terrorist activity. https://arxiv.org/pdf/1703.08429.pdf.

Courrège, Philippe. 1965–1966. Sur la forme intégro-différentielle des opérateurs de C_0^∞ danc C satisfaisant au principe du maximum. *Proc. Séminaire Brelot–Choquet–Deny. Théorie du potentiel*, **10**(1), 1–38.

Crépey, S., Bielecki, T. R., and Brigo, D. 2014. *Counterparty Risk and Funding: A Tale of Two Puzzles*. CRC Press.

Duffie, D., Filipović, D., and Schachermayer, W. 2003. Affine processes and applications in finance. *Ann. Appl. Probab.*, **13**(3), 984–1053.

Dynkin, E. B. 1965. *Markov processes, Vols. I, II*. Translated with the authorization and assistance of the author by J. Fabius, V. Greenberg, A. Maitra, and G. Majone. Academic Press.

Embrechts, Paul, Liniger, Thomas, and Lin, Lu. 2011. Multivariate Hawkes processes: an application to financial data. *J. Appl. Probab.*, **48 A** (New frontiers in applied probability: a Festschrift for Søren Asmussen), 367–378.

Ethier, Stewart N., and Kurtz, Thomas G. 1986. *Markov Processes*. Wiley Series in Probability and Mathematical Statistics, John Wiley & Sons.

Föllmer, Hans, and Protter, Philip. 2011. Local martingales and filtration shrinkage. *ESAIM Probab. Stat.*, **15**, S25–S38 (supplement in honour of Marc Yor).

Gikhman, Iosif I., and Skorokhod, Anatoli V. 2004. *The Theory of Stochastic Processes*. Volume II. Classics in Mathematics. Springer-Verlag. Translated from the Russian by S. Kotz, reprint of the 1975 edition.

Granger, C.W.J. 1969. Investigating causal relations by econometric models and cross-spectral methods. *Econometrica*, **37**(3), 424–438.

Hansen, N. R., Reynaud-Bouret, P., and Rivoirard, V. 2015. Lasso and probabilistic inequalities for multivariate point processes. *Bernoulli*, **21**, 83–143.

Hawkes, Alan G. 1971a. Point spectra of some mutually exciting point processes. *J. Royal Statist. Soc., Series B*, **33**(3), 438–443.

Hawkes, Alan G. 1971b. Spectra of some self-exciting and mutually exciting point processes. *Biometrika*, **58**(1), 83–90.

He, Sheng-Wu, and Wang, Jia-Gang. 1984. Two results on jump processes. Pages 256–267 of *Séminaire de Probabilités XVIII 1982/83*. Springer.

He, Sheng Wu, Wang, Jia Gang, and Yan, Jia An. 1992. *Semimartingale Theory and Stochastic Calculus*. Kexue Chubanshe (Science Press).

Hoh, Walter. 1998. Pseudo differential operators generating Markov processes. Der Fakultat fur Mathematik der Universitat Bielefeld als Habilitationsschrift.

Hoyle, E., and Mengütürk, L. 2013. Archimedean survival processes. *J. Multivariate Anal.*, **115**, 1–15.

Hoyle, E., Hughston, L., and Macrina, A. 2011. Lévy random bridges and the modelling of financial information. *Stochastic Process. Appl.*, **121**(4), 856–884.

Itô, Kiyosi, and McKean, Henry P. Jr. 1974. *Diffusion Processes and Their Sample Paths*. Springer. Second printing, corrected.

Iyengar, Satish. 2001. The analysis of multiple neural spike trains. Pages 507–524 of *Advances in Methodological and Applied Aspects of Probability and Statistics*, N. Balakrishnan (ed.). Taylor and Francis.

Jacob, N. 1998. Characteristic functions and symbols in the theory of Feller processes. *Potential Anal.*, **8**(1), 61–68.

Jacob, N. 2001. *Pseudo Differential Operators and Markov Processes*. Volume I. Imperial College Press.

Jacob, N. 2002. *Pseudo Differential Operators & Markov Processes*. Volume II. Imperial College Press.

Jacob, N. 2005. *Pseudo Differential Operators and Markov Processes*. Volume III. Imperial College Press.

Jacob, Niels, and Schilling, René L. 2001. Lévy-type processes and pseudodifferential operators. Pages 139–168 of *Lévy Processes*. Birkhäuser Boston.

Jacod, Jean. 1974/75. Multivariate point processes: predictable projection, Radon-Nikodým derivatives, representation of martingales. *Z. Wahrscheinlichkeitstheorie und Verw. Gebiete*, **31**, 235–253.

Jacod, Jean, and Shiryaev, Albert N. 2003. *Limit Theorems for Stochastic Processes*, Second Editon. Springer.

Jakubowski, Jacek, and Niewęgłowski, Mariusz. 2008. Pricing bonds and CDS in the model with rating migration induced by a Cox process. Pages 159–182 of Stettner, Łukasz (ed.), *Advances in Mathematics of Finance*. Banach Center Publications.

Jakubowski, Jacek, and Niewęgłowski, Mariusz. 2010. A class of $\mathbb{F}$-doubly stochastic Markov chains. *Electron. J. Probab.*, **15**(56), 1743–1771.

Jakubowski, Jacek, and Pytel, Adam. 2016. The Markov consistency of Archimedean survival processes. *J. Appl. Probab.*, **53**(02), 392–409.

Jeanblanc, Monique, Yor, Marc, and Chesney, Marc. 2009. *Mathematical Methods for Financial Markets*. Springer.

Kallenberg, Olav. 2002. *Foundations of Modern Probability*, Second Edition. Springer.

Kallsen, Jan, and Tankov, Peter. 2006. Characterization of dependence of multidimensional Lévy processes using Lévy copulas. *J. Multivariate Anal.*, **97**(7), 1551–1572.

Kurtz, Thomas G. 1998. Martingale problems for conditional distributions of Markov processes. *Electron. J. Probab.*, **3**, 1–29.

Last, Günter, and Brandt, Andreas. 1995. *Marked Point Processes on the Real Line*. Springer.

Liniger, Thomas Josef. 2009. Multivariate Hawkes processes. Ph.D. thesis, ETH Zurich.

Massoulié, Laurent. 1998. Stability results for a general class of interacting point processes dynamics, and applications. *Stochastic Process. Appl.*, **75**(1), 1–30.

Michalik, Zofia. 2015. Markov chain and Poisson copulae and their applications in finance. M.Phil. thesis, University of Warsaw. In Polish.

Mohler, G. O., Short, M. B., Brantingham, P. J., Schoenberg, F. P., and Tita, G. E. 2011. Self-exciting point process modeling of crime. *J. Amer. Statist. Assoc.*, **106**(493), 100–108.

Nelsen, Roger B. 2006. *An Introduction to Copulas*, Second Edition. Springer.

Oakes, David. 1975. The Markovian self-exciting process. *J. Appl. Probab.*, **12**, 69–77.

Ogata, Y. 1998. Space–time point–process models for earthquake occurrences. *Ann. Inst. Math. Statist.*, **50**, 379–402.

Ogata, Y. 1999. Seismicity analysis through point-process modeling: a review. *Pure Appl. Geophys.*, **155**, 471–507.

Pazy, A. 1983. *Semigroups of Linear Operators and Applications to Partial Differential Equations*. Applied Mathematical Sciences, vol. 44. Springer.

Peszat, S., and Zabczyk, J. 2007. *Stochastic Partial Differential Equations with Lévy Noise*. Encyclopedia of Mathematics and its Applications, vol. 113. Cambridge University Press.

Rao, Murali. 1972. On modification theorems. *Trans. Amer. Math. Soc.*, **167**, 443–450.

Reynaud-Bouret, Patricia, and Schbath, Sophie. 2010. Adaptive estimation for Hawkes processes; application to genome analysis. *Ann. Statist.*, **38**(5), 2781–2822.

Reynaud-Bouret, Patricia, Rivoirard, Vincent, and Tuleau-Malot, Christine. 2013. Inference of functional connectivity in neurosciences via Hawkes processes. Pages 317–320 of *Proc Global Conf. on Signal and Information Processing (GlobalSIP), 2013*. IEEE.

Reynaud-Bouret, Patricia, Rivoirard, Vincent, Grammont, Franck, and Tuleau-Malot, Christine. 2014. Goodness-of-fit tests and nonparametric adaptive estimation for spike train analysis. *J. Math. Neurosci.*, **4**(1), 3.

Richmond, Douglas R. 1995. Issues and problems in "other insurance," multiple insurance, and self-insurance. *Pepperdine Law Rev.*, **22**(4), 1373–1465.

Rogers, L.C.G., and Pitman, J.W. 1981. Markov functions. *Ann. Probab.*, **9**(4), 573–582.

Rogers, L.C.G., and Williams, David. 2000. *Diffusions, Markov Processes, and Martingales*. Volume 2. Cambridge Mathematical Library. Cambridge University Press.

Rüschendorf, Ludger, Schnurr, Alexander, and Wolf, Viktor. 2016. Comparison of time-inhomogeneous Markov processes. *Adv. Appl. Probab.*, **48**(4), 1015–1044.

Scarsini, M. 1989. Copulae of probability measures on product spaces. *J. Multivariate Anal.*, **31**, 201–219.

Schilling, René L. 1998. Growth and Hölder conditions for the sample paths of Feller processes. *Probab. Theory Rel. Fields*, **112**(4), 565–611.

Schnurr, J.A. 2009. The symbol of a Markov semimartingale. Ph.D. thesis, Technischen Universität, Dresden.

Sklar, A. 1959. Fonctions de rèpartition à n dimensions et leurs marges. *Publ. Inst. Statist. Univ. Paris*, **8**, 229–231.

Stricker, C. 1977. Quasimartingales, martingales locales, et filtrations naturelles. *Z. Wahrscheinlichkeitstheorie Verw. Gebiete*, **39**, 55–63.

Stroock, Daniel W. 1975. Diffusion processes associated with Lévy generators. *Z. Wahrscheinlichkeitstheorie und Verw. Gebiete*, **32**(3), 209–244.

Tankov, Peter. 2003. Dependence structure of spectrally positive multidimensional Lévy processes. Technical report.

Vacarescu, A. 2011. Filtering and parameter estimation for partially observed generalized Hawkes processes. Ph.D. thesis, Stanford University.

van Abdel-Hameed, M. 1987. Inspection and maintenance policies of devices subject to deterioration. *Adv. Appl. Probab.*, **19**(4), 917–931.

van Casteren, Jan A. 2011. *Markov Processes, Feller Semigroups and Evolution Equations*. Series on Concrete and Applicable Mathematics, vol. 12. World Scientific.

van Noortwijk, J. 2007. A survey of the application of gamma processes in maintenance. *Reliability Eng. Safety Syst.*, **94**, 2–21.

Watanabe, Shinzo. 1964. On discontinuous additive functionals and Lévy measures of a Markov process. *Japan. J. Math.*, **34**, 53–70.

Wentzell, A. D. 1981. *A Course in the Theory of Stochastic Processes*. McGraw-Hill. Translated from the Russian by S. Chomet, with a foreword by K. L. Chung.

Zhu, L. 2013. Nonlinear Hawkes processes. Ph.D. thesis, New York University.

Notation Index

Subject Index